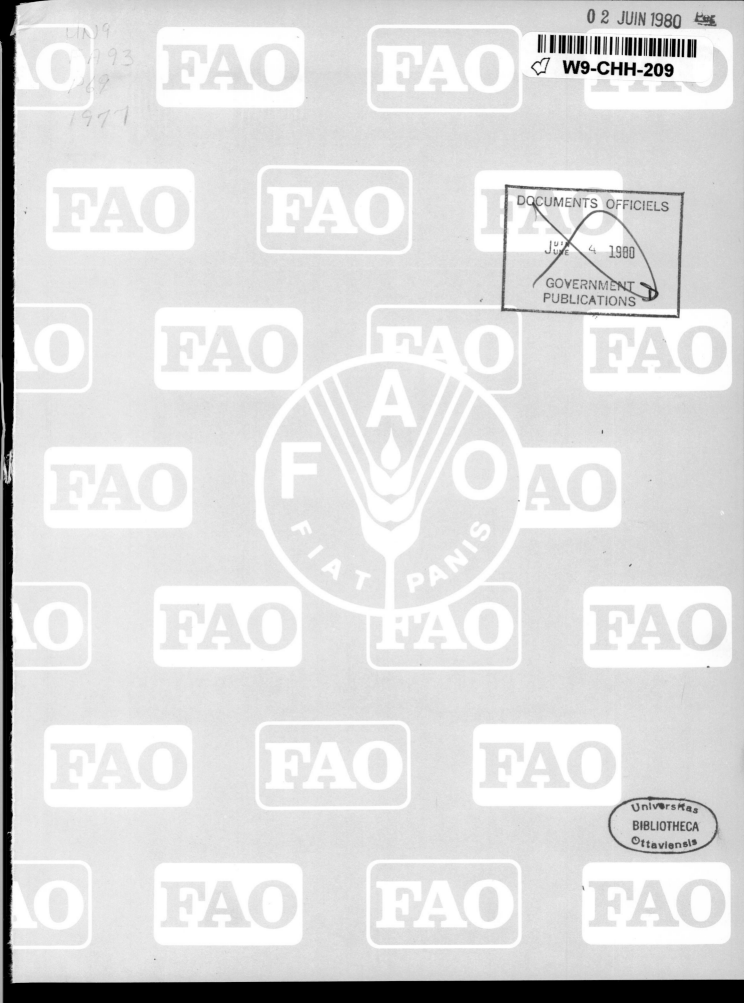

TABLE OF CONTENTS

TABLE DES MATIÈRES

TABLA DE MATERIAS

VI – LIVESTOCK NUMBERS AND PRODUCTS
VI – ÉLEVAGE ET PRODUITS DE L'ÉLEVAGE
VI – GANADO, AVES DE CORRAL Y SUS PRODUCTOS

This is the 31st issue of the *FAO production yearbook*. After the changes made in last year's issue, only three have been introduced this year. In most tables relating to crops, livestock and production means, the five-year average 1961-65 has been substituted by the three-year average 1969-71. Secondly, also in the tables on production index numbers the base period has been shifted from 1961-65 to 1969-71. The third and most important change is the introduction of national average producer prices for the period 1969-71 as weighting coefficients to aggregate the production of each country against the common regional wheat-based price relatives for the period 1961-65 which were used previously (see other details on index numbers in the Notes on the Tables). Also worthy of note is the fact that two other country groups have been introduced in the Yearbook: one, All developed countries, covering developed market economies, eastern Europe and the U.S.S.R.; the other, All developing countries, covering developing market economies and Asian centrally planned economies.

The reader is reminded that certain data on prices and related matters which are no longer published in this Yearbook, can be found in the *FAO Monthly Bulletin of Statistics*, United Nations *Statistical yearbook* and *Monthly Bulletin of Statistics*, International Labour Office *Yearbook of labour statistics* and *Labour Statistics* (monthly).

Data on production, trade, consumption and prices of commercial fertilizers can be found in the FAO *Annual fertilizer review*.

Statistical data on crops and livestock which are no longer shown in this Yearbook are still available in the FAO Statistics Division which is ready to release them on request.

As in past years, this volume also includes a large number of estimates made by FAO on area and production of major crops and on livestock numbers and products, where no official or semiofficial figures were available from the countries themselves. The publication of these estimates gives the countries concerned a chance to examine them, and it is hoped that they will provide FAO with more reliable figures.

For almost all commodity tables, data shown include figures for the year appearing on the cover of the Yearbook. However, several figures for that year are FAO estimates based on unofficial information and are, therefore, preliminary.

As for previous issues, the publication of the Yearbook has been made possible by the cooperation of governments, which have supplied most of the information in the form of replies to annual FAO questionnaires. Collaboration with various agencies has continued in order to achieve conformity in the presentation of international figures. The assistance of governments and agencies is gratefully acknowledged.

EXPLANATORY NOTES

Symbols

Definition of symbols used in the tables:

*   Unofficial figure.

—   None, in negligible quantity (less than one half of the unit indicated), or entry not applicable.

...   Data not available.

F   FAO estimate

NES	Not elsewhere specified or included
HA	Hectare
MT	Metric ton
KG	Kilogramme
HEAD	
NUMBER	
KG/HA	Kilogramme per hectare
KG/AN	Kilogramme per animal

In most of the tables a blank space has the same meaning as the symbols (—) or (...) defined above.

To divide decimals from whole numbers a period (.) is used.

Country and commodity names

In most of the tables the space provided for country and commodity names is limited to 12 and 24 letters, respectively. The commodity names are given in English, French and Spanish; the names of continents, countries and regions in English only. While the abbreviated commodity names are sufficiently clear, the names of countries are at times somewhat obscure, and the reader should consult the List of Countries, Continents, Economic Classes and Regions (page 37), which shows the countries in the order in which they appear in the tables, providing abbreviated names in English and corresponding full names in English, French and Spanish.

Time reference

As initiated in the 1966 edition of the Yearbook, the time reference for statistics on area and production of crops is based on the calendar-year period. That is to say, the data for any particular crop refer to the calendar year in which the entire harvest or the bulk of it took place. This does not necessarily mean however that, for a given commodity, production data are aggregated month by month from January to December, although this is true for certain crops such as tea, sisal, palm kernels, palm oil, rubber, coconuts, and, in certain countries, sugarcane and bananas, which are harvested almost uniformly throughout

the year. The harvest of other crops, however, is generally limited to a few months and even, in certain cases, to a few weeks. Production of these crops is reported by the various countries in different ways: for calendar years, agricultural years, marketing years, etc. Whatever the statistical period used by the countries for presentation of area and production data, these data are allocated commodity by commodity to the calendar year in which the entire harvest or the bulk of it takes place. Obviously, when a crop is harvested at the end of the calendar year, production of this crop will be utilized mostly during the year following the calendar year under which production figures are shown in the tables.

It should be noted that the adoption of a calendar-year time reference period inevitably means that, in a number of cases, crops assigned by countries to a particular split year may appear under two different calendar years in the tables in this Yearbook.

Livestock numbers have been grouped in 12-month periods ending 30 September of the years stated in the tables, that is, animals enumerated in a given country any time between 1 October 1976 and 30 September 1977 are shown under the year 1977.

As regards livestock products, data on meat, milk, and milk products relate to calendar years, with a few exceptions which are mentioned in the Notes on the Tables. Data for other animal products that are produced only in certain periods of the year, for example, honey, wool, etc., are allocated to the calendar year, following a policy similar to that adopted for crops.

For tractors and other agricultural machinery, data refer, as far as possible, to the number in use at the end of the year stated or during the first quarter of the following year.

Data on pesticides are generally for the calendar year.

The FAO index numbers of agricultural production have been calculated on the basis of a calendar-year reference period as defined above.

Crop areas

Figures for crop areas generally refer to harvested areas, although for permanent crops data may refer to total planted area.

Yields per hectare

All yields per hectare, for single countries as well as for continental, regional, and world totals, are given in kilogrammes. In all cases they are computed from detailed area and production data, that is, hectares and metric tons. Data on yields of permanent crops are not as reliable as those for temporary crops, either because most of the area information may relate to planted area, as for grapes, or because of the scarcity and unreliability of the area figures reported by the countries, as for example for cocoa and coffee.

Totals

Continental, regional and world totals are given for data on all commodities except those on pesticides and on milking machines. These totals include only data for the countries shown in the body of the table.

In general these totals reflect adequately the situation in the geographical areas which they represent, with the exception of certain vegetable and fruit crops and certain livestock products.

More details on this subject can be found in the Notes on the Tables.

Notes on the tables and Country notes

For a number of tables and countries the figures require more explanation and qualification than are possible here. Explanation of these points, as well as other elements, including changes in territorial coverage, classification of countries by FAO regions, etc., are given below in the Notes on the Tables and the Country Notes.

As a general rule, data in the Yearbook relate to the country specified with its present *de facto* boundaries. Country names and continental groupings follow, in general, the nomenclature used by the Statistical Office of the United Nations.

NOTES ON THE TABLES

LAND

Land use and irrigation

These tables attempt to bring together all available data on land use and irrigated land throughout the world.

When considering the section on land use it should be borne in mind that definitions used by reporting countries vary considerably and items classified under the same category often relate to greatly differing kinds of land.

Definitions of land-use categories are as follows:

1. *Total area* refers to the total area of the country, including area under inland water bodies.

2. *Land area* refers to total area, excluding area under inland water bodies. The definition of inland water bodies generally includes major rivers and lakes.

3. *Arable land* refers to land under temporary crops (double-cropped areas are counted only once), temporary meadows for mowing or pasture, land under market and kitchen gardens (including cultivation under glass), and land temporarily fallow or lying idle.

4. *Land under permanent crops* refers to land cultivated with crops which occupy the land for long periods and need not be replanted after each harvest, such as cocoa, coffee, and rubber; it includes land under shrubs, fruit trees, nut trees and vines, but excludes land under trees grown for wood or timber.

5. *Permanent meadows and pastures* refers to land used permanently (five years or more) for herbaceous forage crops, either cultivated or growing wild (wild prairie or grazing land).

6. *Forests and woodland* refers to land under natural or planted stands of trees, whether or not productive, and includes land from which forests have been cleared but which will be reforested in the foreseeable future.

7. *Other land* includes unused but potentially productive land, built-on areas, wasteland, parks, ornamental gardens, roads, lanes, barren land, and any other land not specifically listed under items 3 through 6.

Data on irrigation relate to areas purposely provided with water, including land flooded by river water for crop production or pasture improvement, whether this area is irrigated several times or only once during the year stated.

Specific country notes pertaining to land-use categories and irrigation are as follows:

TOTAL AREA

Greenland: Data refer to area free from ice.

Mauritius: Data exclude dependencies.

Namibia: Data include the territory of Walvis Bay.

New Caledonia: Data include dependencies.

South Africa: Data exclude the territory of Walvis Bay.

U.S.S.R.: Data include the White Sea (9 000 000 hectares) and the Azov Sea (3 730 000 hectares).

ARABLE LAND AND LAND UNDER PERMANENT CROPS

Australia: Data on arable land include about 27 000 000 hectares of cultivated grassland.

Cuba: Data refer to the State sector only.

Portugal: Data include about 800 000 hectares of temporary crops grown in association with permanent crops and forests.

PERMANENT MEADOWS AND PASTURES

Australia: Data refer to balance of area of rural holdings.

Egypt: Rough grazing land is included under Other land.

U.S.S.R.: Data exclude pastures for reindeer.

For the following countries, data refer to permanent meadows and pastures on agricultural holdings only: Chile, Dominican Republic, Finland, Guatemala, Kenya, Republic of Korea, Surinam, Trinidad and Tobago, Uruguay.

IRRIGATION

France: Data exclude market and kitchen gardens.

Hungary: Data exclude complementary farm plots and individual farms.

United Kingdom: Data exclude Scotland and Northern Ireland.

For the following countries data refer to land provided with irrigation facilities: Bulgaria, Norway, Romania, Surinam.

For the following countries, data refer to irrigated rice only: Japan, Republic of Korea, Sri Lanka.

POPULATION

Table 3 presents estimates of *total population, agricultural population* and *economically active population*, total and in agriculture, for 1965, 1970, 1975, 1976 and 1977, by country.

The United Nations *Demographic Yearbook* gives time series of estimates on total population. These data are generally provided by the countries, but in some instances the United Nations Statistical Office adjusts the available estimates or prepares new ones. However, for many developing countries further adjustment of available estimates is needed in order to maintain a reasonable degree of consistency. Inconsistency is sometimes internal, within the set of annual estimates themselves, since these estimates fluctuate in a manner that cannot be explained by migration. Sometimes the estimates are inconsistent with data from external sources, mainly from censuses and vital registration. Also there are gaps in the time series of estimates which need to be filled. Therefore, the United Nations Population Division has prepared a complete series of estimates for each country. This series covers a fairly long period and is consistent in its year-to-year variations as well as with other demographic data. [1] The data presented in this table are derived from these series with adjustments to incorporate more recently available official population data, particularly for the developed countries, whose demographic statistics are considered to be reliable.

Data relate generally to the present-in-area (*de facto*) population within the present geographical boundaries.

Agricultural population is defined as all persons depending for their livelihood on agriculture. This comprises all persons actively engaged in agriculture and their non-working dependants.

Economically active population is defined as all persons engaged or seeking employment in an economic activity, whether as employers, own-account workers, salaried employees or unpaid workers assisting in the operation of a family farm or business.

Economically active population in agriculture includes all economically active persons engaged principally in agriculture, forestry, hunting or fishing.

Information on the economically active population and its industrial or occupational breakdowns is available in connection with national population censuses or labour force surveys. The comparability of the data is, however, limited by different country-to-country statistical treatment of, for example, unpaid family workers, particularly housewives. Furthermore, some countries report information on economic activity for persons of all ages, others only for persons of specified ages, e.g., 14 years and over. The ILO has systematically evaluated these data and made them consistent with internationally accepted standard concepts and has prepared estimates of the economically active population broken down by agriculture, industry and services for each fifth year between 1950 and 1970 and projections of the total economically active population for each fifth year between 1970 and 2000. [2] The data in the table are based on these estimates and projections.

Information on agricultural population derived from national population censuses or surveys is scarce. In deriving the series of estimates of agricultural population presented in this table, FAO has largely relied on the close relationship existing between the ratio of economically active population in agriculture to total economically active population (EAA/EA) and the ratio of agricultural population to total population (AP/TP). It was generally assumed that these ratios are equal so that, given EAA/EA, the agricultural population is estimated as a product of this ratio and total population. [3]

Because of the very rough nature of the estimates of the economically active population and of the agricultural population in small countries — generally those with a total population of 250 000 or less in 1970 — these data are not shown separately; the continental, regional and world totals, however, have been adjusted to cover those countries as well.

[1] United Nations, *Single-year population estimates and projections of the major areas, regions and countries of the world, 1950-2000,* mimeographed, ESA/P/WP. 34, New York, 1975.

[2] ILO, *Labour Force, 1950-2000,* Vols. I, II, III, IV, Geneva, 1977.
[3] For details regarding the method and assumptions adopted in arriving at these estimates, see FAO, "Projections of world agricultural population", *Monthly Bulletin of Agricultural Economics and Statistics,* Vol. 21, No. 1, 1972, p. 3-4.

FAO INDEX NUMBERS OF AGRICULTURAL PRODUCTION

These index numbers show the relative level of the aggregate volume of agricultural production for each year in comparison with the base period 1969-71. They are based on the sum of price-weighted quantities of different agricultural commodities produced after deductions of quantities used as seed and feed weighted in a similar manner. The resulting aggregate represents, therefore, disposable production for any use except as seed and feed.

Deductions for seed (in the case of eggs, for hatching) and for livestock and poultry feed apply to both domestically produced and imported commodities. They cover primary agricultural products used as such (maize, potatoes, pumpkins, etc.) and semi-processed feeds, such as bran, oilcakes, meals, molasses, dry whey, etc. The reason for subtracting amounts of seed and feed from the production data is because the FAO index numbers, based on the concept of world agriculture as one farm, measure agricultural production by avoiding double counting of seed and feed (which are already counted in the production data) and the crops and livestock produced from them.

The country indices are calculated by the Laspeyres formula. Production quantities of each commodity are weighted by average national producer prices for the period 1969-71 and summed for each year. The aggregate for a given year is divided by the average aggregate for the base period 1969-71 to obtain the index number.

Indices for continents, regions and the world are computed by summing up the country aggregates, which for that purpose are converted into US dollars. The exchange rates of national currencies for US dollars are those published by the International Monetary Fund, with a few exceptions: eastern European countries, the U.S.S.R., China and other centrally planned Asian economies for which the exchange

rates were derived from non-commercial, tourist and other rates which are not basic (fixed) rates.

The commodities covered in the computation of index numbers of agricultural production are all crops and livestock products originating in each country on which information is available. The index numbers of food production include commodities which are considered edible and contain nutrients. Accordingly, together with inedible commodities coffee and tea are excluded because, although edible, they have practically no nutritive value.

The index numbers are based on production data presented on the basis of a calendar-year time reference.

The differences between the indices presented in this issue of the Yearbook and the indices shown in previous issues are basically three: 1. The production data used for the computation of index numbers are now, with very few exceptions, primary commodities. Therefore, production of sugar, olive oil, copra, raisins and wine used earlier for the computation of index numbers has been substituted by production of sugarcane, sugar beet, olives, coconuts and grapes. The processed commodities that still remain are palm oil, palm kernels, cottonseed and cotton lint. Production of palm nuts could not be used for lack of data on production and prices, and seed cotton could not be used because it is a mixed food and non-food product. 2. The base period has been shifted from 1961-65 average to 1969-71 average. 3. The most important change is the use of *national* average producer prices (1969-71) as weights for computation of aggregate production of each country instead of *regional* wheat-based price relatives (1961-65 averages) applied to each country of the region.

The FAO index numbers may differ from those produced by the countries themselves because of differences in concepts of production, coverage, weights, time reference of data, and methods of calculation.

CROPS

Statistical summary of world and regional agricultural production

Figures for production of *apples* exclude those for the U.S.S.R. Data on *oil crops* represent the *total production* of oilseeds, oil nuts, and other oil crops harvested in the year indicated and *expressed in terms of oil equivalent*. That is to say, these figures do not relate to the actual production of vegetable oils, but to the potential production, on the hypothesis that the total amounts produced from all oil crops were processed into oil in producing countries in the same year in which they were harvested. Naturally, the total production of oil crops is never processed into oil in its entirety, since, depending on the various crops, important quantities are used also for seed, feed and food. On the other hand, although oil extraction rates vary from country to country, in this table for each crop the same extraction rate has been applied for all countries. Moreover, it should be borne in mind that the crops harvested during the latter months of the year are generally processed into oil during the following year. In spite of these deficiencies in coverage, extraction rates and time reference, the data reported here are useful as they provide a valid indication of year-to-year changes in the size of total oil crop production. The actual production of vegetable oils in the world is about 80% of the production reported here. *Meat* production figures relate to total production from animals slaughtered in each country, irrespective of origin.

Cereals

Area and production data on cereals relate to crops harvested for grain only. Cereal crops harvested for hay, green feed and silage or used for grazing are therefore excluded. Area data relate to harvested area. Some countries report sown or cultivated area only; however, in these countries the sown or cultivated area does not differ significantly in normal years from the area actually harvested, either because practically the whole area sown is harvested or because the area surveys are conducted around the harvest period.

Cereals, total. Data shown in this table also include other cereals which are no longer published in this Yearbook, such as mixed grains, buckwheat, etc.

Wheat and spelt. Available data for spelt are included with those for wheat except for the U.S.S.R.

Millet and sorghum. Millet and sorghum are grown chiefly as feed for livestock and poultry in Europe and North America, but are used to a large extent as food in Asia, Africa and the U.S.S.R. Wherever possible, statistics are given separately for millet and sorghum, but many countries, especially in Africa, make no distinction between the two grains in their reports; in such cases, combined figures are given in the table on millet.

Root crops

Roots and tubers, total. Data shown in this table also include other root crops which are no longer published in the Yearbook, such as taros, yams, etc.

Cassava. Cassava (*Manihot esculenta*, Crantz) is a root crop commonly divided into two groups: bitter and sweet cassava, sometimes considered as two different species with the botanical names of *M. utilissima* and *M. dulcis* or *aipi*. In the table bitter and sweet cassava are reported together.

Pulses

Pulses, total. Data shown in this table also include other pulses which are no longer published in the Yearbook, such as cowpeas, vetch, etc.

As far as can be ascertained the tables show production of crops harvested for dry grain only, whether used for food or feed.

Dry beans. For a few European countries, data on area and production of dry beans grown mixed with other crops have been aggregated to those for beans grown alone. Area data for these countries are overestimated and consequently yields per hectare appear extremely low.

Oilseeds, oil nuts, and oil kernels

Rapeseed. Sweden's production is given with an 18 percent water content. Figures for a few countries, such as India and Pakistan, also include mustard seed.

Linseed. Area data for the U.S.S.R. and a few minor producing countries relate to crops grown for both seed and fibre.

Cottonseed. Direct production figures for cottonseed are reported by countries accounting for about 60 percent of world production. Data for the remainder are derived from ginned cotton production, according to ratios obtained from earlier years in these countries or from countries with similar conditions.

Olive oil. With few exceptions, data relate to total oil production, including oil extracted from olive residues.

Coconuts. Data shown relate to total production of coconuts whether ripe or unripe, whether consumed fresh or processed into copra or desiccated coconut. Production is expressed in terms of weight of the whole nut, excluding only the fibrous outer husk.

Vegetables and melons

Data shown in the tables relate to vegetable crops grown mainly for human consumption. Crops such as cabbages, pumpkins, carrots, etc., when explicitly cultivated for animal feed, are therefore excluded. Statistics on vegetables are not available in many countries, and the coverage of the reported data differs from country to country. In general, it appears that the estimates refer to crops grown in fields and market gardens mainly for sale, thus excluding crops cultivated in kitchen gardens or small family gardens mainly for household consumption. In Austria reported data relate to field crops only; in Cuba they refer to procurement from state and private farms; in the U.S.S.R. data refer to collective and state farms. Production

from family and other small gardens not included in current statistical surveys and consequently not included in the tables of the Yearbook constitutes quite an important part of the estimated total production in certain countries: for example in Austria and in the Federal Republic of Germany, over 40 percent; in Italy over 15 percent.

For the reasons mentioned above, continental, regional and world totals are far from representative of the total area and production of the different kinds of vegetables. Production data shown in Table 40 include data published in the individual tables on vegetables presented in this Yearbook, as well as data on all other kinds of vegetables. They also include estimates for non-reporting countries and, when available, production of non-commercial crops for countries reporting only production for sale.

Cabbages. The main varieties of cabbages included in the table are: red, white, and savoy cabbages, China cabbages, Brussels sprouts, green kale, and sprouting broccoli.

Tomatoes. Data for Belgium, Denmark, Finland, Norway, Sweden and the United Kingdom relate to crops grown mainly or totally under glass. This explains the larger yields per hectare obtained in these countries.

Cauliflowers. Where possible, data on heading broccoli are also included in the figures.

Cucumbers and gherkins. About half of the European producing countries, Canada and New Zealand cultivate their crops totally or partially under glass, and yields per hectare reflect this.

Green beans. Data refer to beans (*Phaseolus* and *Dolichos* species only) harvested green; they exclude data on snap or string beans, at least for countries, such as France and the United States, which publish separate statistics for green-shell beans and string beans. Data on green beans for processing, originally reported by a few countries in shelled weight, have been converted into beans in the shell at about 200 percent.

Green peas. Data refer to peas (*Pisum sativum* and *P. arvense*) harvested green. Data for a few countries, originally reported in shelled weight, have been converted to peas in the shell at 225-250 percent.

Cantaloupes and other melons. Data for Romania include watermelons.

Watermelons. Data for Algeria, Bulgaria, Turkey and Yugoslavia include melons; for the U.S.S.R., data include melons (about 18 percent) and pumpkins and squash (about 30 percent).

Grapes and wine

Grapes. Certain countries, such as Algeria, Austria, Chile, France, the Federal Republic of Germany, and Iran — to mention the important producers — do not publish data on total grape production. Estimates for these countries shown in the table are based on information available on the production of table grapes, raisins and wine. Area data for Italy include area under grapes grown mixed with other crops; 23.5 percent of this mixed area is included in the total area under grapes.

Wine. In most of the major wine-growing countries, wine production is estimated from the quantity of grapes crushed at harvest time; consequently, it corresponds to the amount of " grapes for wine " for the same crop year and represents total output at wine presses, irrespective of whether it is finally consumed as wine, vinegar, or distilling material. Unfortunately, it has not yet been possible to obtain statistics on this basis from all countries, and gaps have been filled by using tax returns or trade estimates. Some countries do not publish statistics on wine production or they give unreliable data, either because they do not include total wine production or because they include mixtures of wine and fruit juices. Wine production for these countries has been estimated on the basis of quantities of grapes crushed for wine when such information was available.

Sugarcane, sugar beets and sugar

Sugarcane and sugar beets. Area and production data on sugarcane and sugar beets generally cover all crops harvested, except the crops grown explicitly for feed. Most of the crop is used for the production of centrifugal and non-centrifugal sugar; however, in several countries important quantities of sugarcane are used also for seed, feed, fresh consumption, the manufacture of alcohol and other uses; some sugar beet production is used for feed and alcohol.

Centrifugal sugar. Data include both cane and beet sugar and are shown, as far as possible, in terms of raw value as reported by the countries. It is not certain, however, whether all countries report raw sugar in terms of 96° polarization as requested in the FAO questionnaires. Australia, for instance, reports sugar production at 94° net titre. Figures for two countries, Haiti and Indonesia, are given as *tel quel*, which is the actual physical weight of all centrifugal sugar produced. Data reported by countries as refined sugar have been converted to a raw basis at 108.7 percent.

Non-centrifugal sugar. This table includes any sugar produced from sugarcane which has not undergone centrifugation. Practically all non-centrifugal sugar is used for local consumption.

Fruits and berries

Data refer to total production of fresh fruit, whether finally used for direct consumption for food or feed, or processed into different products: dry fruit, juice, jam, alcohol, etc.

Statistics on fruit, especially tropical fruit, are unavailable in many countries, and the coverage of the reporting countries suffers from lack of uniformity. Generally, production data relate to plantation crops or orchard crops grown mainly for sale. Data on production from scattered trees used mainly for home consumption are not usually collected. Production from wild plants, particularly berries, which is of some importance in certain countries, is generally disregarded by national statistical services. Therefore, the figures published in this Yearbook for the various fruits and berries are rather incomplete, particularly in respect of regions other than Europe, North America, Australia and New Zealand. However, in most of the tables, the totals shown — though relating to the limited number of countries listed — are nevertheless believed to give a reliable indication of these crops insofar as they influence international trade. The totals, in any case, provide an indication of annual changes in the size of the crops.

Production data shown in Table 40 include data published in the individual tables on fruits and berries, as well as data on all other kinds of fruits and berries. Dates, plantains and total grapes are also included in the total fruit figures, while olives are excluded. Figures in this table are more complete than those published for the single commodities because they include estimates for most of the non-reporting countries as well as data for countries, such as the U.S.S.R., reporting total production of fruits in a single figure without specification by kinds.

Oranges. Data for Guinea, Sierra Leone, Swaziland, the U.S.S.R. and a few other minor producing countries relate to production of all citrus fruit. Data for a few other countries may include tangerines.

Tangerines, mandarines, clementines and satsumas. Figures for the United States include tangelos, a tangerine/grapefruit hybrid, and temples, a sweet orange/tangerine hybrid.

Lemons and limes. Figures for Bolivia and Peru include grapefruit.

Citrus fruit n.e.s. Data for Japan include grapefruit and lemons. Figures for other countries generally refer to production of total or unidentified citrus crops.

Bananas and plantains. Figures on bananas refer, as far as possible, to all edible fruit-bearing species of the genus *Musa* except *M. paradisiaca*, commonly known as plantain. Unfortunately, several countries make no distinction in their statistics between bananas and plantains and publish only overall estimates. When this occurs and there is some indication or assumption that the data reported relate mainly to bananas, the data are included in the table. It is worthy of note that none of the countries excluded from the table are significant exporters. Comparability of production data on bananas and plantains, as reported by the various countries, is also rendered difficult by the fact that a number of countries report in terms of bunches, which generally means that the stalk is included in the weight.

Raspberries. Some data on raspberries (*Rubus idaeus*) appear to include other berries of the genus *Rubus*, such as blackberries, loganberries, and dewberries.

Currants. Data on currants include *Ribes rubrum*, *album* and *nigrum*.

Nuts

Production of nuts (including chestnuts) relates to nuts in the shell or in the husk. Statistics are very scanty and generally refer only to crops for sale.

In addition to the six kinds of nuts shown in the separate tables in the Yearbook, production data in Table 40 include all other nuts mainly used as dessert or table nuts, such as Brazil nuts, pili nuts, sapucaia nuts, macadamia nuts, etc. Nuts mainly used for flavouring beverages are excluded from the table, as are masticatory and stimulant nuts and nuts used mainly for the extraction of oil or butter: areca/betel nuts, cola nuts, illipe nuts, karite nuts, coconuts, tung nuts, oil-palm nuts, etc.

Coffee, cocoa and tea

Production figures for coffee relate to green beans. Data for a few countries reporting in terms of cherries or as parchment coffee have been converted into clean coffee by using appropriate conversion factors.

For cocoa, production data relate to cocoa beans fermented and dried. Official area statistics for coffee and cocoa are available only for certain countries and are not always reliable. Yields per hectare are therefore not very meaningful.

Production data on coffee and cocoa shown for Brazil are the official figures published in the Brazilian statistical yearbook. Data on coffee reported in terms of dry cherries have been converted into green coffee at 50 percent.

As regards tea, production figures relate to made tea. For Indonesia, however, about one third of the production shown is given in green-leaf weight. Burma, each year, produces about 45 000 tons of tea leaves. This, however, is not included in the table since practically the entire production is consumed fresh as a vegetable.

Hops

Production data relate to dried-cone weight except for Spain, whose data relate to green weight.

Tobacco

Production figures refer, as far as could be determined, to farm sales weight. Data available on a dry weight basis have therefore been converted into farm sales weight at about 90 parts to 100.

Fibre crops

Flax fibre. Data shown refer generally to scutched and hackled flax and include tow. Figures for countries reporting production in terms of straw or retted flax were converted into flax fibre and tow to make them comparable with data for other countries.

Hemp fibre. As for flax, data on hemp refer to scutched fibre and include tow. Figures for Bangladesh, India and Pakistan relate to sunn hemp (*Crotalaria juncea*), while data for other countries refer to true hemp (*Cannabis sativa*).

Jute and jute-like fibres. Jute fibres are obtained from *Corchorus capsularis* and *C. olitorius*. Jute-like fibres include a number of jute substitutes, the main ones being kenaf or mesta and roselle (*Hibiscus* spp.) and Congo jute or paka (*Urena lobata*).

Sisal. Data on sisal comprise fibres and waste of *Agave sisalana*. Area data are usually — even when official — very rough estimates.

Cotton (lint). The table on cotton lint was prepared in cooperation with the International Cotton Advisory Committee. For most countries, the production figures are those officially reported as lint and do not include cotton linters. In a few cases where production was reported in terms of unginned cotton, and where no specific conversion factor for lint was known, the lint equivalent was taken to be one third.

Fibre crops n.e.s. The main vegetable fibres shown in this table are Mauritius fibre (*Furcraea gigantea*), New Zealand flax (*Phormium tenax*), fique (*Furcraea macrophylla*), caroa (*Neoglazovia variegata*), istle (*Samuela carnerosana*), ramie and rhea (*Boehmeria* spp.), kapok (*Ceiba pentandra*), coir (a fibre contained in the husk of coconuts). Agave fibres and abaca are excluded.

Natural rubber

The table on rubber was prepared in cooperation with the International Rubber Study Group, which defines natural rubber (*Hevea* spp.) as including the dry content weight of latex and excluding balata, gutta-percha, and all rubber-allied gums, as well as scrap rubber. The latter are regarded as products which generally go into entirely different uses from those to which natural rubber is put.

LIVESTOCK NUMBERS AND PRODUCTS

Livestock numbers

The data on livestock numbers in this Yearbook are intended to cover all domestic animals irrespective of their age and place or purpose of their breeding. Estimates have been made for non-reporting countries as well as for countries reporting partial coverage. However, in certain countries, data for chickens, ducks, and turkeys do not seem yet to represent the total number of these birds. Certain other countries give a single figure for all poultry. Data for these countries are shown under chickens.

Livestock products

Slaughterings, carcass weight, production of meat. Tables 83 to 87 present, for major species, the number of animals slaughtered, the average dressed carcass weight and the corresponding production of meat. Data shown in these tables relate to *animals slaughtered* within national boundaries irrespective of their origin. Similarly, the data on production of horse meat, poultry meat and total meat shown in Table 88 relate to animals slaughtered in the country concerned regardless of the origin of the animal.

The concept of production of meat reported in Table 89 is different. Here the production figures relate to *indigenous animals*, i.e., they include the meat equivalent of exported live animals and exclude the meat equivalent of imported live animals.

All data shown relate to total meat production, that is, from both commercial and farm slaughter. Data are given in terms of dressed carcass weight, excluding offal and slaughter fats. Production of beef and buffalo meat includes veal; mutton and goat meat includes meat from lambs and kids; pig meat includes bacon and ham in fresh equivalent.

Poultry meat includes meat from all domestic birds and refers to dressed weight. Data for countries reporting in terms of live weight or on a ready-to-cook basis have been converted into dressed weight. Total meat production includes the data reported in Tables 83 to 87 (meat from animals slaughtered in countries, irrespective of their origin), horse meat, poultry meat, and from all other domestic or wild animals such as camels, rabbits, reindeer, game animals, etc.

Although most countries report data on a calendar-year basis, there are a few exceptions. Israel and New Zealand, for example, give data for years ending 30 September, Australia for years ending 30 June.

Cow milk, milking cows, milk yield and milk production. Data on cow milk production relate to total production of whole fresh milk, excluding the milk sucked by young animals but including amounts fed to livestock. However, Czechoslovakia, France, the Federal Republic of Germany, the German Democratic Republic, Hungary, Italy and Romania report production including milk sucked by young animals. Official statistics on cow milk production are available for most countries; where they were not available, estimates based on food consumption surveys and other indicators have been used. For a few countries where statistics on milking cows were not available, data shown in the table have been estimated on the basis of milk production and on the actual or presumed yield per cow. Yield per cow shown in the table is the result of dividing the production of milk by the number of milking cows. Milk production data shown for Israel refer to years ending 30 September; for Australia, to years ending 30 June; for New Zealand, to years ending 31 May. Also a few other minor producing countries report data for periods other than the calendar year.

Buffalo, sheep and goat milk. The concept of production reported in this table is the same as for cow milk; however, the coverage is probably less adequate.

Dairy products. Data shown for the commodities in this group refer, generally, to total production whether manufactured at milk plants or on farms. No data are available for certain countries, and the coverage reported by other countries may be underestimated, particularly as regards farm production. Naturally, continental, regional and world totals reflect the limited coverage of the data.

Data on *cheese* published in the Yearbook relate to all kinds of cheese produced: from full-fat cheese to fatless cheese, hard and soft cheese, ripe and fresh cheese, cottage cheese and curd.

Data on *butter* include ghee, which is liquid butter clarified by boiling, produced chiefly in countries of the Far East.

Eggs. Some countries have no statistics on egg production, and estimates had to be derived from such related data as chicken or total poultry numbers and reported or assumed rates of egg laying.

Most of the countries having statistics on egg production report either the total weight of eggs or numbers of eggs produced; data on numbers have been converted into weight, using official conversion factors wherever possible.

Data generally refer to total production, including eggs for hatching, in both agricultural and non-agricultural sectors.

Honey. The data presented in the table are still incomplete, particularly with regard to African and Asian countries.

Raw silk. Although data for a few producing countries are missing, data for the countries listed in the table adequately represent total world production.

Wool. Wool production statistics are generally given on a greasy basis. Such wool, however, contains from 30 to 65 percent of impurities. In order to make figures comparable, data are given also on a clean (scoured) basis.

Hides and skins. All figures refer to fresh weight of hides and skins. Data for countries reporting production in numbers or expressed in dry, cured, or salted weight have been converted into fresh weight using appropriate conversion factors. Where no official data were available, estimates based on slaughterings and on other information have been given.

FOOD SUPPLY

Since 1971, FAO has been developing an integrated and computerized system of compiling and maintaining, in the form of supply/utilization accounts (SUAs), current agricultural statistics covering about 240 primary food and agricultural commodities and 290 processed products derived therefrom, for 215 countries and territories, with data series from 1961 through 1977.

The total quantity of foodstuffs produced in a country, added to the total quantity imported and adjusted to any change in stocks that may have occurred since the beginning of the reference period, gives the *supply* available during that period. On the *utilization* side, a distinction is made between the quantities exported, fed to livestock, used for seed, put to industrial and other non-food uses, or lost during storage and transportation, and food supplies available for human consumption at the retail level, i.e., as the food leaves the retail shop or otherwise enters the household. The per caput supply of each food item available for human consumption is then obtained by dividing the food supplies available for human consumption by the related data on the population actually partaking of it. Data on per caput food supplies are expressed in terms of quantity and also, by applying appropriate food composition factors, in terms of nutrient elements (calories, proteins and fats). In the Yearbook

only nutrient elements are given to meet the immediate requirements of many users in the review and evaluation of the food and nutritional situation of the countries shown.

It is important to note that the quantities of food available relate to the quantities of food *reaching the consumer* but not necessarily to the amounts of food actually *consumed*, which may be lower than the quantity shown depending on the extent of losses of edible food and nutrients in the household, e.g., during storage, preparation and cooking, plate-waste or quantities fed to domestic animals and pets, or thrown away.

The figures shown do not give any indication of the differences that may exist in the diet consumed by different population groups, e.g., different socio-economic groups, ecological zones and geographical areas within a country, nor do they provide information on seasonal variations in the total food supply. They represent only the average supply for the population as a whole and do not indicate what is actually consumed by individuals.

Provisional detailed food balance sheets for the average period 1972-74, together with long-term series of per caput food supply by major food groups in terms of calories, protein and fat for the average period 1961-63 and for the individual years 1964 to 1974, have been prepared and sent to the countries for comments.

MEANS OF PRODUCTION

Agricultural machinery

Tractors. Data generally refer to total wheel and crawler tractors (excluding garden tractors) used in agriculture.

Milking machines. This table shows the number of milking machines or, for certain countries, the number of farms with milking machines. Data for the United Kingdom exclude milking machines in Scotland and Northern Ireland. Data for Australia refer to number of units (cow capacity).

Pesticides

Data refer generally to quantities of pesticides used in, or sold to, agriculture. (Data for Mexico refer to production; data for Nicaragua refer to imports; data for the United States refer to sales except where otherwise stated.) They are shown in terms of active ingredients except for the following countries where data refer to formulation weight: Botswana, Chile, Colombia, Costa Rica, Cuba, Ecuador, Egypt, France, German Democratic Republic, Guadeloupe, Guatemala, Hungary, Italy,

Jordan, Kuwait, Martinique, Mauritius, Morocco, Nicaragua, Nigeria, Puerto Rico, Reunion, Samoa, South Africa, Sweden, Trinidad and Tobago, Uganda, Virgin Islands (U.S.) and Yugoslavia. Formulation weight usually includes active ingredients plus diluents and adjuvants; for this reason, and because of incomplete coverage, it has not been possible to calculate totals.

When reading the tables on the products mentioned below, the following points should be kept in mind.

DDT. Data for Argentina include exports.

BHC. Data for Austria include lindane.

Lindane. Data for Hungary include BHC.

Aldrin and others. Data for Austria include toxaphene. Up to 1974, data for the United States include toxaphene.

Other chlorinated hydrocarbons. Data for Puerto Rico include all chlorinated hydrocarbons except DDT and chlordane. Data for Cuba, Ecuador, Hong Kong, Israel, New Zealand, Norway and South Africa refer to all chlorinated hydrocarbons.

Parathion. Data for Sweden refer to all organophosphorus.

Other organophosphorus. Data for Cuba, Hong Kong, Iceland, Israel, Jordan, Norway, Puerto Rico and South Africa refer to all organophosphorus.

Pyrethrum. Data for the United States refer to imports.

Other botanical insecticides. Data for Norway include all botanical insecticides. Data for the United States refer to imports.

Other insecticides. Data for Colombia, Denmark, Guadeloupe, Guatemala, Martinique, Mauritius, Morocco, Reunion, Trinidad and Tobago, Uganda and Yugoslavia include all insecticides. Data for Puerto Rico include all insecticides except those shown separately. Data for Bahrain include all insecticides except parathion.

Other fungicides. Data for Bahrain, Colombia, Cuba, Ecuador, Guadeloupe, Guatemala, Hong Kong, Jordan, Mauritius, Morocco, Puerto Rico, Reunion, South Africa, Trinidad and Tobago, Uganda and Yugoslavia include all fungicides. Data for Martinique include all fungicides but exclude copper compounds, which are shown separately.

2,4-D. Data for Canada and Norway include mixtures of 2,4-D and 2,4,5-T. Data for India include other herbicides.

Other herbicides. Data for Bahrain, Colombia, Cuba, Denmark, Ecuador, German Democratic Republic, Guadeloupe, Guatemala, Hong Kong, Martinique, Morocco, Puerto Rico, Reunion, Switzerland, Uganda and Yugoslavia include all herbicides. Data for New Zealand refer to all herbicides excluding 2,4,5-T. Data for Sweden include all herbicides except those shown separately.

Bromides. Data for Israel include fumigant carbamates.

Other fumigants. Data for Canada, the Federal Republic of Germany, Guatemala and South Africa refer to all fumigants.

Other rodenticides. Data for Denmark, Mauritius and Puerto Rico include all rodenticides. Data for the United States refer to imports.

Other pesticides. Data for Costa Rica include all insecticides and fungicides. Data for the German Democratic Republic refer to all pesticides except herbicides. Data for New Zealand refer to all insecticides (except chlorinated hydrocarbons), fungicides and fumigants.

PRICES

As a step toward the use of a single set of conversion factors in various FAO statistical publications, weighted average trade conversion factors, as published in *International Financial Statistics*, have been used as exchange rates in converting data from national currencies to U.S. dollars, thus bringing these conversions into line with the practices used for trade values.

The price series are obtained from many sources: special questionnaires, international commodity groups, official and non-official literature, and — in some cases — special correspondents. The "List of Sources for Price Tables" published in the 1960 *Production Yearbook* for the price series in that issue may still be referred to as a general guide, although it does not parallel the presentation of tables in this volume.

The local quantity units for which original prices were quoted are expressed in terms of their equivalent metric or British standard units.

COUNTRY NOTES

Berlin

The data which relate to the Federal Republic of Germany and the German Democratic Republic include the relevant data relating to Berlin, for which separate data are not available for most items.

China

Data for China generally include those for Taiwan Province. Crop and livestock production estimates for China present one of the main difficulties in preparing the *FAO production yearbook*. China is the most populated country in the world and its agricultural production, representing apparently almost 12 percent of the world total, follows that of the United States and that of the U.S.S.R.

Adequate official statistics, however, have not been published by China for the last 19 years. Consequently, the estimates shown are based on fragmentary official information reported by the Chinese mass media, and on evaluations of qualified persons who live in or visit China or neighbouring countries and territories. Trade statistics derived from publications of trading partners of China have also been taken into account, as has available information on meteorological phenomena, irrigation facilities, mechanization of agriculture, consumption of fertilizers, etc.

Cyprus

Due to the present situation of Cyprus, data for recent years might not cover the whole country.

East Timor

Data for recent years are generally included in those of Indonesia.

India and Pakistan

Data relating to Kashmir-Jammu, the final status of which has not yet been determined, are generally included in figures for India and excluded from those of Pakistan.

Data for Sikkim are included with those of India.

Jordan

Except for Table 1 (Land Use and Irrigation), data from 1967 on refer to the East Bank only.

CLASSIFICATION OF COUNTRIES BY ECONOMIC CLASSES AND REGIONS

Regional totals are given in all tables in which continental totals are shown. They appear even when identical with continental totals or easily derivable from them, in order to avoid any possible ambiguity.

The Economic Classes and Regions into which the world is divided for the purposes of FAO's analytical studies are given below.

CLASS I: DEVELOPED MARKET ECONOMIES

Region A — North America: Canada, United States.

Region B — Western Europe: Andorra, Austria, Belgium-Luxembourg, Denmark, Faeroe Islands, Finland, France, Federal Republic of Germany (incl. West Berlin), Gibraltar, Greece, Holy See, Iceland, Ireland, Italy, Liechtenstein, Malta, Monaco, Netherlands, Norway, Portugal (incl. Azores and Madeira), San Marino, Spain, Sweden, Switzerland, United Kingdom (incl. Channel Islands and Isle of Man), Yugoslavia.

Region C — Oceania: Australia, New Zealand.

Region D — Other developed market economies: Israel, Japan (incl. Bonin and Ryukyu Is.), South Africa.

CLASS II: DEVELOPING MARKET ECONOMIES

Region A — Africa: Algeria, Angola, Benin, Botswana, British Indian Ocean Territory, Burundi, Cameroon, Cape Verde, Central African Empire, Chad, Comoros, Congo, Djibouti, Equatorial Guinea, Ethiopia, Gabon, Gambia, Ghana, Guinea, Guinea-Bissau, Ivory Coast, Kenya, Lesotho, Liberia, Madagascar, Malawi, Mali, Mauritania, Mauritius, Morocco, Mozambique, Namibia, Niger, Nigeria, Reunion, Rhodesia, Rwanda, St. Helena, São Tomé and Principe, Senegal, Seychelles, Sierra Leone, Somalia, Spanish North Africa, Swaziland, Tanzania, Togo, Tunisia, Uganda, Upper Volta, Western Sahara, Zaire, Zambia.

Region B — Latin America: Antigua, Argentina, Bahamas, Barbados, Belize, Bolivia, Brazil, Cayman Islands, Chile, Colombia, Costa Rica, Cuba, Dominica, Dominican Republic, Ecuador (incl. Galapagos Islands), El Salvador, Falkland Islands (Malvinas), French Guiana, Grenada, Guadeloupe, Guatemala, Guyana, Haiti, Honduras, Jamaica, Martinique, Mexico, Montserrat, Netherlands Antilles, Nicaragua, Panama, Panama Canal Zone, Paraguay, Peru, Puerto Rico, St. Kitts-Nevis-Anguilla, St. Lucia, St. Vincent, Surinam, Trinidad and Tobago, Turks and Caicos Islands, Uruguay, Venezuela, Virgin Islands (U.K.), Virgin Islands (U.S.).

Region C — Near East: Africa: Egypt, Libya, Sudan. *Asia:* Afghanistan, Bahrain, Cyprus, Gaza Strip (Palestine), Iran, Iraq, Jordan, Kuwait, Lebanon, Oman, Qatar, Saudi Arabia, Syria, Turkey, United Arab Emirates, Yemen Arab Republic, Democratic Yemen.

Region D — Far East: Bangladesh, Bhutan, Brunei, Burma, East Timor, Hong Kong, India, Indonesia, Republic of Korea, Lao, Macau, Malaysia (Peninsular Malaysia, Sabah, Sarawak), Maldives, Nepal, Pakistan, Philippines, Singapore, Sri Lanka, Thailand.

Region E — Other developing market economies: America: Bermuda, Greenland, St. Pierre and Miquelon. *Oceania:* American Samoa, Canton and Enderbury Islands, Christmas Island (Aust.), Cocos (Keeling) Islands, Cook Islands, Fiji, French Polynesia, Gilbert Islands, Guam, Johnston Island, Midway Islands, Nauru, New Caledonia, New Hebrides, Niue Island, Norfolk Island, Pacific Islands (Trust Territ.), Papua New Guinea, Pitcairn Island, Samoa, Solomon Islands, Tokelau, Tonga, Tuvalu, Wake Island, Wallis and Futuna Islands.

CLASS III: CENTRALLY PLANNED ECONOMIES

Region A — Asia: China, Democratic Kampuchea, Democratic People's Republic of Korea, Mongolia, Viet Nam.

Region B — Eastern Europe and U.S.S.R: Albania, Bulgaria, Czechoslovakia, German Democratic Republic (incl. East Berlin), Hungary, Poland, Romania, U.S.S.R.

ALL DEVELOPED COUNTRIES

Includes developed market economies and Region B of centrally planned economies.

ALL DEVELOPING COUNTRIES

Includes developing market economies and Region A of centrally planned economies.

INTRODUCTION

Le présent volume constitue la trente et unième édition de l'*Annuaire FAO de la production*. De nombreuses innovations ont été apportées au volume précédent; trois nouveaux changements seulement sont introduits dans la présente édition. Premièrement, dans la plupart des tableaux relatifs aux cultures, à l'élevage et aux moyens de production, la moyenne quinquennale 1961-65 a été remplacée par la moyenne de trois ans 1969-71. Deuxièmement, dans les tableaux donnant les nombres-indices de la production, on a de même utilisé 1969-71 comme période de référence au lieu de 1961-65. La troisième nouveauté, la plus importante, porte également sur les indices; les coefficients de pondération utilisés pour calculer les agrégats de la production de chaque pays sont désormais des moyennes nationales des prix à la production pour la période 1969-71 et non plus les prix régionaux relatifs par rapport au blé pour la période 1961-65 comme par le passé (pour d'autres détails sur les indices, voir les Notes sur les tableaux). On notera aussi que deux nouveaux groupes de pays ont été introduits dans l'Annuaire: l'un, Tous les pays développés, comprend les pays développés à économies de marché, l'Europe orientale et l'U.R.S.S.; l'autre, Tous les pays en développement, comprend les pays en développement à économies de marché et les pays d'Asie à économies centralement planifiées.

On se rappellera que certaines données relatives aux prix et à des questions connexes qui ne sont plus publiées dans l'Annuaire peuvent être trouvées dans le *Bulletin mensuel FAO de statistiques*, l'*Annuaire statistique* et le *Bulletin mensuel de statistique* des Nations Unies, l'*Annuaire des statistiques du travail* et les *Statistiques du travail* [mensuelles] du Bureau international du travail.

On trouvera des données sur la production, le commerce et le prix des engrais commerciaux dans le *Rapport annuel sur les engrais* de la FAO.

Les statistiques relatives aux cultures et à l'élevage qui ne sont plus publiées dans l'Annuaire restent disponibles dans les dossiers de la Division de la statistique de la FAO, qui est prête à les communiquer sur demande.

Comme dans les éditions précédentes, ce volume contient aussi un grand nombre d'estimations que la FAO a établies en ce qui concerne la superficie et la production des principales cultures, ainsi que l'élevage et ses produits, lorsqu'elle ne disposait pas de sources officielles ou semi-officielles dans les pays eux-mêmes. La publication de ces estimations donne aux pays intéressés l'occasion de les examiner, et il faut espérer qu'à l'avenir ces pays fourniront à la FAO des données meilleures.

Dans presque tous les tableaux relatifs aux produits, les données se réfèrent à l'année indiquée sur la couverture de l'Annuaire. Cependant, il s'agit, dans certains cas, d'estimations de la FAO fondées sur des renseignements officieux et ayant par conséquent un caractère provisoire.

Comme par le passé, la publication de l'Annuaire a été rendue possible par le concours apporté par les gouvernements qui fournissent la plupart des renseignements en répondant aux questionnaires annuels de la FAO. Afin de normaliser davantage la présentation des statistiques internationales, la FAO continue de collaborer avec divers organismes qu'elle tient à remercier ici de leur concours.

Signes conventionnels

Les signes conventionnels sont les suivants:

*	Renseignement non officiel
—	Néant, négligeable (inférieur à la moitié de l'unité visée), ou n'ayant pas lieu de figurer
...	Renseignement non disponible
F	Estimation de la FAO
NDA	Non désigné ailleurs
HA	Hectare
MT	Tonne métrique
KG	Kilogramme
HEAD	Tête
NUMBER	Nombre
KG/HA	Kilogramme par hectare
KG/AN	Kilogramme par animal

Dans la plupart des tableaux, un espace blanc a la même signification que les signes conventionnels (—) ou (...) dont la définition vient d'être donnée.

Un point (.) sépare la partie fractionnaire du nombre entier.

Noms de pays et de produits

Dans la plupart des tableaux, les noms de pays et de produits doivent être inscrits dans 12 et 24 espaces au maximum, respectivement. Les noms des produits sont indiqués en anglais, français et espagnol; ceux des continents, pays et régions en anglais seulement. Si les noms abrégés des produits sont faciles à comprendre, ceux des pays sont bien moins clairs dans certains cas. Le lecteur peut alors consulter la Liste des pays, continents, catégories économiques et régions (page 37), où les noms apparaissent dans le même ordre que dans les tableaux. Cette liste donne les noms abrégés en anglais et les noms complets en anglais, français et espagnol.

Période de référence

Comme on a commencé à le faire dans l'édition de 1966 de l'Annuaire, les périodes de référence pour les statistiques des superficies et de la production des cultures se fondent sur l'année civile. En d'autres termes, les données relatives à une culture se rapportent à l'année civile au cours de laquelle l'intégralité ou le gros de la récolte s'effectue. Si les chiffres relatifs à la superficie et à la production des cultures se réfèrent à l'année civile, il ne s'ensuit pas nécessairement que les chiffres relatifs à la production d'un produit déterminé constituent la somme des chiffres établis mois par mois de janvier à décembre, bien qu'il en soit ainsi pour certains produits comme le thé, le sisal, les palmistes, l'huile de palme, le caoutchouc, la noix de coco et, dans certains pays, la canne à sucre et la banane, dont la récolte se poursuit presque uniformément pendant toute l'année. En revanche, la récolte d'autres produits est limitée à quelques mois, voire quelques semaines. Les pays communiquent la production de ces cultures sur des bases diverses: année civile, campagne agricole, campagne commerciale, etc.

Quelle que soit l'année statistique appliquée par les pays pour présenter les chiffres de superficie et de production, ces chiffres sont attribués, produit par produit, à l'année civile au cours de laquelle l'intégralité ou le gros de la récolte s'effectue. De toute évidence, lorsqu'un produit est récolté à la fin de l'année civile, cette récolte est utilisée principalement pendant l'année qui suit l'année civile pour laquelle les chiffres de production sont classés dans les tableaux.

Il convient de noter que l'adoption de l'année civile comme période de référence implique inévitablement que, dans un certain nombre de cas, les récoltes indiquées par le pays comme ayant été effectuées au cours d'une campagne donnée peuvent figurer sous deux années civiles différentes dans les tableaux du présent Annuaire.

Les chiffres concernant les effectifs du cheptel sont groupés en périodes de 12 mois se terminant le 30 septembre de l'année indiquée dans les tableaux; ainsi, les animaux recensés dans un pays déterminé à une date quelconque comprise entre le 1er octobre 1976 et le 30 septembre 1977 sont rapportés à l'année 1977.

En ce qui concerne les produits de l'élevage, les chiffres relatifs à la viande, au lait et aux produits laitiers se réfèrent aux années civiles, sous réserve de quelques exceptions indiquées dans les Notes sur les tableaux. Les chiffres concernant d'autres produits animaux dont la production est limitée à certaines périodes de l'année, comme le miel ou la laine, sont rapportés à l'année civile suivant les mêmes modalités que les chiffres relatifs aux cultures.

En ce qui concerne les tracteurs et les autres machines agricoles, les chiffres se rapportent autant que possible au nombre d'unités en service à la fin de l'année indiquée ou pendant le premier trimestre de l'année suivante.

Les chiffres relatifs aux produits antiparasitaires se réfèrent généralement à l'année civile.

Les nombres-indices FAO de la production agricole ont été calculés sur la base de l'année civile selon la méthode indiquée ci-dessus.

Superficie des cultures

Il s'agit en général de la superficie récoltée mais, pour les cultures permanentes, il peut s'agir de la superficie plantée totale.

Rendements à l'hectare

Les rendements à l'hectare — nationaux, continentaux, régionaux et mondiaux — sont tous exprimés en kilogrammes. Ils sont toujours calculés sur la base de chiffres détaillés de superficie et de production exprimés en hectares et en tonnes. Les rendements indiqués pour les cultures permanentes ne sont pas aussi sûrs que ceux des cultures temporaires, soit parce que la plupart des chiffres de superficie peuvent se rapporter à la superficie plantée, comme pour la vigne, soit parce que les chiffres de superficie communiqués par les pays sont peu abondants et peu fiables, comme cela est le cas pour le cacao et le café.

Totaux

On a donné les totaux continentaux, régionaux et mondiaux pour tous les produits, à l'exception des produits antiparasitaires et des trayeuses. Ces totaux ne comprennent que les chiffres qui se rapportent aux pays indiqués dans les tableaux mêmes.

En règle générale, ces totaux traduisent bien la situation des zones géographiques auxquelles ils se rapportent, à l'exception de ceux qui concernent certains fruits et légumes et certains produits de l'élevage.

On trouvera des observations plus détaillées à ce sujet dans les Notes sur les pays.

Notes sur les tableaux et sur les pays

Dans plusieurs cas, les données appellent des explications et des réserves qu'il n'était pas possible d'indiquer ici. Les détails nécessaires, ainsi que des renseignements sur les changements territoriaux, la classification des pays par région selon la FAO, etc., sont donnés plus loin dans les Notes sur les tableaux et les Notes sur les pays.

En règle générale, les données figurant dans l'Annuaire s'appliquent aux pays indiqués dans les limites de leurs frontières de fait actuelles. Pour les noms de pays et le groupement par continent, on a suivi d'une manière générale la nomenclature utilisée par le Bureau de statistique des Nations Unies.

NOTES SUR LES TABLEAUX

TERRES

Utilisation des terres et irrigation

On s'est efforcé de réunir dans ces tableaux des données sur l'utilisation des terres et sur les terres irriguées dans le monde. On ne perdra pas de vue en étudiant le chapitre sur l'utilisation des terres que les définitions dont se servent les pays varient beaucoup, et que, selon les pays, une même rubrique englobe des catégories de terres très différentes.

Les catégories de terres sont définies comme suit:

1. *Superficie totale*: celle du pays, y compris les eaux intérieures.

2. *Superficie des terres*: comme au paragraphe 1, moins la superficie des eaux intérieures. L'expression « eaux intérieures » s'entend en général des principaux cours d'eau et lacs.

3. *Terres arables*: terres affectées aux cultures temporaires (les superficies récoltées deux fois n'étant comptées qu'une fois), prairies temporaires à faucher ou à pâturer, jardins maraîchers ou potagers (y compris les cultures sous verre) et terres en jachères temporaires ou incultes.

4. *Cultures permanentes*: terres consacrées à des cultures qui occupent le terrain pendant de longues périodes et ne doivent pas être replantées après chaque récolte, comme le cacao, le café et le caoutchouc. Cette rubrique comprend les superficies couvertes d'arbustes, d'arbres fruitiers et de vignes, mais non les terres plantées en arbres destinés à la production de bois ou de grumes.

5. *Prairies et pâturages permanents*: terres consacrées de façon permanente (cinq ans au minimum) aux herbacées fourragères, cultivées ou sauvages (prairies sauvages ou pâturages).

6. *Forêts et terrains boisés*: toutes terres portant des peuplements naturels ou artificiels, qu'ils soient productifs ou non. Cette rubrique comprend les terres déboisées, mais dont le reboisement est envisagé dans un proche avenir.

7. *Autres terres*: terres non utilisées mais potentiellement productives, terrains bâtis, terres inutilisables, parcs ou jardins d'ornement, routes ou chemins, terres improductives et toutes autres terres n'entrant pas spécifiquement dans la définition des paragraphes 3 à 6 inclus.

Les données sur l'irrigation se réfèrent aux superficies irriguées volontairement — y compris les terres couvertes par les crues — à des fins de culture ou pour améliorer les pâturages, qu'elles aient été irriguées plusieurs fois ou une seule fois dans l'année.

En ce qui concerne les catégories relatives à l'utilisation des terres et à l'irrigation, les notes se rapportant à des pays particuliers sont les suivantes:

SUPERFICIE TOTALE

Groenland: les données se rapportent aux superficies non couvertes de glace.

Maurice: les données ne comprennent pas les dépendances

Namibie: les données comprennent le territoire de Walvis Bay.

Nouvelle-Calédonie: les données comprennent les dépendances.

Afrique du Sud: les données ne comprennent pas le territoire de Walvis Bay.

U.R.S.S.: les données comprennent la mer Blanche (9 000 000 d'hectares) et la mer d'Azov (3 730 000 hectares).

TERRES ARABLES ET CULTURES PERMANENTES

Australie: les données sur les terres arables comprennent environ 27 000 000 d'hectares d'herbages cultivés.

Cuba: les données se rapportent uniquement au secteur d'Etat.

Portugal: les données comprennent environ 800 000 hectares de cultures temporaires associées à des cultures permanentes ou à des forêts.

PRAIRIES ET PÂTURAGES PERMANENTS

Australie: les données se rapportent à la superficie des autres exploitations rurales.

Egypte: les pâturages sauvages sont inclus dans « Autres terres ».

U.R.S.S.: les données ne comprennent pas les pâturages pour les rennes.

Pour les pays suivants, les données se rapportent uniquement aux prairies et pâturages permanents des exploitations agricoles: Chili, République Dominicaine, Finlande, Guatemala, Kenya, République de Corée, Surinam, Trinité-et-Tobago, Uruguay.

IRRIGATION

France: les données ne comprennent pas les jardins maraîchers et potagers.

Hongrie: les données ne comprennent pas les parcelles complémentaires et les exploitations individuelles.

Royaume-Uni: les données ne comprennent pas l'Ecosse et l'Irlande du Nord.

Pour les pays suivants, les données se rapportent aux terres où il existe des installations d'irrigation: Bulgarie, Norvège, Roumanie, Surinam.

Pour les pays suivants, les données se rapportent uniquement au riz irrigué: Japon, République de Corée, Sri Lanka.

POPULATION

Le tableau 3 fournit, pour la *population totale*, la *population agricole* et la *population active* (totale et agricole) de chaque pays, des estimations se rapportant aux années 1965, 1970, 1975, 1976 et 1977.

L'*Annuaire démographique* des Nations Unies donne des séries chronologiques d'estimations de la population totale. Ces données proviennent généralement des pays mais, dans certains cas, le Bureau de statistique des Nations Unies ajuste les estimations dont il dispose ou en prépare de nouvelles. Pour nombre de pays en développement, il faut toutefois procéder à un nouvel ajustement des estimations disponibles pour assurer une certaine uniformité. Il arrive que la série d'estimations annuelles accuse elle-même des écarts, celles-ci étant soumises à des fluctuations que l'on ne peut expliquer par la migration. Quelquefois, les estimations ne correspondent pas aux données provenant de sources extérieures, c'est-à-dire principalement des recensements et des statistiques démographiques. Les séries chronologiques d'estimations présentent aussi des lacunes qu'il est nécessaire de combler. Aussi la Division de la population des Nations Unies a-t-elle préparé une série complète d'estimations pour chaque pays. Cette série, qui couvre une période assez longue, est uniforme dans ses variations d'année en année et cadre avec d'autres données démographiques [1]. Les données présentées dans ce tableau sont tirées de cette série et assorties des ajustements voulus pour incorporer les chiffres officiels les plus récents sur la population, notamment pour les pays développés dont les statistiques démographiques sont jugées sûres.

Les statistiques se rapportent en général à la population dénombrée dans le pays ou territoire à l'intérieur des limites géographiques actuelles (de facto).

Par population agricole, on entend toutes les personnes dont l'agriculture constitue le moyen d'existence. Celle-ci comprend toutes les personnes qui se livrent effectivement à l'agriculture ainsi que les personnes à leur charge qui ne travaillent pas.

Par population active, on entend toutes les personnes qui se livrent à une activité économique ou cherchent un emploi: employeurs, personnes travaillant pour leur propre compte, salariés ou aides familiaux non rémunérés.

La population agricole active comprend l'ensemble des personnes occupées économiquement et de manière principale dans l'agriculture, la foresterie, la chasse et la pêche.

Des renseignements relatifs à la population active et à ses ventilations par industrie ou occupation sont disponibles à partir des recensements nationaux de la population et des enquêtes sur la main-d'œuvre. La comparabilité des données est toutefois réduite du fait que les statistiques nationales ne traitent pas toutes de la même manière certaines catégories, par exemple les aides familiaux non rémunérés et surtout les femmes au foyer. En outre, les statistiques de l'emploi se rapportent parfois aux personnes de tous âges; d'autres fois elles se limitent à un groupe d'âge déterminé, par exemple au-dessus de 14 ans. Le BIT a systématiquement évalué ces données qu'il a normalisées, compte tenu des concepts adoptés sur le plan international, et a établi des estimations de la population active par secteur: agriculture, industrie et services, pour toute cinquième année de la période 1950 à 1970, et des projections de la population active totale pour toute cinquième année de la période 1970 à 2 000 [2]. Les données contenues dans le tableau sont fondées sur ces estimations et projections.

Les statistiques sur la population agricole établies d'après les recensements et enquêtes des différents pays sont peu nombreuses. Pour obtenir les estimations de la population agricole présentées dans ce tableau, la FAO s'est largement fondée sur le lien étroit existant entre le rapport population agricole active/population active totale et le rapport population agricole/population totale. On a en général retenu l'hypothèse que ces rapports sont équivalents, de sorte que, si l'on connaît le rapport population agricole active/population active totale, la population agricole estimée est le produit de ce rapport et de la population totale [3].

Etant donné que les estimations de la population active et de la population agricole dans les petits pays — c'est-à-dire en général les pays dont la population totale s'élevait au maximum à 250 000 personnes en 1970 — sont très sommaires, ces données ne sont pas indiquées séparément; toutefois, les totaux continentaux, régionaux et mondiaux ont été ajustés de façon à englober également ces pays.

---

[1] Nations Unies, *Single-year population estimates and projections of the major areas, regions and countries of the world, 1950-2000*, multicopié, ESA/P/WP.34, New York, 1975.

[2] BIT, *Labour Force, 1950-2000*, Vol. I, II, III, IV, Genève, 1977.
[3] Pour les détails concernant les méthodes et hypothèses ayant servi à l'élaboration de ces estimations, voir FAO. Projections de la population agricole mondiale, *Bulletin mensuel: économie et statistique agricoles*, vol. 21, n° 1, 1972, p. 3-4.

NOMBRES-INDICES FAO DE LA PRODUCTION AGRICOLE

Les nombres-indices indiquent le niveau relatif du volume global de la production agricole pour chaque année, par rapport à la période de référence 1969-71. Ils sont calculés en faisant la somme pondérée par le prix de la production des différents produits agricoles, après déduction des quantités utilisées comme semences ou pour l'alimentation animale, pondérées de même. L'agrégat ainsi obtenu représente donc la production disponible pour toutes les utilisations, sauf comme semences ou aliments pour les animaux.

Les quantités utilisées comme semences (ou, dans le cas des œufs, les œufs à couver) et comme aliment pour le bétail ou la volaille sont déduites aussi bien de la production nationale que des importations. Il s'agit de produits agricoles primaires utilisés tels quels (maïs, pommes de terre, courges, etc.) et d'aliments semi-transformés (son, tourteaux d'oléagineux, farine, mélasse, poudre de lactosérum, etc.). Les quantités utilisées comme semences et pour l'alimentation animale sont déduites des chiffres de production parce que les indices de la FAO étant calculés comme si l'agriculture mondiale était une exploitation unique, en évaluant la production agricole, on évite autant que possible de compter deux fois les semences ou les aliments pour animaux, une fois comme production primaire et une fois avec les cultures et le bétail qui en dérivent.

Les nombres-indices relatifs aux pays sont calculés au moyen de la formule de Laspeyres. Les quantités de chaque produit sont pondérées par les prix moyens nationaux à la production pour la période 1969-71 et additionnées pour chaque année. L'agrégat pour une année déterminée est divisé par l'agrégat moyen de la période 1969-71 pour obtenir le nombre-indice. Les indices relatifs aux continents, aux régions et au monde sont calculés en faisant la somme des agrégats des pays, convertis à cet effet en dollars des Etats-Unis. Les taux de change des monnaies nationales en dollars sont, à quelques exceptions près, ceux qui sont publiés par le Fonds monétaire international. Ces exceptions sont les pays d'Europe orientale, l'U.R.S.S., la Chine et les autres pays d'Asie à économies centralement planifiées, pour lesquels les taux de change sont dérivés des taux non commerciaux, touristiques et autres cours non fixés.

Les produits qui entrent dans le calcul des nombres-indices de la production agricole sont tous les produits de l'agriculture et de l'élevage obtenus dans tous les pays pour lesquels on possède des informations. Les nombres-indices de la production alimentaire comprennent les produits considérés comme comestibles et contenant des éléments nutritifs. En conséquence, outre les produits non comestibles, ils excluent le café et le thé car, bien que ces produits entrent dans l'alimentation humaine, ils n'ont pratiquement aucune valeur nutritive.

Les nombres-indices sont calculés à partir des données de la production présentées sur la base de l'année civile.

Dans le présent volume de l'Annuaire, les indices présentés diffèrent de ceux des éditions précédentes principalement à trois égards: 1. Les chiffres de production utilisés pour calculer les nombres-indices portent désormais, à de très rares exceptions près, sur les produits primaires. En conséquence, au lieu de la production de sucre, d'huile d'olive, de coprah, de raisins secs et de vin qui entrait autrefois dans le calcul des nombres-indices, on a utilisé la production de canne à sucre, de betteraves à sucre, d'olives, de noix de coco et de raisins. Comme produits transformés, on a conservé l'huile de palme, les palmistes, les graines de coton et les fibres de coton. La production d'amandes de palme n'a pas pu être utilisée parce que l'on manquait de chiffres tant sur la production que sur les prix, non plus que celle de coton à graines parce qu'il s'agit d'un produit à la fois alimentaire et non alimentaire. 2. Comme période de référence, on a utilisé la moyenne 1969-71 au lieu de la moyenne 1961-65. 3. La principale innovation est l'utilisation comme coefficient de pondération pour calculer la production globale de chaque pays des prix moyens *nationaux* à la production (1969-71) au lieu des prix moyens *régionaux* relatifs par rapport au blé (1961-65) dans chaque pays de la région.

Les indices de la FAO peuvent être différents de ceux qui sont publiés par les pays eux-mêmes en raison de divergences dans la définition de la production, du champ d'application, dans les coefficients de pondération, dans la période de référence des données et dans les méthodes de calcul.

CULTURES

Sommaire statistique de la production agricole mondiale et régionale

Les chiffres relatifs à la production de *pommes* ne comprennent pas la production de l'U.R.S.S. Pour les *cultures oléagineuses*, les chiffres donnés représentent la *production totale* de graines, noix et autres oléagineux récoltés pendant les années indiquées et sont *exprimés en équivalent d'huile*. En d'autres termes, ces chiffres ne se rapportent pas à la production effective mais à la production potentielle d'huiles végétales, dans l'hypothèse où les volumes totaux de produits provenant de toutes les cultures d'oléagineux seraient transformés en huile dans les pays producteurs l'année même où ils ont été récoltés. Bien entendu, la production totale d'oléagineux n'est jamais transformée intégralement en huile, car des quantités importantes, qui varient suivant les cultures, sont également utilisées pour les semailles, l'alimentation animale et l'alimentation humaine. D'un autre côté, bien que les taux d'extraction d'huile varient selon les pays, on a appliqué dans ce tableau le même taux à tous les pays pour chaque oléagineux. En outre, il ne faut pas oublier que les produits récoltés au cours des derniers mois de l'année sont généralement transformés en huile dans le courant de l'année suivante. En dépit de ces imperfections qui concernent le champ d'application, les taux d'extraction et les

références chronologiques, les chiffres présentés ici sont utiles, car ils donnent une indication valable des variations de volume que la production totale d'oléagineux enregistre d'une année à l'autre. La production mondiale effective d'huiles végétales atteint 80 pour cent environ de la production indiquée ici. Les chiffres relatifs à la production de *viande* correspondent à la production totale des animaux abattus dans les pays, quelle que soit leur origine.

Céréales

Les données sur la superficie et la production se rapportent uniquement aux céréales récoltées pour le grain; celles cultivées pour le foin, le fourrage vert et l'ensilage ou le pâturage en sont exclues. Les statistiques de superficies s'entendent des superficies récoltées. Bien que certains pays fassent seulement état des superficies ensemencées ou cultivées, celles-ci n'y diffèrent pas sensiblement, pendant les années normales, des superficies effectivement récoltées, soit parce que la quasi-totalité de la superficie ensemencée est récoltée, soit parce que les enquêtes sur les superficies s'effectuent au moment de la moisson.

Céréales, total. Les données figurant dans ce tableau comprennent également les statistiques relatives à d'autres céréales qui ne sont plus publiées dans l'Annuaire, par exemple: mélanges de céréales, sarrasin, etc.

Blé et épeautre. Les données disponibles sur l'épeautre sont comprises avec le blé, sauf pour l'U.R.S.S.

Millet et sorgho. Le millet et le sorgho, qui sont surtout des aliments du bétail et de la volaille en Europe et en Amérique du Nord, s'emploient beaucoup dans l'alimentation humaine en Asie, en Afrique et en U.R.S.S.

Chaque fois que possible, on a donné des chiffres distincts pour le millet et pour le sorgho, mais de nombreux pays (notamment en Afrique) groupent ces deux céréales dans leurs rapports, auquel cas on a indiqué un chiffre global.

Racines et tubercules

Racines et tubercules, total. Les données figurant dans ce tableau comprennent également les statistiques relatives à d'autres racines et tubercules qui ne sont plus publiées dans l'Annuaire, par exemple: taros, ignames, etc.

Manioc. Le manioc (*Manihot esculenta*, Crantz) est une plante racine communément divisée en deux variétés — l'une amère, l'autre douce — parfois considérées comme deux espèces différentes sous les noms de *M. utilissima* et *M. dulcis* ou *aipi*. Les données figurant dans le tableau se rapportent à l'ensemble des deux variétés.

Légumineuses

Légumineuses sèches, total. Les données figurant dans ce tableau comprennent également les statistiques relatives à d'autres légumineuses qui ne sont plus publiées dans l'Annuaire, par exemple: pois à vache, vesces, etc.

Pour autant qu'on sache, les tableaux indiquent la production totale des cultures récoltées en grains secs, sans considérer si elle était utilisée pour l'alimentation humaine ou pour celle du bétail.

Haricots secs. Pour quelques pays européens, les données des superficies et de la production des haricots secs cultivés en association avec d'autres plantes ont été groupées avec celles concernant les haricots en monoculture. Les données des superficies intéressant ces pays ont été surestimées et c'est pourquoi les rendements par hectare apparaissent extrêmement faibles.

Graines, noix et amandes oléagineuses

Graines de colza. La production de la Suède est indiquée avec une teneur en eau de 18 pour cent. Dans le cas de quelques pays, dont l'Inde et le Pakistan, les chiffres fournis englobent aussi les graines de moutarde.

Graines de lin. Pour l'U.R.S.S. et quelques pays dont la production est peu importante, les statistiques des superficies concernent le lin cultivé à la fois pour la graine et la fibre.

Graines de coton. Un certain nombre de pays, dont la production représente environ 60 pour cent de la production mondiale, communiquent des statistiques directes de la production de graines de coton. Pour le reste, les chiffres ont été calculés à partir de la production de coton égrené, à laquelle on a appliqué des coefficients obtenus antérieurement, soit dans lesdits pays, soit dans les pays où les conditions étaient comparables.

Huile d'olive. A de rares exceptions près, les données concernant les pays européens se rapportent à la production totale d'huile, y compris celle extraite de résidus d'olives.

Noix de coco. Les données fournies portent sur la production totale de noix de coco, qu'elles soient mûres ou non, consommées fraîches ou transformées en coprah ou en noix séchées. La production est exprimée en poids de la noix entière, à l'exclusion uniquement de la coque extérieure fibreuse.

Légumes et melons

Les données figurant aux tableaux se rapportent aux légumes cultivés principalement pour la consommation humaine. Les légumes comme les choux, les potirons, les carottes, etc., cultivés expressément pour l'alimentation animale, sont donc exclus. Peu de pays disposent de statistiques sur les légumes et ces statistiques couvrent des champs divers selon les pays. D'une manière générale, il semble que les estimations se rapportent à la production de plein champ et maraîchère destinée surtout à la vente; en sont donc exclues les cultures pratiquées dans les potagers et les petits jardins familiaux et principalement destinées à la consommation familiale. Les chiffres fournis par l'Autriche concernent uniquement les cultures de plein champ, tandis que ceux communiqués par Cuba se rapportent exclusivement à la production des fermes d'Etat et privées, et qu'en U.R.S.S. ils s'appliquent aux fermes collectives et d'Etat. La production des jardins familiaux et autres petits potagers, non relevée dans les enquêtes statistiques actuelles et exclue par conséquent des tableaux de l'Annuaire, représente une partie importante de la production totale estimée dans certains pays: par exemple environ 40 pour cent en Autriche et en République fédérale d'Allemagne; plus de 15 pour cent en Italie.

Aussi, les totaux continentaux, régionaux et mondiaux figurant dans le présent Annuaire sont-ils loin d'être représentatifs de la superficie et de la production totales des diverses sortes de légumes. Les statistiques de la production figurant au tableau 40 comprennent les données publiées dans les tableaux consacrés par le présent volume aux divers légumes, ainsi que les statistiques sur toutes autres espèces de légumes. Ils comportent aussi des estimations sur des pays ne fournissant pas d'information et, le cas échéant, la production des cultures non commerciales pour les pays faisant uniquement état de la production destinée à la vente.

Choux. Sont incluses dans le tableau les principales variétés de choux ci-après: chou rouge, chou blanc, chou frisé de Milan, chou de Chine, chou de Bruxelles, chou frisé vert et brocoli branchu.

Tomates. Les données pour la Belgique, le Danemark, la Finlande, la Norvège, la Suède et le Royaume-Uni se rapportent aux tomates cultivées principalement ou entièrement en serres. Ceci explique les plus forts rendements à l'hectare obtenus dans ces pays.

Choux-fleurs. Lorsque cela a été possible, on a également inclus dans les données des chiffres concernant les brocolis pommés.

Concombres et cornichons. Près de la moitié des pays producteurs européens, ainsi que le Canada et la Nouvelle-Zélande, cultivent tout ou partie de ces produits en serres, ce qui explique les rendements à l'hectare.

Haricots récoltés verts. Les données concernent les haricots (*Phaseolus* et *Dolichos* seulement) récoltés verts mais ne comprennent pas les haricots mange-tout, au moins pour les pays tels que la France et les Etats-Unis qui publient des statistiques séparées pour les haricots à écosser et les haricots mange-tout. Les données sur les haricots récoltés verts aux fins de transformation, communiquées par quelques pays en poids écossé, ont été converties en poids non écossé à raison de 200 pour cent environ.

Petits pois récoltés verts. Ces données se rapportent aux petits pois (*Pisum sativum* et *P. arvense*) récoltés verts. Pour quelques pays, les données exprimées à l'origine en poids écossé ont été converties en poids non écossé à raison de 225-250 pour cent.

Cantaloups et autres melons. Pour la Roumanie, les données comprennent les pastèques.

Pastèques. Les données pour l'Algérie, la Bulgarie, la Turquie et la Yougoslavie comprennent les melons; pour l'U.R.S.S., les melons (environ 18 pour cent), ainsi que les potirons et les courges (30 pour cent environ).

Raisins de table et vin

Raisins. Certains pays, dont l'Algérie, l'Autriche, le Chili, la France, la République fédérale d'Allemagne et l'Iran, pour ne mentionner que les producteurs principaux, ne publient pas de données sur la production totale de raisins. Les estimations fournies par le tableau pour ces pays se fondent donc sur les informations disponibles concernant la production de raisins de table, de raisins secs et de vin. Dans le cas de l'Italie, les statistiques des superficies englobent les superficies cultivées en raisins de table en association avec d'autres cultures. La superficie sous culture mixte représente 23,5 pour cent de la superficie totale plantée en vignes.

Vin. Dans la plupart des pays grands producteurs de vin, la production est dérivée des quantités de raisins pressés à la vendange; elle correspond donc à la récolte de « raisins à cuve » de la même campagne et représente la production totale au pressoir, indépendamment de son usage ultérieur (consommation comme vin ou vinaigre, ou distillation). On n'a malheureusement pas pu obtenir de tous les pays des statistiques établies sur cette base et il a fallu combler les lacunes en se servant des déclarations fiscales ou des estimations commerciales. Certains pays ne publient aucune statistique sur la production de vin ou fournissent des données peu sûres, soit qu'elles n'englobent pas la production totale de vin, soit qu'elles comprennent aussi des mélanges de vins et de jus de fruits. En pareil cas, on a estimé la production de vin à partir des quantités de raisins à cuve envoyées au pressoir, lorsque l'on disposait de ces informations.

Canne à sucre, betteraves à sucre et sucre

Canne à sucre et betteraves à sucre. Pour la canne à sucre et les betteraves à sucre, les statistiques des superficies et de la production se rapportent en général à toutes les cultures récoltées, sauf dans le cas où les cultures sont expressément destinées à l'alimentation animale. La plupart de la production sert à la fabrication de sucre centrifugé et non centrifugé. Cependant, dans plusieurs pays, d'importantes quantités de canne à sucre servent également aux usages suivants: semences, aliments du bétail, consommation à l'état frais, fabrication d'alcool et autres utilisations; les betteraves à sucre sont utilisées comme aliments du bétail et pour la fabrication d'alcool.

Sucre centrifugé. Le sucre de betterave ou de canne centrifugé est indiqué autant que possible en équivalent de sucre brut, comme le signalent les pays. Il n'est pas certain toutefois que tous les pays rendent compte de leur production de sucre brut sur la base d'un taux de raffinage de 96°, comme demandé dans les questionnaires de la FAO. C'est ainsi que l'Australie fait état de sa production de sucre sur la base d'un taux de raffinage de 94°. Pour deux pays, à savoir Haïti et Indonésie, on a donné une valeur de sucre telle quelle, c'est-à-dire le poids effectif de tout le sucre centrifugé produit. Les données communiquées par les pays en termes de sucre raffiné ont été converties en sucre brut à raison de 108,7 pour cent.

Sucre non centrifugé. Ce tableau fait état de tout le sucre tiré de la canne n'ayant pas subi le processus de centrifugation. La quasi-totalité du sucre non centrifugé sert à la consommation locale.

Fruits et baies

Les données concernent la production totale de fruits frais utilisés, soit directement pour la consommation humaine ou animale, soit pour la fabrication de divers produits: fruits séchés, jus de fruits, confitures, alcools, etc.

Les statistiques sur les fruits, notamment les fruits tropicaux, manquent dans nombre de pays, tandis que celles fournies par d'autres ne sont pas uniformes. De manière générale, les données de production se rapportent aux cultures pratiquées en plein champ et dans les vergers, principalement en vue de la vente. Les pays ne relèvent généralement pas la production des arbres disséminés essentiellement destinée à la consommation familiale, pas plus que les services statistiques nationaux ne recueillent de données sur la production des plantes sauvages, bien que, dans certains pays, celle-ci ait quelque importance. Aussi les chiffres fournis dans le présent Annuaire pour les divers fruits et baies sont-ils assez incomplets, notamment en ce qui concerne les régions autres que l'Europe, l'Amérique du Nord, l'Australie et la Nouvelle-Zélande. Bien que les totaux figurant dans la plupart des tableaux ne se rapportent qu'au nombre limité de pays énumérés, ils n'en donnent pas moins une indication valable sur ces cultures, dans la mesure où elles influent sur le commerce international. De toute façon, ces totaux donnent une idée des fluctuations annuelles auxquelles est soumise la production de ces cultures.

Les statistiques de la production figurant au tableau 40 englobent les données publiées dans les tableaux individuels sur les fruits et les baies, ainsi que des statistiques sur toutes autres espèces de fruits et baies. La production de dattes, de bananes plantains et de raisins de tous types est également incluse dans les chiffres totaux pour tous fruits, alors que la production d'olives en est exclue. Les chiffres indiqués dans ce tableau sont plus complets que ceux publiés pour les produits respectifs, car ils comprennent des estimations pour la plupart des pays ne fournissant pas de données, ainsi que des renseignements sur les pays qui, comme l'U.R.S.S. notamment, signalent leur production totale de fruits par un seul chiffre, sans préciser les espèces.

Oranges. Pour la Guinée, la Sierra Leone, le Swaziland, l'U.R.S.S. et quelques autres pays petits producteurs, les données se rapportent à la production de tous les agrumes. Il se peut que, dans le cas de quelques autres pays, les données englobent les tangerines.

Tangerines, mandarines, clémentines et satsoumas. Dans le cas des Etats-Unis, les chiffres donnés englobent les « tangelos », hybrides de tangerines et de pamplemousses, et les « temples », hybrides d'oranges douces et de tangerines.

Citrons et limes. Pour la Bolivie et le Pérou, les chiffres indiqués englobent les pamplemousses.

Agrumes n.d.a. Les données pour le Japon comprennent les pamplemousses et les citrons. Pour d'autres pays, les chiffres se rapportent le plus souvent à la production de tous les agrumes ou agrumes non identifiés.

Bananes et plantains. Les données sur les bananes se rapportent autant que possible à toutes les variétés fructifères comestibles du genre *Musa*, à l'exception des fruits de *Musa paradisiaca*, communément dite banane plantain. Malheureusement, plusieurs pays ne font pas de distinction dans leurs statistiques entre la banane proprement dite et la banane plantain et publient seulement des estimations globales pour les deux espèces. Dans les cas de ce genre, lorsqu'il existe certaines indications ou que l'on suppose que les statistiques fournies se rapportent principalement aux bananes, celles-ci sont incluses dans le tableau. Il convient de noter qu'aucun pays exclu du tableau n'est un exportateur important. Il est également difficile de comparer, pour les bananes et les plantains, les données de production fournies par divers pays, car un certain nombre d'entre eux donnent le chiffre des régimes, ce qui signifie généralement que la tige est comprise dans le poids.

Framboises. Certaines statistiques sur les framboises (*Rubus idaeus*) semblent englober d'autres baies du genre *Rubus*, comme la mûre, la ronce-framboise et la mûre de ronce.

Groseilles. Les statistiques sur les groseilles englobent les fruits *Ribes rubrum*, *album* et *nigrum*.

Fruits à coque

La production de fruits à coque (y compris les châtaignes) se rapporte aux fruits à coque non écalés. Les statistiques sur ces fruits sont très limitées et s'appliquent le plus souvent aux seuls fruits à coque cultivés pour la vente.

Outre les six espèces de fruits à coque figurant dans les tableaux distincts de l'Annuaire, les statistiques de la production présentées au tableau 40 englobent tous les autres fruits à coque consommés principalement comme dessert ou fruits de table, telles les noix du Brésil, de pili, de sapucaia, de macadamia, etc. Les noix utilisées principalement pour aromatiser des boissons sont donc exclues de ce tableau, comme le sont les noix masticatoires et toniques et celles utilisées principalement pour l'extraction de l'huile ou de beurre, par exemple noix d'arec/bétel, cola, illipe, karité, noix de coco, noix d'abrasin, graine d'éléis, etc.

Café, cacao et thé

Les chiffres de la production de café se rapportent aux grains verts. Pour quelques pays fournissant des données exprimées en cerises ou en café en parche, les chiffres ont été convertis en café nettoyé à l'aide de coefficients de conversion appropriés.

Pour le cacao, les statistiques de la production concernent les fèves de cacao fermentées et séchées. Pour le café et le cacao, les statistiques des superficies sont disponibles seulement pour certains pays et ne sont pas toujours fiables. Aussi les données sur les rendements à l'hectare ne sont-elles pas très significatives.

Dans le cas du Brésil, les données sur la production de café et de cacao représentent les chiffres officiels publiés dans l'Annuaire statistique de ce pays. Les données sur le café communiquées en termes de cerises ont été converties en café vert à raison de 50 pour cent.

Les chiffres de production de thé concernent le thé manufacturé. Toutefois, pour l'Indonésie, un tiers environ de la production est exprimé en poids de feuilles vertes. Chaque année, la Birmanie produit environ 45 000 tonnes de feuilles de thé. Cependant, on n'a pas fait figurer ces quantités dans le tableau étant donné que la presque totalité de la production est consommée fraîche comme légume.

Houblon

Les données de la production se rapportent au poids de cônes séchés, sauf pour l'Espagne où elles s'appliquent au poids vert.

Tabac

Dans la mesure où l'on a pu le déterminer, les chiffres de la production concernent le poids des ventes à la ferme. Les chiffres disponibles sur la base du poids sec ont donc été convertis en poids des ventes à la ferme à raison de 90 pour cent environ.

Fibres

Lin (fibre). Les statistiques se rapportent généralement au lin teillé et peigné et elles comprennent l'étoupe. Pour les pays indiquant leur production en termes de paille ou de lin roui, on a converti ces chiffres en fibre et étoupe pour pouvoir les comparer avec les données émanant d'autres pays.

Chanvre (fibre). Comme pour le lin, les données sur le chanvre se rapportent au chanvre teillé et comprennent l'étoupe. Dans le cas du Bangladesh, de l'Inde et du Pakistan, les chiffres s'appliquent au chanvre sunn (Crotalaria juncea), alors que pour les autres pays ils s'appliquent au véritable chanvre (Cannabis sativa).

Jute et substituts. Les fibres de jute s'obtiennent de Corchorus capsularis et C. olitorius. Parmi les fibres apparentées au jute figurent un certain nombre de produits de remplacement du jute, les principaux étant le kénaf ou mesta et la roselle (Hibiscus spp.) et le jute du Congo ou paka (Urena lobata).

Sisal. Les données sur le sisal englobent les fibres et les déchets d'Agave sisalana. Les statistiques sur les superficies sont le plus souvent très approximatives, même lorsqu'elles sont officielles.

Coton (fibre). Le tableau consacré au coton a été établi en coopération avec le Comité consultatif international du coton. Pour la plupart des pays, les chiffres sont ceux qui ont été communiqués officiellement comme représentant la production de coton égrené et comprennent les linters. Lorsque les renseignements reçus concernaient le coton non égrené et que l'on ne disposait pas d'un coefficient spécifique du rendement en coton égrené, on a considéré que ce rendement était un tiers.

Fibres n.d.a. Les principales fibres végétales figurant dans ce tableau sont les suivantes: Aloès (Furcraea gigantea), lin de Nouvelle-Zélande (Phormium tenax), fique (Furcraea macrophylla), caroa (Neoglazovia variegata), tampico (Samuela carnerosana), ramie et rhea (Boehmeria spp.), kapok (Ceiba pentandra), fibre de coco (fibre contenue dans la coque des noix de coco), etc. Sont exclues les fibres d'agaves et d'abaca (chanvre de Manille).

Caoutchouc naturel

Le tableau sur le caoutchouc a été préparé en coopération avec le Groupe international d'étude du caoutchouc, d'après lequel le caoutchouc naturel (Hevea spp.) doit s'entendre du poids séché de latex, à l'exclusion du balata, de la gutta-percha et des diverses gommes analogues au caoutchouc, ainsi que du caoutchouc de récupération. Ces derniers produits servent en général à des usages complètement différents de ceux du caoutchouc naturel.

ÉLEVAGE ET PRODUITS DE L'ÉLEVAGE

Effectifs du cheptel

Les statistiques sur les effectifs du cheptel figurant au présent Annuaire s'entendent de tous les animaux domestiques, quel que soit leur âge, leur emplacement ou le but de leur élevage. Des estimations ont été établies pour les pays n'ayant pas communiqué de renseignements comme pour ceux ayant fourni des données partielles. Dans certains pays toutefois, les statistiques sur les poules, les canards et les dindons ne représentent pas encore, semble-t-il, l'effectif total de ces volatiles. D'autres pays donnent un seul chiffre pour toute la volaille. Les statistiques pour ces pays figurent à la rubrique « Poules ».

Produits de l'élevage

Abattages, poids carcasse, production de viande. Les tableaux 83 à 87 indiquent, pour les principales espèces, le nombre des animaux abattus, le poids carcasse parée moyen et la production correspondante de viande. Les données contenues dans ces tableaux se rapportent aux animaux abattus sur un territoire national, indépendamment de leur origine. Les données du tableau 88 concernant la production de viande équine, de viande de volaille et des viandes en général se rapportent aux animaux abattus dans les divers pays, abstraction faite de leur provenance.

Dans le tableau 89, en revanche, les chiffres relatifs à la production de viande se rapportent aux animaux indigènes, c'est-à-dire qu'ils comprennent l'équivalent en viande des animaux exportés sur pied mais non l'équivalent en viande des animaux importés sur pied.

Toutes les statistiques se rapportent à la production totale de viande, provenant tant des animaux abattus commercialement que des animaux sacrifiés à la ferme. Les statistiques sont données en termes de poids carcasse parée, après enlèvement des abats et de la graisse d'abattage. Les statistiques de la production comprennent, pour la viande de bœuf et de buffle, celle de veau, pour la viande de mouton et de chèvre, celle des agneaux et chevreaux et, pour la viande de porc, le bacon et le jambon en équivalent de viande fraîche.

La viande de volaille comprend la viande de tous les volatiles de basse-cour et est exprimée en poids paré. Pour les pays fournissant des statistiques en termes de poids vif ou sur la base de « prêt à cuire », on a converti les chiffres en poids paré. La production totale de viande comprend les chiffres donnés aux tableaux 83 à 87 (viande fournie par des animaux abattus dans le pays, sans considération d'origine), ainsi que la viande de cheval, la viande de volaille et celle de tous autres animaux domestiques ou sauvages (chameaux, lapins, rennes, gibier, etc.).

La plupart des pays communiquent leurs données sur la base de l'année civile, mais il existe quelques exceptions. Les statistiques d'Israël et de la Nouvelle-Zélande, par exemple, portent sur des campagnes se terminant le 30 septembre et celles de l'Australie sur des campagnes se terminant le 30 juin.

Lait de vache, vaches laitières, rendement laitier et production de lait. Par rendement laitier on entend la production totale par vache laitière de lait frais, y compris le lait donné aux jeunes animaux, mais non celui qu'ils ont tété. Toutefois, quelques pays, dont la Tchécoslovaquie, la France, la République fédérale d'Allemagne, la République démocratique allemande, la Hongrie, l'Italie et la Roumanie, font figurer dans leur production laitière le lait tété au pis par les jeunes animaux. La plupart des pays disposent de statistiques officielles sur la production de lait de vache; dans les cas où ces statistiques font défaut, on a inclus des estimations fondées sur des enquêtes de consommation alimentaire ainsi que d'autres indicateurs. Pour les rares pays ne relevant pas de statistiques sur les vaches laitières, les données figurant au tableau ont été dégagées en se fondant sur la production laitière et sur le rendement effectif ou supposé par vache. Le rendement par vache indiqué dans le tableau a été obtenu en divisant la production de lait par le nombre de vaches laitières.

Les statistiques de production laitière se rapportent pour Israël aux années se terminant le 30 septembre, pour l'Australie aux années se terminant le 30 juin et pour la Nouvelle-Zélande aux années se terminant le 31 mai. Quelques autres pays petits producteurs fournissent également des données s'appliquant à des années autres que l'année civile.

Lait de bufflonne, de brebis et de chèvre. Dans ce tableau, on a adopté la même conception de la production que pour le lait de vache; il est probable toutefois que les statistiques sont dans ce cas moins précises.

Produits laitiers. Les statistiques relatives aux produits de ce groupe se rapportent généralement à la production totale, qu'il s'agisse de celle produite en usine ou dans les exploitations. Pour certains pays, les données font absolument défaut, tandis que, pour d'autres, les données communiquées risquent d'être sous-estimées, notamment en ce qui concerne la production sur l'exploitation. Aussi les totaux continentaux, régionaux et mondiaux reflètent-ils ces lacunes.

Les données sur le *fromage* publiées dans l'Annuaire se rapportent à toutes les sortes de fromages: fromage tout gras, fromage maigre, fromage à pâte dure et molle, fromage fermenté et frais, « cottage cheese » et caillé.

Les statistiques sur le *beurre* englobent le ghee, beurre liquide qui est clarifié par ébullition et est fabriqué essentiellement dans les pays d'Extrême-Orient.

Œufs. Certains pays n'ont aucune statistique de la production d'œufs et il a fallu procéder à des estimations à partir de données telles que les effectifs des poules ou de la volaille en général, les taux de ponte connus ou supposés, etc.

La plupart des pays qui établissent des statistiques expriment la production, soit en poids total, soit en nombre d'œufs. Dans ce dernier cas, les données fournies ont été converties en poids, en utilisant, chaque fois que possible, les coefficients de conversion officiels.

Les chiffres s'entendent en général de la production totale, y compris les œufs à couver, tant dans le secteur agricole que dans le secteur non agricole.

Miel. Les statistiques figurant au tableau sont encore incomplètes, notamment en ce qui concerne les pays africains et asiatiques.

Soie grège. Bien que l'on manque de données pour quelques pays producteurs, les statistiques relatives aux pays énumérés dans le tableau représentent convenablement la production mondiale totale.

Laine. En général, les statistiques de la production lainière sont exprimées en poids de laine en suint, qui contient de 30 à 65 pour cent d'impuretés. Afin de rendre les chiffres comparables, les données sont également fournies en poids de laine lavée.

Cuirs et peaux. Toutes les données s'entendent en poids frais de cuirs et peaux. Pour les pays indiquant leur production en nombre ou en poids séché, tanné ou salé, les données ont été converties en poids frais en appliquant des facteurs de conversion appropriés. Lorsqu'on ne disposait pas de données officielles, on a incorporé des estimations fondées sur l'abattage et d'autres indicateurs.

DISPONIBILITÉS ALIMENTAIRES

Depuis 1971, la FAO travaille à mettre au point un système intégré de rassemblement et mise à jour sur ordinateur des statistiques agricoles courantes — sous forme de bilans disponibilités/utilisation (SUAs) —, qui englobe environ 240 produits alimentaires et agricoles primaires et 290 produits transformés tirés des premiers, pour 215 pays et territoires, avec des séries chronologiques allant de 1961 à 1977.

Les quantités totales de denrées alimentaires produites dans un pays, ajoutées aux importations totales et ajustées pour tenir compte de toute variation des stocks survenue depuis le début de la période de référence, donnent les *disponibilités* pour la période considérée. En ce qui concerne l'*utilisation*, on fait une distinction entre les quantités exportées, utilisées comme aliments du bétail ou comme semences, ou consacrées à des utilisations industrielles et à d'autres fins non alimentaires, ou perdues durant le stockage et le transport, d'une part, et les quantités disponibles pour la consommation humaine au niveau du détail d'autre part, c'est-à-dire au moment où les aliments quittent le magasin de détail ou entrent dans les ménages. Pour chaque denrée alimentaire disponible pour la consommation humaine, on obtient alors les disponibilités par habitant en divisant les quantités disponibles pour la consommation humaine par le nombre correspondant d'individus qui, en fait, se les partagent. Les données sur les disponibilités alimentaires par habitant peuvent être exprimées quantitativement et aussi, en utilisant les facteurs appropriés relatifs à la composition des aliments, en termes d'éléments nutritifs (calories, protéines et lipides). Seuls les éléments nutritifs figurent dans l'Annuaire, pour répondre aux besoins immédiats de nombreux utilisateurs qui désirent analyser et évaluer la situation alimentaire et nutritionnelle des pays indiqués.

Il convient de noter que les quantités d'aliments disponibles se rapportent uniquement aux quantités d'aliments qui arrivent au consommateur et pas obligatoirement aux quantités d'aliments réellement consommés, qui peuvent être plus faibles que la quantité indiquée selon l'importance des pertes d'aliments comestibles et de nutriments qui surviennent dans les ménages, à savoir pendant le stockage, la préparation et la cuisson, sans tenir compte des déchets laissés sur les assiettes, des quantités données aux animaux domestiques ou familiers ou des quantités jetées.

Les chiffres ne fournissent aucune indication sur les différences qui peuvent exister entre le régime alimentaire des divers groupes de population, c'est-à-dire en fonction des différents groupes socio-économiques, des zones écologiques et des régions géographiques à l'intérieur d'un pays donné. Ils ne donnent également aucune indication sur les variations saisonnières des disponibilités alimentaires totales. Ils ne représentent que les disponibilités moyennes pour l'ensemble de la population et n'indiquent pas ce que consomment en réalité les individus.

Les bilans alimentaires détaillés provisoires moyens pour la période 1972-74, ainsi que des séries à long terme des disponibilités alimentaires par habitant pour les grands groupes d'aliments, exprimés en calories, en protéines et en matières grasses (moyenne pour la période 1961-63 et années 1964 à 1974), ont été préparés et envoyés aux pays pour observations.

MOYENS DE PRODUCTION

Machines agricoles

Tracteurs. Les données comprennent généralement tous les tracteurs, à pneus ou à chenilles, utilisés dans l'agriculture, à l'exclusion des motoculteurs.

Trayeuses. Ce tableau indique soit le nombre de trayeuses soit, pour certains pays, le nombre d'exploitations possédant de telles machines. Les données concernant le Royaume-Uni ne comprennent pas les trayeuses dont disposent l'Ecosse et l'Irlande du Nord. Pour l'Australie, les données indiquent le nombre d'unités (capacité en nombre de vaches).

Produits antiparasitaires

Les données se rapportent généralement aux quantités de pesticides utilisées ou achetées par les agriculteurs. (Les chiffres du Mexique concernent la production; ceux du Nicaragua, les importations; ceux des Etats-Unis, les ventes, sauf indications contraires). Les données sont exprimées en équivalent d'éléments actifs, sauf pour les pays suivants, où ils se rapportent au poids du produit préparé: Botswana, Chili, Colombie, Costa Rica, Cuba, Equateur, Egypte, France, République démocratique allemande, Guadeloupe, Guatemala, Hongrie, Italie, Jordanie, Koweït, Martinique, Maurice, Maroc, Nicaragua, Nigéria, Porto Rico, Réunion, Samoa, Afrique du Sud, Suède, Trinité-et-Tobago, Ouganda, îles Vierges (E.-U.) et Yougoslavie. Le poids du produit préparé comprend généralement le composant actif plus les diluants et les adjuvants; pour cette raison, et du fait également que les statistiques sont incomplètes, il n'a pas été possible de calculer de totaux.

En consultant les tableaux relatifs aux produits mentionnés ci-dessous, il convient de tenir compte des observations suivantes:

DDT. Pour l'Argentine, les données comprennent les exportations.

HCH. Pour l'Autriche, les données comprennent le lindane.

Lindane. Les données relatives à la Hongrie comprennent l'HCH.

Aldrine et autres produits. Pour l'Autriche, ainsi que pour les Etats-Unis jusqu'en 1974, les données comprennent le toxaphène.

Autres hydrocarbures chlorurés. Les données relatives à Porto Rico comprennent tous les hydrocarbures chlorurés à l'exception du DDT et du chlordane. Celles de Cuba, de l'Equateur, de Hong-kong, d'Israël, de la Nouvelle-Zélande, de la Norvège et de l'Afrique du Sud se rapportent à tous les hydrocarbures chlorurés.

Parathion. Pour la Suède, les données concernent tous les composés organiques phosphorés.

Autres composés organiques phosphorés. Pour Cuba, Hong-kong, l'Islande, Israël, la Jordanie, la Norvège, Porto Rico et l'Afrique du Sud, les données concernent tous les composés organiques phosphorés.

Pyrèthre. Les données relatives aux Etats-Unis portent sur les importations.

Autres insecticides botaniques. Les données relatives à la Norvège comprennent tous les insecticides botaniques. Celles relatives aux Etats-Unis portent sur les importations.

Autres insecticides. Pour la Colombie, le Danemark, la Guadeloupe, le Guatemala, la Martinique, Maurice, le Maroc, la Réunion, Trinité-et-Tobago, l'Ouganda et la Yougoslavie, les données comprennent tous les insecticides. Les données pour Porto Rico couvrent tous les insecticides à l'exception de ceux qui sont indiqués séparément. Pour Bahreïn, les données comprennent tous les insecticides à l'exception du parathion.

Autres fongicides. Les chiffres comprennent tous les fongicides pour les pays suivants: Bahreïn, Colombie, Cuba, Equateur, Guadeloupe, Guatemala, Hong-kong, Jordanie, Maurice, Maroc, Porto Rico, Réunion, Afrique du Sud, Trinité-et-Tobago, Ouganda et Yougoslavie. Les chiffres relatifs à la Martinique comprennent tous les fongicides à l'exclusion des composés cupriques, qui figurent séparément.

2,4-D. Les données relatives au Canada et à la Norvège comprennent des mélanges de 2,4-D et de 2,4,5-T. Les données relatives à l'Inde comprennent les autres herbicides.

Autres herbicides. Les chiffres relatifs aux pays suivants comprennent tous les herbicides: Bahreïn, Colombie, Cuba, Danemark, Equateur, République démocratique allemande, Guadeloupe, Guatemala, Hong-kong, Martinique, Maroc, Porto Rico, Réunion, Suisse, Ouganda et Yougoslavie. Les chiffres relatifs à la Nouvelle-Zélande comprennent tous les herbicides à l'exclusion de 2,4,5-T. Les chiffres relatifs à la Suède comprennent tous les herbicides à l'exception de ceux qui figurent séparément.

Bromures. Les chiffres relatifs à Israël comprennent les carbamates pour fumigation.

Autres produits pour fumigation. Les données relatives au Canada, à la République fédérale d'Allemagne, au Guatemala et à l'Afrique du Sud se rapportent à la totalité des produits de cette nature.

Autres rodenticides. Les données relatives au Danemark, à Maurice et à Porto Rico couvrent tous les produits de cette nature. Les données relatives aux Etats-Unis se rapportent aux importations.

Autres pesticides. Pour le Costa Rica, les données comprennent tous les insecticides et fongicides. Pour la Nouvelle-Zélande, elles couvrent tous les insecticides, à l'exception des hydrocarbures chlorurés, fongicides et produits pour fumigation.

PRIX

Comme première mesure d'uniformisation des facteurs de conversion employés dans diverses publications statistiques de la FAO, on a utilisé les facteurs commerciaux de conversion moyens pondérés, tels que publiés par le Fonds monétaire international dans *International Financial Statistics,* comme taux de change pour la conversion en dollars des Etats-Unis des données exprimées en monnaies nationales, de sorte que ces conversions s'effectuent selon la pratique suivie pour les valeurs commerciales.

Les séries de prix sont de provenances diverses: question-

naires spéciaux, groupes internationaux s'occupant de produits, publications officielles et non officielles et — dans certains cas — correspondants spéciaux. La « Liste des sources pour les tableaux des prix » publiée dans l'Annuaire de 1960 pour les séries de prix figurant dans cette édition peut encore servir de guide général, bien qu'elle ne suive pas exactement la présentation des tableaux adoptée dans ce volume. Les unités locales de quantité, dans lesquelles les prix étaient originellement exprimés, ont été ramenées aux unités métriques ou britanniques.

NOTES SUR LES PAYS

Berlin

Les données relatives à la République fédérale d'Allemagne et à la République démocratique allemande incluent, le cas échéant, des chiffres ayant trait à Berlin, pour lequel des données séparées ne sont pas disponibles dans la plupart des cas.

Chine

Les données relatives à la Chine comprennent en général les chiffres pour la Province de Taïwan.

Une des principales difficultés lors de l'élaboration de l'*Annuaire FAO de la production* a été l'estimation de la production végétale et animale de la Chine. Ce pays est le plus peuplé du monde et sa production agricole, qui représente apparemment près de 12 pour cent du total mondial, vient après celle des Etats-Unis et avant celle de l'U.R.S.S.

Toutefois, la Chine n'a publié aucune statistique officielle adéquate depuis 19 ans. En conséquence, les estimations paraissant dans le présent Annuaire pour ce pays sont fondées sur des informations officielles fragmentaires transmises par les media du pays, et sur l'évaluation de personnes qualifiées qui vivent ou se rendent en Chine, ou dans des pays et territoires voisins. Toutes les statistiques commerciales tirées des publications des partenaires commerciaux de la Chine ont été prises en considération, ainsi que toute information disponible sur les conditions météorologiques, les systèmes d'irrigation, la mécanisation agricole, la consommation d'engrais, etc.

Chypre

En raison de la situation actuelle, les données relatives à Chypre sont peut-être partielles.

Timor oriental

Les données des années récentes sont généralement comprises dans celles de l'Indonésie.

Inde et Pakistan

Les données se rapportant au Cachemire-Jammu, dont le statut final n'a pas encore été déterminé, figurent en général dans les chiffres relatifs à l'Inde et ne sont pas comprises dans celles relatives au Pakistan.

Les données relatives au Sikkim sont comprises dans celles de l'Inde.

Jordanie

A l'exception du tableau 1 (Utilisation des terres et irrigation), depuis 1967, les données se rapportent seulement à la rive orientale.

CLASSIFICATION DES PAYS PAR CATÉGORIES ÉCONOMIQUES ET RÉGIONS

Tous les tableaux qui indiquent des totaux continentaux contiennent également des totaux régionaux qui figurent même s'ils sont identiques aux totaux continentaux ou s'il est facile de les en déduire.

Pour procéder à ses études analytiques, la FAO a divisé le monde selon les catégories économiques et régions ci-après; les noms des pays sont donnés ici dans l'ordre où ils apparaissent dans les tableaux:

CATÉGORIE I: PAYS DÉVELOPPÉS A ÉCONOMIES DE MARCHÉ

Région A — Amérique du Nord: Canada, Etats-Unis.

Région B — Europe occidentale: Andorre, Autriche, Belgique-Luxembourg, Danemark, îles Féroé, Finlande, France, République fédérale d'Allemagne (y compris Berlin Ouest), Gibraltar, Grèce, Saint-Siège, Islande, Irlande, Italie, Liechtenstein, Malte, Monaco, Pays-Bas, Norvège, Portugal (y compris les Açores et Madère), Saint-Marin, Espagne, Suède, Suisse, Royaume-Uni (y compris les îles Anglo-Normandes et l'île de Man), Yougoslavie.

Région C — Océanie: Australie, Nouvelle-Zélande.

Région D — Autres pays développés à économies de marché: Israël, Japon (y compris les îles Bonin et Ryu-kyu), Afrique du Sud.

CATÉGORIE II: PAYS EN DÉVELOPPEMENT A ÉCONOMIES DE MARCHÉ

Région A — Afrique: Algérie, Angola, Bénin, Botswana, Territoire britannique de l'océan Indien, Burundi, Cameroun, Cap-Vert, Empire centrafricain, Tchad, Comores, Congo, Djibouti, Guinée équatoriale, Ethiopie, Gabon, Gambie, Ghana, Guinée, Guinée-Bissau, Côte-d'Ivoire, Kenya, Lesotho, Libéria, Madagascar, Malawi, Mali, Mauritanie, Maurice, Maroc, Mozambique, Namibie, Niger, Nigéria, Réunion, Rhodésie, Rwanda, Sainte-Hélène, Saint-Thomas et Prince, Sénégal, Seychelles, Sierra Leone, Somalie, Afrique du Nord espagnole, Swaziland, Tanzanie, Togo, Tunisie, Ouganda, Haute-Volta, Sahara occidental, Zaïre, Zambie.

Région B — Amérique latine: Antigoa, Argentine, Bahama, Barbade, Belize, Bolivie, Brésil, îles Caïmanes, Chili, Colombie, Costa Rica, Cuba, Dominique, République Dominicaine, Equateur (y compris les îles Galapagos), El Salvador, îles Falkland (Malvinas), Guyane française, Grenade, Guadeloupe, Guatemala, Guyane, Haïti, Honduras, Jamaïque, Martinique, Mexique, Montserrat, Antilles néerlandaises, Nicaragua, Panama, Zone du canal de Panama, Paraguay, Pérou, Porto Rico, Saint-Christophe-Nevis-Anguilla, Sainte-Lucie, Saint-Vincent, Surinam, Trinité-et-Tobago, îles Turques et Caïques, Uruguay, Venezuela, îles Vierges (R.-U.), îles Vierges (E.-U.).

Région C — Proche-Orient: Afrique: Egypte, Libye, Soudan. *Asie:* Afghanistan, Bahreïn, Chypre, Zone de Gaza (Palestine), Iran, Irak, Jordanie, Koweït, Liban, Oman, Katar, Arabie saoudite, Syrie, Turquie, Emirats arabes unis, République arabe du Yémen, Yémen démocratique.

Région D — Extrême-Orient: Bangladesh, Bhoutan, Brunéi, Birmanie, Timor oriental, Hong-kong, Inde, Indonésie, République de Corée, Lao, Macao, Malaisie, (Malaisie péninsulaire, Sabah, Sarawak), Maldives, Népal, Pakistan, Philippines, Singapour, Sri Lanka, Thaïlande.

Région E — Autres pays en développement à économies de marché: Amérique: Bermudes, Groenland, Saint-Pierre-et-Miquelon. *Océanie:* Samoa américaines, îles Canton et Enderbury, île Christmas (aust.), îles Cocos (Keeling), îles Cook, Fidji, Polynésie française, îles Gilbert, Guam, île Johnston, îles Midway, Nauru, Nouvelle-Calédonie, Nouvelles-Hébrides, île Nioué, île Norfolk, îles du Pacifique (Territ. sous tutelle), Papouasie Nouvelle-Guinée, île Pitcairn, îles Salomon, Samoa, Tokélaou, Tonga, Tuvalu, île de Wake, îles Wallis et Futuna.

CATÉGORIE III: PAYS A ÉCONOMIES CENTRALEMENT PLANIFIÉES

Région A — Asie: Chine, Kampuchea démocratique, République populaire démocratique de Corée, Mongolie, Viet Nam.

Région B — Europe orientale et U.R.S.S.: Albanie, Bulgarie, Tchécoslovaquie, République démocratique allemande (y compris Berlin Est), Hongrie, Pologne, Roumanie, U.R.S.S.

TOUS LES PAYS DÉVELOPPÉS

Pays développés à économies de marché et Région B des économies centralement planifiées.

TOUS LES PAYS EN DÉVELOPPEMENT

Pays en développement à économies de marché et Région A des économies centralement planifiées.

INTRODUCCION

Este volumen constituye la 31ª edición del *Anuario FAO de Producción*. Después de los cambios efectuados en la edición correspondiente al año pasado, este año se han introducido sólo tres cambios. En primer lugar, en la mayoría de los cuadros relativos a cultivos, ganado y medios de producción, el promedio del quinquenio 1961-65 se ha sustituido por el promedio del trienio 1969-71. En segundo lugar, también en los cuadros relativos a los números índices de la producción, se ha cambiado el período base de 1961-65 a 1969-71. El tercero y más importante de los cambios se refiere asimismo a los números índices, y consiste en la introducción de precios medios nacionales al productor para el período 1969-71 como coeficientes de ponderación para obtener la producción global de cada país, en comparación con los precios regionales comunes referidos al trigo para el período 1961-65, que se utilizaban anteriormente (véanse otros detalles referentes a los números índices en las Notas sobre los cuadros). Cabe asimismo señalar que se han introducido otros dos grupos de países en el Anuario: uno, Todos los países desarrollados, que abarca las economías de mercado desarrolladas, Europa oriental y la U.R.S.S.; el otro, Todos los países en desarrollo, abarca las economías de mercado en desarrollo y las economías asiáticas de planificación centralizada.

Se recuerda al lector que algunos datos sobre precios y asuntos afines, que ya no se publican en este Anuario, pueden encontrarse en el *Boletín mensual FAO de estadísticas*, el *Anuario de estadística* y el *Boletín mensual de estadística* de las Naciones Unidas, el *Anuario de estadísticas del trabajo* y *Estadísticas del trabajo* (mensual) de la Oficina Internacional del Trabajo.

Los datos relativos a la producción, el comercio, el consumo y los precios de los fertilizantes comerciales pueden encontrarse en el *Informe anual sobre los fertilizantes*, de la FAO.

Los datos estadísticos referentes a los cultivos y al ganado, que ya no se consignan en este Anuario, se encuentran en la Dirección de Estadística de la FAO, que los facilitará a los interesados previa solicitud.

Igual que en años precedentes, este volumen comprende también un gran número de estimaciones de la FAO sobre la superficie y producción de los cultivos principales y sobre el número de cabezas de ganado y productos pecuarios, cuando no se ha podido disponer de cifras oficiales u oficiosas de los países mismos.

La publicación de estas estimaciones da a los países interesados la posibilidad de examinarlas, y es de esperar que tales países proporcionen a la FAO cifras más fidedignas.

Los datos consignados en casi todos los cuadros de productos incluyen cifras relativas al año que aparece en la tapa del Anuario. Sin embargo, algunas de las cifras relativas a ese año son estimaciones de la FAO basadas en informes extraoficiales y, por consiguiente, se trata de cifras provisionales.

En cuanto a los números anteriores, la publicación del Anuario ha sido posible gracias a la cooperación de los gobiernos, que han suministrado casi toda la información en forma de respuestas a los cuestionarios anuales de la FAO. Ha continuado la colaboración con diversas organizaciones, con el fin de conseguir uniformidad en la presentación de las cifras internacionales. Se reconoce con agradecimiento la ayuda prestada por los gobiernos y las organizaciones.

Símbolos

Definición de los símbolos utilizados en los cuadros:

*	Cifras extraoficiales
—	Nada, cantidad insignificante (menos de la mitad de la unidad indicada) o partida no aplicable
...	No se dispone de datos
F	Estimación de la FAO
NEP	No especificado ni incluido en ninguna otra parte
HA	Hectárea
MT	Tonelada métrica
KG	Kilogramo
HEAD	Cabeza
NUMBER	Número
KG/HA	Kilogramo por hectárea
KG/AN	Kilogramo por animal

En la mayoría de los cuadros, un espacio en blanco tiene el mismo significado que los símbolos (—) o (...) antes definidos.
Los decimales se separan de los enteros por un punto (.).

Nombres de países y productos

En la mayoría de los cuadros, el espacio destinado a los nombres de países y productos está limitado a 12 y a 24 letras respectivamente. Los nombres de los productos aparecen en inglés, francés y español; los de los continentes, países y regiones sólo en inglés. Aunque los nombres abreviados de los productos son suficientemente claros, no siempre sucede lo mismo con los de los países y el lector debe consultar la Lista de Países, Continentes, Clases Económicas y Regiones, en la página 37, que indica los países en el orden en que aparecen en los cuadros. Esta lista contiene las abreviaturas en inglés y los nombres completos en inglés, francés y español.

Período de referencia

Como se empezó a hacer en la edición de 1966 del Anuario, el período de referencia en cuanto a las estadísticas de superficie y producción de cultivos se basa en el año civil, es decir, los datos correspondientes a un determinado cultivo se refieren al año civil en que se recogió la totalidad o la parte principal de la cosecha, lo que, sin embargo, no significa necesariamente que los datos de producción de un producto determinado se acumulen mes por mes de enero a diciembre, si bien suceda así en el caso de ciertos productos tales como el té, el sisal, la almendra de palma, el aceite de palma, el caucho, el coco y, en ciertos países, la caña de azúcar y los bananos, que se cosechan casi uniformemente durante todo el año. Pero la cosecha de otros cultivos se limita generalmente a unos cuantos meses y, en algunos casos, a unas cuantas semanas. La producción de estos cultivos la notifican los diversos países de diferente modo: por años civiles, años agrícolas, años comerciales, etc. Sea cual fuere el año estadístico que utilicen los países para la presentación de los datos sobre superficie y producción, los datos se asignan, producto por producto, al año civil en que se recoge la totalidad o el grueso de la cosecha. Naturalmente, cuando una cosecha se recoge a fines del año civil, la producción del cultivo se utilizará en su mayor parte durante el año siguiente al año civil correspondiente a las cifras de producción en los cuadros.

Debe advertirse que la adopción del año civil como período de referencia significa inevitablemente que, en una serie de casos, las cosechas asignadas por los países a un año emergente determinado pueden aparecer bajo dos años civiles distintos en los cuadros de este Anuario.

La población pecuaria se ha agrupado en períodos de 12 meses que terminan el 30 de septiembre de los años que se indican en los cuadros, es decir, los animales enumerados en un país determinado en cualquier momento entre el 1 de octubre de 1976 y el 30 de septiembre de 1977 se consignan en el año 1977.

Por lo que respecta a los productos pecuarios, los datos sobre carne, leche y productos lácteos se refieren a años civiles propiamente dichos, salvo unas cuantas excepciones que se mencionan en las Notas sobre los cuadros. Los datos relativos a otros productos pecuarios cuya producción se limita a ciertos períodos del año, por ejemplo, miel, lana, etc., se asignan al año civil, siguiendo un sistema análogo al adoptado para los cultivos.

Los datos relativos a tractores y a otra maquinaria agrícola representan, hasta donde es posible, el número que estaba en uso al terminar el año indicado o durante el primer trimestre del siguiente año.

Los datos sobre plaguicidas corresponden por lo general al año civil.

Los números índices de la FAO relativos a la producción agrícola se han calculado ajustándose al período de referencia del año civil según antes se define.

Superficies de cultivo

Las cifras relativas a superficies de cultivo se refieren generalmente a la superficie cosechada, aunque las correspondientes a cultivos permanentes pueden referirse a la superficie total plantada.

Rendimiento por hectárea

Todos los rendimientos por hectárea, lo mismo los nacionales que los totales continentales, regionales y mundiales, se expresan en kilogramos. En todos los casos dichos rendimientos se han calculado a partir de los datos detallados de superficie y producción, es decir, hectáreas y toneladas. Los datos sobre rendimientos de los cultivos permanentes no son tan fidedignos como los de los cultivos temporales, bien porque la mayoría de los datos sobre superficie pueden referirse a la superficie plantada, como en el caso de las uvas, o bien por la escasez y la inseguridad de las cifras sobre superficie notificadas por los países, como sucede, por ejemplo, con el cacao y el café.

Totales

Se dan totales continentales, regionales y mundiales respecto a todos los productos, con excepción de los plaguicidas y de las ordeñadoras mecánicas. Dichos totales sólo incluyen datos relativos a los países, los cuales aparecen en el cuerpo del cuadro.

En general, estos totales reflejan adecuadamente la situación en las zonas geográficas que representan, con excepción de algunos cultivos de hortalizas y frutas, así como de ciertos productos pecuarios.

En las Notas sobre los cuadros se dan mayores detalles a este respecto.

Notas sobre cuadros y países

Las cifras de algunos cuadros y países requieren más aclaraciones y reservas de las que es posible dar en esta sección. En las Notas sobre los cuadros y las Notas sobre países, que aparecen a continuación, se exponen estos puntos, así como los demás elementos, incluso cambios en la extensión territorial, clasificación de los países por regiones de la FAO, etc.

Como regla general, los datos del Anuario se refieren al país especificado con sus actuales fronteras de facto. Para los nombres de los países y las agrupaciones continentales se sigue, en general, la nomenclatura empleada por la Oficina de Estadística de las Naciones Unidas.

NOTAS SOBRE LOS CUADROS

TIERRAS

Aprovechamiento de tierras y riego

Estos cuadros constituyen un intento de reunir todos los datos disponibles sobre aprovechamiento de tierras y tierras de regadío en todo el mundo.

Cuando se examine la sección sobre aprovechamiento de tierras deberá tenerse presente que las definiciones que emplean los países al comunicar la información estadística varían considerablemente y que a veces dentro de una misma categoría se incluyen muchas tierras destinadas a diferentes usos.

A continuación figuran las definiciones de las categorías relativas al aprovechamiento de las tierras:

1. *Superficie total:* Se refiere a la extensión del país en su totalidad, incluyendo la superficie comprendida por las masas de agua interiores.

2. *Superficie terrestre:* Se refiere a la extensión total de las tierras, sin incluir las aguas interiores. La definición de las aguas interiores comprende en general los ríos y lagos principales.

3. *Tierras arables o de labranza:* Comprenden las tierras bajo cultivos temporales (las que dan dos cosechas se toman en cuenta sólo una vez), las praderas temporales para corte o pastoreo, las tierras dedicadas a huertas comerciales o huertos (incluidos los cultivos de invernadero), y las tierras temporalmente en barbecho o no cultivadas.

4. *Tierras destinadas a cultivos permanentes:* Se refieren a la tierra cultivada con cosechas que ocupan el terreno durante largos períodos y no necesitan ser replantadas después de cada cosecha, tales como el cacao, el café y el caucho; incluye tierras ocupadas por arbustos, árboles frutales, nogales y otros

árboles de fruto seco, vides y otras plantas trepadoras, pero excluye la tierra dedicada a árboles para la producción de leña o de madera.

5. *Praderas y pastos permanentes:* Se refieren al terreno utilizado permanentemente (cinco años o más) para forrajes herbáceos, ya sean cultivados o silvestres (praderas o tierras de pastoreo silvestres).

6. *Terrenos forestales y montes abiertos:* Se refieren a las tierras con masas de árboles naturales o plantadas, sean productivas o no. Incluyen los terrenos de los que se han talado los bosques, pero que serán repoblados con árboles en un futuro previsible.

7. *Otras tierras:* Comprenden las tierras no utilizadas, pero potencialmente productivas, superficies edificadas, terrenos baldíos, parques, jardines ornamentales, carreteras, caminos, tierras incultas, y cualesquiera otras tierras que no se hayan incluido en los párrafos 3 a 6.

Los datos sobre riego se refieren a las superficies a las que voluntariamente se proporciona agua, incluidos los terrenos inundados por las aguas fluviales, y destinados a la producción de cultivos o al mejoramiento de pastos, independientemente de si estas superficies son regadas varias veces o solamente una vez durante el año respectivo.

Las notas sobre determinados países que figuran a continuación se refieren a las categorías relativas al aprovechamiento de las tierras y al riego:

SUPERFICIE TOTAL

Groenlandia: Los datos se refieren a la zona exenta de hielo.

Mauricio: Los datos no incluyen a las zonas dependientes.

Namibia: Los datos incluyen el territorio de Walvis Bay.

Nueva Caledonia: Los datos incluyen las zonas dependientes.

Sudáfrica: Los datos no incluyen el territorio de Walvis Bay.

U.R.S.S.: Los datos incluyen el Mar Blanco (9 000 000 de hectáreas) y el Mar de Azov (3 730 000 hectáreas).

TIERRAS ARABLES O DE LABRANZA, Y TIERRAS DESTINADAS A CULTIVOS PERMANENTES

Australia: Los datos sobre tierras arables incluyen alrededor de 27 000 000 de hectáreas de tierras cultivadas destinadas a pastizales.

Cuba: Los datos se refieren solamente al sector perteneciente al Estado.

Portugal: Los datos incluyen alrededor de 800 000 hectáreas de cultivos temporales sembrados conjuntamente con cultivos permanentes y bosques.

PRADERAS Y PASTOS PERMANENTES

Australia: Los datos se refieren al resto de la superficie comprendida en las explotaciones rurales.

Egipto: Las tierras accidentadas de pastoreo han sido incluidas en Otras tierras.

U.R.S.S.: Los datos no incluyen los pastizales destinados a los renos.

En el caso de los siguientes países, los datos se refieren solamente a las praderas y pastos permanentes existentes en las explotaciones agrícolas: Corea (República de), Chile, Finlandia, Guatemala, Kenya, República Dominicana, Surinam, Trinidad y Tabago y Uruguay.

RIEGO

Francia: Los datos no incluyen las huertas comerciales ni las domésticas.

Hungría: Los datos no incluyen las parcelas agrícolas complementarias ni las explotaciones agrícolas particulares.

Reino Unido: Los datos no incluyen Escocia ni Irlanda del Norte.

En el caso de los países que se indican a continuación, los datos se refieren a las tierras dotadas de instalaciones de riego: Bulgaria, Noruega, Rumania y Surinam.

En el caso del Japón, la República de Corea y Sri Lanka, los datos se refieren solamente al arroz que se cultiva con riego.

POBLACION

El Cuadro 3 presenta estimaciones de *la población total, la población agrícola* y *la población económicamente activa*, total y en la agricultura, correspondientes a 1965, 1970, 1975, 1976 y 1977, por países.

El *Anuario Demográfico* de las Naciones Unidas da series cronológicas de estimaciones sobre la población total. Por lo general, estos datos son facilitados por los países, pero en algunos casos la Oficina de Estadística de las Naciones Unidas ajusta las estimaciones disponibles o prepara unas nuevas. Por lo que respecta a muchos de los países en desarrollo se necesitan ulteriores ajustes de las estimaciones disponibles con objeto de mantener un grado razonable de coherencia. La incoherencia está a veces en una serie de las propias estimaciones anuales, puesto que éstas fluctúan de una manera que la migración no puede explicar. Algunas veces las estimaciones están en desacuerdo con los datos procedentes de fuentes externas, sobre todo los derivados de censos y del registro civil. En las series cronológicas de estimaciones también hay lagunas que necesitan llenarse. Por consiguiente, la División de Población de las Naciones Unidas ha preparado para cada país una serie completa de estimaciones que abarca un período bastante grande y es coherente en sus variaciones de año en año, así como con otros datos demográficos[1]. Los datos contenidos en este cuadro proceden de esas series, reajustados para incorporar los datos oficiales sobre población disponibles más recientemente, en particular de los países desarrollados, cuyas estadísticas demográficas se consideran fidedignas.

Los datos se refieren, en general, a la población presente (*de facto*) en las zonas respectivas, dentro de las actuales fronteras geográficas.

La población agrícola se define como todas las personas que dependen de la agricultura para su subsistencia. Comprende todas las personas activamente ocupadas en la agricultura y sus familiares a cargo que no trabajan.

La población económicamente activa se define como todas las personas dedicadas a una actividad económica o que buscan empleo en ella, bien sea en calidad de patronos, como trabajadores por cuenta propia, empleados asalariados o bien trabajadores no remunerados que colaboran en la explotación de una finca o empresa familiar.

La población económicamente activa en la agricultura comprende todas las personas económicamente activas que se dedican principalmente a actividades agrícolas, forestales, venatorias o pesqueras.

Se dispone de información sobre la población económicamente activa agrupada por industrias u oficios en relación con los censos demográficos o las encuestas sobre mano de obra nacionales. La comparabilidad de los datos está limitada, sin embargo, por las diferencias existentes en el tratamiento estadístico que cada país da, por ejemplo, a los miembros de la familia que trabajan sin remuneración, particularmente las amas de casa. Además, algunos países envían datos sobre la actividad económica de personas de todas las edades, y otros los dan sólo en cuanto a personas de edades determinadas, por ejemplo, 14 años o más. La OIT ha evaluado sistemáticamente esos datos ajustándolos a conceptos uniformes internacionalmente aceptados y ha elaborado estimaciones de la población económicamente activa distribuida entre la agricultura, la industria y los servicios para cada quinto año entre 1950 y 1970, y proyecciones de la población total económicamente activa para cada quinto año entre 1970 y 2 000[2]. Los datos del cuadro se basan en esas estimaciones y proyecciones.

Es escasa la información que se ha obtenido sobre la población agrícola con los censos o encuestas demográficas nacionales. Para las estimaciones de la población agrícola que figuran en este cuadro, la FAO ha recurrido en gran parte a la estrecha relación existente entre la razón población económicamente activa en la agricultura-población económicamente activa total (EAA/EA) y la razón población agrícola-población total (PA/PT).

[1] Naciones Unidas, *Single-year population estimates and projections of the major areas, regions and countries of the world, 1950-2000,* mimeografiado, ESA/P/WP.34, Nueva York, 1975.

[2] OIT, *Labour Force, 1950-2000,* volúmenes I, II, III, IV, Ginebra, 1977.

Se ha supuesto, en general, que estas razones son iguales, por lo que, dada la EAA/EA, la población agrícola se estima como un producto de esta razón y de la población total[3].

[3] Para detalles relativos al método y las hipótesis adoptados para llegar a estas estimaciones, véanse las Proyecciones de la población agrícola mundial, FAO, *Boletín mensual de economía y estadística agrícolas*, Vol. 21, N° 1, 1972, págs. 2-4.

A causa de la naturaleza muy aproximativa de las estimaciones de la población económicamente activa y de la población agrícola en los países pequeños (generalmente los que tenían una población total de 250 000 habitantes o menos en 1970), estos datos no se muestran separadamente; sin embargo, los totales continentales, regionales y mundiales se han reajustado para abarcar también esos países.

NUMEROS INDICES DE LA FAO DE LA PRODUCCION AGRICOLA

Estos números índices ponen de manifiesto la cuantía relativa del volumen global de la producción agrícola para cada uno de los años en comparación con el período base 1969-71. Están basados en la suma de cantidades a precios ponderados de diversos productos agrícolas obtenida después de haber deducido las cantidades utilizadas como semilla y piensos ponderadas en forma semejante. El volumen global resultante representa, por tanto, la producción disponible para cualquier uso, salvo la destinada a semilla y a piensos.

Las deducciones de la producción destinada a semilla (en el caso de los huevos para empollar) y a la alimentación del ganado y las aves de corral se aplican tanto a los productos obtenidos internamente como a los importados. Entre ellas figuran productos agrícolas primarios utilizados en su forma original (maíz, patatas, calabazas, etc.) y alimentos semielaborados, tales como salvado, tortas oleaginosas, harinas, melaza, suero deshidratado, etc. La razón por la cual las cantidades destinadas a semillas y a piensos se deducen de los datos de producción es que los números índices de la FAO, basados en el concepto de considerar a la agricultura mundial como una sola explotación agrícola, miden tal producción evitando el contar dos veces las semillas y los piensos, ya incluidos en los datos de producción, y los cultivos y el ganado obtenidos a partir de ellos.

Los índices de los países se calculan mediante la fórmula de Laspeyres. Las cantidades obtenidas de cada producto se ponderan con arreglo a la media de los precios nacionales al productor correspondientes al período 1969-71, y se suman para cada año. Para obtener el número índice, se divide la suma global de un determinado año entre la suma global media correspondiente al período base 1969-71. Los índices relativos a los continentes, las regiones y el mundo se calculan sumando los totales globales de los países, que para ese fin se han convertido en dólares de los Estados Unidos. Los tipos de cambio de las monedas nacionales con relación al dólar son, salvo pocas excepciones, los publicados por el Fondo Monetario Internacional, con unas pocas excepciones: países de la Europa oriental, la U.R.S.S., China y otros países asiáticos de planificación económica centralizada, para los cuales los tipos de cambio se derivaron de tipos no comerciales, turísticos y otros tipos que no son básicos (fijos).

Los productos incluidos en el cálculo de los números índices de la producción agrícola son todos los cultivos y productos pecuarios producidos en cada país respecto a los cuales se dispone de datos. Los números índices de la producción de alimentos comprenden productos que se consideran comestibles y que contienen nutrientes. Por consiguiente, al igual que los productos no comestibles, se excluyen el café y el té, porque, a pesar de que son comestibles, no tienen prácticamente ningún valor nutritivo. Los números índices se basan en datos de producción consignados con arreglo al período de referencia basado en el año civil.

Las tres diferencias fundamentales entre los índices que se presentan en esta edición del Anuario y los índices presentados en ediciones anteriores son: 1. Los datos de producción utilizados para calcular los números índices se refieren actualmente, salvo pocas excepciones, a productos primarios. Por tanto, la producción de azúcar, aceite de oliva, copra, pasas y vino utilizada anteriormente para calcular los números índices se ha sustituido por la producción de caña de azúcar, remolacha azucarera, aceitunas, cocos y uvas. Los productos elaborados que han permanecido son el aceite de palma, las almendras de palma, la semilla de algodón y el algodón fibra. La producción de nueces de palma no pudo utilizarse por falta de datos sobre la producción y los precios, y el algodón sin desmotar no pudo utilizarse porque se trata de un producto mixto que es, a la vez, alimenticio y no alimenticio. 2. El período base se ha cambiado del promedio de 1961-65 al promedio de 1969-71. 3. El cambio más importante consiste en la utilización de los precios medios *nacionales* al productor (1969-71) como coeficientes de ponderación para estimar la producción global de cada país, en lugar de los precios *regionales* referidos al trigo (promedios de 1961-65) aplicados a cada país de la región.

Quizá los números índices de la FAO difieran de los obtenidos por los países mismos por ser distintos los conceptos de producción, alcance estadístico, coeficientes de ponderación, período de referencia de los datos y métodos de cálculo.

CULTIVOS

Resumen estadístico de la producción agrícola mundial y regional

De las cifras de producción de *manzanas*, se excluyen las correspondientes a la U.R.S.S. Los datos sobre *cultivos oleaginosos* representan la *producción total* de semillas oleaginosas, nueces oleaginosas y otras plantas oleaginosas de los años que se indican, *expresada en equivalente en aceite*. En otras palabras, estas cifras no corresponden a la producción real, sino a la producción potencial de aceites vegetales, por basarse en la hipótesis de que la producción total de todos los cultivos oleaginosos se transforma en aceite en los países productores en el año mismo de la cosecha. Naturalmente, la producción total de las plantas oleaginosas nunca se transforma en aceite por entero, ya que, con variaciones según los cultivos, se utilizan cantidades importantes para semillas, piensos y alimentos. Por otro lado, aunque el índice de extracción de aceite varía de un país a otro, en el presente cuadro se ha aplicado a cada cultivo oleaginoso el mismo índice de extracción en todos los países. Además, hay que tener presente que las plantas oleaginosas cosechadas durante los últimos meses del año se transforman de ordinario en aceite al año siguiente. A pesar de estas deficiencias en el ámbito de aplicación, los índices de extracción y la comprensión cronológica, los datos que aquí se dan son útiles, ya que ofrecen una indicación válida de los cambios anuales del volumen de la producción total de los cultivos oleaginosos. La producción real mundial de aceites vegetales es alrededor del 80% de la cifra aquí citada. Las cifras de la producción de *carne* se refieren a la producción total de los animales sacrificados en los países, sin consideración de su origen.

Cereales

Los datos de superficie y producción de cereales se refieren solamente a los cosechados en grano. Los cosechados para heno, forraje verde y ensilaje o que se utilizan para apacentamiento están por consiguiente excluidos. Los datos de superficie se refieren a la superficie cosechada. Aunque algunos países sólo facilitan datos referentes a la superficie sembrada o cultivada, la superficie no difiere notablemente en años normales de la realmente cosechada, bien sea porque se cosecha prácticamente toda la superficie sembrada o porque las encuestas sobre superficie se efectúan más o menos en el período de recolección.

Cereales, total. Los datos contenidos en el presente cuadro comprenden también otros cereales que ya no se publican en este Anuario, como cereales mixtos, alforfón común, etc.

Trigo y escanda. Los datos disponibles sobre la escanda se han incluido con los del trigo, excepto en el caso de la U.R.S.S.

Mijo y sorgo. El mijo y el sorgo se cultivan en Europa y en América del Norte, principalmente para alimentación del ganado y las aves de corral, pero en Asia, Africa y la U.R.S.S. se emplean mucho como alimento humano.

Siempre que ha sido posible se dan por separado las estadísticas del mijo y del sorgo, pero como muchos países, en particular los de Africa, no hacen distinciones sobre estos dos granos en sus informes, se ofrecen, en tales casos, las cifras conjuntas en el cuadro sobre el mijo.

Raíces y tubérculos

Raíces y tubérculos, total. Los datos comprendidos en este cuadro comprenden también otros cultivos de raíces que ya no se publican en el Anuario, como taro, ñame, etc.

Yuca (mandioca). La yuca (*Manihot esculenta*, Crantz) es una raíz que suele tener dos variedades, la yuca amarga y la dulce, que a veces se consideran como dos especies botánicas distintas denominadas respectivamente *M. utilissima* y *M. dulcis* o *aipi*. En el cuadro figuran juntas la yuca amarga y la dulce.

Legumbres

Legumbres secas, total. Los datos contenidos en este cuadro comprenden también otras legumbres que ya no se publican en el Anuario, como caupíes, vezas, etc.

En cuanto se ha podido determinar, las cifras se refieren a la producción total de los cultivos, cosechados para secar el grano, bien se destine a alimentación humana o a pienso.

Frijoles secos. Los datos sobre superficie y producción de frijoles secos producidos en cultivos mixtos se han unido a los de frijoles que se producen en monocultivo, con referencia a algunos países europeos. Las cifras dadas para la superficie cultivada en estos países resultan exageradas y, por consiguiente, los rendimientos por hectárea parecen en extremo bajos.

Semillas, almendras y nueces oleaginosas

Semilla de colza. La producción de Suecia se da con un 18 por ciento de contenido en agua. Las cifras para algunos países,

tales como India y Pakistán, incluyen también la semilla de mostaza.

Linaza. Las cifras de superficie correspondientes a la U.R.S.S. y a algunos países productores secundarios se refieren tanto a la producción de semilla como de fibra.

Semilla de algodón. Comunican cifras directas de producción países a los que corresponde alrededor del 60 por ciento de la cosecha mundial de semillas de algodón. Las cifras correspondientes al resto de los países cultivadores se han calculado a base de la producción de algodón desmotado, usando coeficientes obtenidos en años anteriores en el mismo país o en otros de condiciones similares.

Aceite de oliva. Con pocas excepciones, los datos se refieren a la producción total de aceite, incluido el extraído de los residuos de la oliva.

Cocos. Los datos consignados se refieren a la producción total de cocos, maduros o no, para el consumo en fresco o transformados en copra o secos. La producción se expresa en función del peso de todo el coco, excluyendo sólo la capa fibrosa exterior.

Hortalizas y melones

Los datos consignados en los cuadros se refieren a los cultivos hortícolas producidos principalmente para consumo humano. Se excluyen productos como col, calabaza, zanahoria, etc., cuando se emplean como forraje o pienso. En muchos países no se dispone de estadísticas correspondientes a las hortalizas, y el alcance de los datos comunicados difiere de un país a otro. En general, parece ser que las estimaciones se refieren a lo que se cultiva en campos y huertas principalmente para la venta, excluyéndose por tanto lo cultivado en pequeños huertos familiares, principalmente para el consumo doméstico. En Austria los datos comunicados se refieren sólo a cultivos de campo; en Cuba, al acopio del sector estatal y del sector privado; en la U.R.S.S., a las granjas colectivas y estatales. La producción en huertos familiares y otros pequeños no incluida en las encuestas estadísticas actuales no, por consiguiente, en los cuadros del Anuario, constituye una parte bastante importante de la producción total estimada en ciertos países, por ejemplo, en Austria y la República Federal de Alemania, más del 40 por ciento; en Italia, más del 15 por ciento.

Por las razones antes citadas, los totales continentales, regionales y mundiales que se dan en este Anuario distan mucho de ser representativos de la superficie y la producción totales de las diferentes clases de hortalizas. Los datos de producción que se dan en el Cuadro 40 incluyen datos publicados en los distintos cuadros sobre hortalizas de este Anuario, así como los de todas las demás hortalizas. También incluyen estimaciones de países que no han enviado datos y, cuando se dispone de ellos, de producción de cultivos no comerciales de países que han notificado sólo la producción para venta.

Coles. Las principales variedades de coles comprendidas en el cuadro son: rojas (lombardas), blancas y berzas de Saboya, col de la China, coles de Bruselas, col rizada y brécol romano o de brotes.

Tomates. Los datos correspondientes a Bélgica, Dinamarca, Finlandia, Noruega, Reino Unido y Suecia se refieren a cultivos producidos principal o totalmente en invernadero, lo que explica que sean mayores los rendimientos por hectárea que se obtienen en estos países.

Coliflores. Cuando ha sido posible también se han incluido en las cifras datos sobre el brécol.

Pepinos y pepinillos. En más o menos la mitad de los países europeos productores, Canadá y Nueva Zelandia, los cultivos son total o parcialmente en invernadero. Lo demuestran los rendimientos por hectárea.

Frijoles verdes. Los datos se refieren a los frijoles recolectados en verde (sólo de las especies *Phaseolus* y *Dolichos*), excluyendo los relativos a los frijoles para consumo en verde, al menos en cuanto a países como Francia y los Estados Unidos, que publican estadísticas separadas para los de desgranar y los de consumo en verde. Los datos sobre frijoles verdes para elaboración, originalmente notificados por algunos países en peso del fríjol desvainado, han sido convertidos en frijol en vaina a razón de 200 por ciento aproximadamente.

Guisantes verdes. Los datos se refieren a guisantes (*Pisum sativum* y *P. arvense*) recolectados en verde. Los datos relativos a algunos países, originalmente notificados en peso del guisante desvainado, se han convertido a guisantes en vaina a razón de 225-250 por ciento.

Cantalupos y otros melones. Los datos para Rumania incluyen sandías.

Sandías. Los datos para Argelia, Bulgaria, Turquía y Yugoslavia incluyen melones; los correspondientes a la U.R.S.S. incluyen melones (alrededor del 18 por ciento) y calabazas y calabacines (alrededor del 30 por ciento).

Uvas y vino

Uvas. Ciertos países, entre los que figuran importantes productores tales como Argelia, Austria, Chile, Francia, Irán y la República Federal de Alemania, no publican datos sobre producción total de uva. Las estimaciones que figuran en el cuadro referentes a estos países se basan en la información disponible sobre la producción de uvas de mesa, pasas y vino. Los datos sobre superficie de Italia incluyen la superficie donde se cultiva la vid y otras plantas. De esta superficie mixta, el 23,5 por ciento está incluido en la superficie total de viñedos.

Vino. Casi todos los grandes países vinícolas calculan la producción del vino a partir de la cantidad de uvas prensadas durante la vendimia; por tanto, sus datos corresponden a la cantidad de mosto del mismo año agrícola y representan la producción total de las prensas, independientemente de que el producto se haya consumido como vino o vinagre, o se haya destilado. Desgraciadamente, no ha sido posible obtener estadísticas sobre esa base para todos los países y se han llenado los vacíos empleando datos sobre recaudaciones tributarias o estimaciones comerciales. Algunos países no publican estadísticas sobre producción de vino o dan datos poco dignos de confianza, ya sea porque no incluyen la producción total de vino o porque incluyen también mezclas de vino y jugos de fruta. La producción de vino de estos países se ha estimado a partir de las cantidades de uva prensada para vino, cuando se dispuso de dicha información.

Caña de azúcar, remolacha azucarera y azúcar

Caña de azúcar y remolacha azucarera. Los datos de superficie y de producción de caña de azúcar y remolacha azucarera comprenden en general todo lo cosechado, excepto lo cultivado expresamente para pienso. Con casi toda la producción se fabrica azúcar centrifugada y no centrifugada; en varios países se emplean importantes cantidades de caña de azúcar para semilla, piensos, consumo en fresco, fabricación de alcohol y otros usos; parte de la producción de remolacha azucarera se emplea para piensos y fabricación de alcohol.

Azúcar centrifugada. Los datos incluyen tanto el azúcar de caña como el de remolacha y se han expresado, en lo posible, en su equivalente en azúcar sin refinar, como lo comunican los países. No se tiene la certeza de que todos los países comuniquen los datos sobre azúcar sin refinar en términos de 96° de polarización, como se solicita en el cuestionario de la FAO. Australia, por ejemplo, da la producción de azúcar a título neto 94°. Respecto a Haití e Indonesia, las cifras se dan « tel quel », que es el peso real de todo el azúcar centrifugada fabricada. Las cifras comunicadas por los países como azúcar refinada se han convertido a azúcar sin refinar a razón de 108,7 por ciento.

Azúcar no centrifugada. Este cuadro incluye todo el azúcar de caña que no ha sido sometido a centrifugación. Prácticamente, todo el azúcar no centrifugada se consume en las zonas donde se produce.

Frutas y bayas

Los datos se refieren a la producción total de fruta fresca, ya se consuma directamente como alimento o pienso, o se prepare en forma de diferentes productos: frutas secas, zumos, conservas, alcohol, etc.

En muchos países no hay estadísticas de la fruta, especialmente la tropical, y no son uniformes los datos de los países que los comunican. En general, los datos sobre producción se refieren a la de plantaciones o huertas que se destina principalmente a la venta. Los países no suelen reunir datos sobre la producción de árboles diseminados y destinada principalmente a consumo doméstico. En general, la producción de plantas silvestres, que en algunos países reviste cierta importancia — particularmente en lo que se refiere a bayas — no la tienen en cuenta los servicios nacionales de estadística. Por consiguiente, la amplitud de las cifras publicadas en este Anuario relativas a diversos frutos y bayas es más bien incompleta, particularmente con respecto a regiones fuera de Europa, América del Norte, Australia y Nueva Zelandia. Se considera que, en la mayoría de los cuadros, los totales que se indican — aunque se refieran al limitado número de países comprendidos — dan una indicación fidedigna de estos cultivos por lo que respecta a su influencia en el comercio internacional. En todo caso, los totales constituyen un indicio de los cambios anuales en la magnitud de los cultivos.

Los datos de producción que figuran en el Cuadro 40 incluyen los datos publicados en los distintos cuadros sobre frutas y bayas, así como los relativos a todas las demás clases de frutas y bayas. La producción de dátiles, plátanos y todas las uvas se incluye en las cifras totales de la fruta, pero se excluye la de aceitunas. Las cifras en este cuadro son más completas que las publicadas para cada producto por separado, debido a que incluyen estimaciones para la mayoría de los países que no han enviado notificaciones, así como datos para los países, tales como la U.R.S.S., que han comunicado la producción total de frutas en una cifra global sin mencionar clases.

Naranjas. Los datos de Guinea, Sierra Leona, Swazilandia, la U.R.S.S. y algunos otros países que producen poco se refieren a todos los frutos cítricos. Los datos de algunos otros países pueden incluir tangerinas.

Tangerinas, mandarinas, clementinas y satsumas. Las cifras de los Estados Unidos incluyen « tangelos », un híbrido de tangerina/toronja, y « temples », un híbrido de naranja dulce/tangerina.

Limones y limas. Las cifras de Bolivia y Perú incluyen toronjas.

Frutos cítricos n.e.p. Las cifras de Japón incluyen toronjas y limones. Las de otros países se refieren generalmente a la producción total de cítricos o a cultivos cítricos sin detallar.

Bananos y plátanos. Las cifras sobre bananos se refieren, en lo posible, a las especies de frutos comestibles del género *Musa*, exceptuando la especie *Musa paradisiaca*, comúnmente conocida con el nombre de plátano. Lamentablemente, hay varios países que no establecen distinción alguna en sus estadísticas entre los bananos y los plátanos y sólo publican estimaciones globales que comprenden ambos. En tales casos, cuando se cree que los datos consignados se refieren principalmente a bananos, los datos se incluyen en el cuadro. Debe notarse que ninguno de los países excluidos del cuadro son exportadores de consideración. Es difícil comparar los datos de producción de bananos y plátanos porque algunos países la expresan en racimos, lo que generalmente suele significar que el tallo va incluido en el peso.

Frambuesas. Algunos datos relativos a las frambuesas (*Rubus idaeus*) parecen incluir otras bayas del género *Rubus*, tales como zarzamoras, moras de logan y « dewberries ».

Grosellas. Los datos sobre grosellas incluyen *Ribes rubrum, album* y *nigrum.*

Nueces de todas clases

La producción de nueces de todas clases (comprendidas las castañas) se refiere a las nueces con cáscara. Las estadísticas sobre nueces son muy escasas y generalmente se refieren únicamente a las destinadas a la venta.

Además de las seis clases de nueces que se indican en los cuadros separados del Anuario, los datos de producción contenidos en el Cuadro 40 comprenden las demás nueces utilizadas principalmente como nueces de mesa, tales como nueces de Brasil, de pili, sapucaia, macadamia, etc. Las nueces destinadas principalmente para aromatizar bebidas están excluidas del cuadro, así como las masticatorias y estimulantes y las utilizadas principalmente para la extracción de aceite o mantequilla, por ejemplo, las de areca/betel, cola, illipe, karite, coco, tung, palma oleaginosa, etc.

Café, cacao y té

Las cifras son de producción de café en grano verde. Los datos de unos cuantos países que informan respecto a la producción de café en cerezas o en pergamino han sido convertidos a café limpio aplicando los oportunos coeficientes.

En el caso del cacao, los datos de producción se refieren al cacao en grano fermentado y seco. Sólo en el caso de ciertos países se dispone de estadísticas oficiales de superficie de café y cacao y no siempre son fidedignas. Los rendimientos por hectárea no son, por consiguiente, muy significativos.

Los datos de producción de café y cacao de Brasil son los oficiales publicados en el Anuario Estadístico de ese país. Los datos sobre el café notificados como cereza seca han sido convertidos a café verde a razón del 50 por ciento.

En lo que respecta al té, las cifras de producción son del elaborado. Para Indonesia, sin embargo, alrededor de una tercera parte de la producción se expresa en peso de hoja verde. Birmania produce cada año alrededor de 45 000 toneladas de hojas de té. Sin embargo, esta cantidad no se ha incluido en el cuadro debido a que casi toda la producción se consume en su estado fresco, como hortaliza.

Lúpulos

Los datos de producción son del peso del cono seco, salvo en el caso de España en el que son del peso en verde.

Tabaco

Las cifras de producción se refieren, hasta donde ha sido posible determinarlo, al peso de venta en la explotación. Los datos disponibles calculados en peso en seco se han convertido, por tanto, en peso de venta en granja, a razón de unas 90 partes por 100.

Cultivos de fibras

Lino (fibra). Los datos se refieren en general al lino agramado y rastrillado e incluyen la estopa. Las cifras para los países que informan de la producción en equivalente en paja o lino enriado se convirtieron en fibra de lino y estopa para hacerlas comparables con los datos para otros países.

Cáñamo (fibra). Lo mismo que para el lino, los datos sobre cáñamo se refieren a la fibra agramada e incluyen la estopa. Las cifras para Bangladesh, India y Pakistán se refieren al cáñamo de Bengala (*Crotalaria juncea*), en tanto que los datos correspondientes a los otros países se refieren al cáñamo verdadero (*Cannabis sativa*).

Yute y otras fibras afines. Las fibras de yute se obtienen de la *Corchorus capsularis* y *C. olitorius.* Las afines al yute incluyen diversos sustitutos, entre los cuales los principales son kenaf o mesta y cáñamo de Guinea (*Hibiscus* spp.) y yute del Congo o paka (*Urena lobata*).

Sisal. Los datos sobre el sisal comprenden fibras y desechos del *Agave sisalana.* Las cifras de superficie son aproximadas en la mayoría de los casos, incluso tratándose de datos oficiales.

Algodón (fibra). El cuadro sobre fibra de algodón se ha preparado en cooperación con el Comité Consultivo Internacional del Algodón. Las cifras dadas para los productos de la mayoría de los países son las que éstos han comunicado oficialmente como correspondientes a la fibra y no incluyen la borra. En unos cuantos casos en que las cifras comunicadas se referían al algodón sin desmotar, y cuando se ignoraba el factor específico de conversión correspondiente al algodón desmotado, el equivalente en fibra se calculó en un tercio.

Fibras n.e.p. Las principales fibras vegetales que se indican en este cuadro son: cáñamo de Mauricio (*Furcraea gigantea*), lino de Nueva Zelandia, formio chileno o maulan (*Phormium tenax*), fique (*Furcraea macrophylla*), caroa (*Neoglazovia variegata*), ixtle (*Samuela carnerosana*), ramio (*Boehmeria* spp.), kapok (*Ceiba pentandra*), bonote (filamento de la corteza del coco). Se excluyen las fibras de agave y abacá.

Caucho natural

El Grupo Internacional de Estudios sobre el Caucho, entidad que colaboró en la preparación de este cuadro, define el caucho natural (*Hevea* spp.) como el peso seco del látex, pero no incluye el de la balata, gutapercha y demás gomas análogas, ni tampoco el de los desechos. Estos últimos se consideran como productos diferentes y, por lo general, se utilizan en forma distinta del caucho natural.

GANADO, AVES DE CORRAL Y SUS PRODUCTOS

Número de cabezas

Los datos sobre número de cabezas que se presentan en este Anuario comprenden todos los animales domésticos sea cual fuere su edad, lugar, o finalidad de su cría. Se han hecho estimaciones para los países que no han comunicado datos, así como para los que envían estadísticas parciales. Hay países en los que los datos de gallinas, patos y pavos todavía no parecen representar el total de estas aves. Ciertos países incluyen a todas las aves de corral en una sola cifra. Los datos correspondientes a estos países se presentan bajo el concepto de Gallinas.

Productos pecuarios

Sacrificios, peso en canal, producción de carne. En los cuadros 83 a 87 se facilitan, refiriéndose a las principales especies, el número de reses sacrificadas, el promedio del peso de la canal limpia y la correspondiente producción de carne. Los datos contenidos en estos cuadros se refieren a *animales sacrificados* en los países, independientemente de su origen. También los datos acerca de la producción de carne de caballo, de aves de corral y carne total, que aparecen en el Cuadro 88, se refieren a animales sacrificados en los países, independientemente de su origen.

Es diferente el concepto de producción de carne que se expone en el Cuadro 89. En este caso, los datos de la producción se refieren a *animales indígenas*, es decir, que incluyen el equivalente en carne de los animales exportados en pie y excluyen el equivalente en carne de los animales importados vivos.

En todos los casos los datos se refieren a la producción total de carne, es decir de la matanza industrial y privada, y representan el peso de la canal limpia, excluidos los despojos y las grasas de matadero. En producción de carne de vacuno y de búfalo está incluida la de ternera; en la carne de carnero y de cabra está la de corderos y cabritos; en la de cerdo, el bacon y jamón, en equivalente en fresco.

La carne de aves de corral incluye la de todas las aves domésticas y se refiere al peso en limpio. Los datos para los países que han notificado el peso vivo o el listo-para-cocinar se han convertido en peso en limpio. La producción total de carne incluye los datos de los cuadros 83 a 87 (carne de animales sacrificados en los países, independientemente de su origen) y comprende: carne de caballo, de aves de corral y de cualquier otro animal doméstico o silvestre, como camellos, conejos, renos, caza, etc.

Si bien los datos comunicados por la mayoría de los países corresponden al año civil, hay unas pocas excepciones. Por ejemplo, Israel y Nueva Zelandia dan datos para años que terminan el 30 de septiembre, y Australia para los que concluyen el 30 de junio.

Leche de vaca, vacas lecheras, rendimiento y producción de leche. Los datos sobre producción de leche de vaca se refieren a la total de leche fresca entera, omitiendo la mamada por el ternero, pero incluyendo la que se usa para alimentar a otros animales. Checoslovaquia, Francia, Hungría, Italia, República Democrática Alemana, República Federal de Alemania y Rumania, incluyen en los datos de producción la mamada por los terneros. Se tienen de casi todos los países estadísticas oficiales de producción de leche de vaca; de no disponerse de ellas, se han empleado estimaciones basadas en las encuestas de consumo de alimentos y otros indicadores. Con respecto a algunos países sin estadísticas de vacas lecheras, los datos expresados en el cuadro han sido estimados a partir de la producción de leche y del rendimiento efectivo o supuesto por vaca. El rendimiento por vaca que se expresa en el cuadro fue obtenido dividiendo la producción de leche por el número de vacas lecheras. Las cifras de producción de leche correspondientes a Israel se refieren a años que terminan el 30 de septiembre;

las de Australia, a años que terminan el 30 de junio; las de Nueva Zelandia, a años que terminan el 31 de mayo. Hay un pequeño número de otros países productores secundarios que comunican datos que no se refieren al año civil.

Leche de búfala, oveja y cabra. El concepto de producción en este cuadro es el mismo que para las vacas lecheras, pero las estadísticas son probablemente menos completas.

Productos lácteos. Las cifras que se dan para los productos de este grupo se refieren, en general, a la producción total, ya se trate de los elaborados en centrales lecheras o en las granjas. Para ciertos países se carece totalmente de datos, y los datos comunicados por otros países pueden ser subestimaciones, particularmente en lo que se refiere a la producción en la granja. Naturalmente, los totales continentales, regionales y mundiales acusan la escasez de datos.

Los datos sobre *queso* publicados en el Anuario se refieren a todas las clases de quesos que se producen: queso con y sin grasa, duro y blando, maduro y fresco, requesón y cuajada.

Los datos sobre *mantequilla* incluyen « ghee », que es la líquida clarificada por ebullición, producida principalmente en países del Lejano Oriente.

Huevos de gallina. Algunos países no tienen estadísticas sobre la producción de huevos y los cálculos han tenido que fundarse en datos indirectos, como el número de gallinas o el número total de aves de corral y la cuantía declarada o hipotética de la puesta de huevos.

La mayoría de los países que tienen estadísticas de producción de huevos comunican el peso total o el número de huevos. En este último caso, el número se ha convertido a su equivalente en peso, empleando siempre que ha sido posible los coeficientes oficiales de conversión.

Los datos se refieren generalmente a la producción total, incluidos los huevos destinados a la incubación, tanto en los sectores agrícolas como no agrícolas.

Miel. Los datos del cuadro son todavía incompletos, particularmente por lo que respecta a países africanos y asiáticos.

Seda cruda. Aunque se carece de información de algunos países productores, los datos para los enumerados en el cuadro representan adecuadamente la producción total mundial.

Lana. En las estadísticas de producción las cifras representan, por lo general, el equivalente en lana grasienta, la que contiene de un 30 a un 65 por ciento de impureza. Con el fin de poder comparar los datos, se incluyen también los equivalentes en lana limpia (desgrasada).

Cueros y pieles. Todas las cifras se refieren al peso en fresco de los cueros y pieles. Los datos de los países que han comunicado la producción en número o en peso de los cueros y pieles, secos, curados, o salados, se han convertido al equivalente en peso fresco empleando los oportunos coeficientes de conversión. Cuando no se ha dispuesto de datos oficiales, se dan estimaciones basadas en los sacrificios y en otra información.

DISPONIBILIDAD DE ALIMENTOS

Desde 1971, la FAO se ha dedicado a crear un sistema integrado y elaborado en ordenadores para acopiar y mantener, en forma de cuentas de suministro y utilización (SUAs), estadísticas agrícolas actuales que abarcan unos 240 productos alimenticios y agrícolas primarios, y 290 productos elaborados a partir de éstos, para 215 países y territorios, con series de datos correspondientes al período de 1961 a 1977 inclusive.

La cantidad total de alimentos producidos en un país, añadida a la cantidad total importada, y reajustada en atención a todos los cambios que pueda haber en las existencias desde principios del período de referencia, da el *suministro* disponible durante ese período. Desde el punto de vista de la *utilización*, se distingue entre las cantidades exportadas, las dadas como alimento al ganado, las utilizadas para semilla, las destinadas a usos industriales y otros no alimentarios o las perdidas durante el alma-

cenamiento y el transporte y los suministros de alimentos disponibles para el consumo humano al por menor, es decir, en la forma en que el alimento sale de la tienda de venta al por menor o bien entra en el hogar. El suministro de cada producto alimenticio disponible per cápita para el consumo humano se obtiene dividiendo las existencias alimenticias disponibles para ese consumo por los datos afines de la población que realmente participa en él. Los datos sobre suministros alimentarios por persona se expresan en términos de cantidad y también mediante la aplicación de coeficientes apropiados de composición de los alimentos, en términos de elementos nutrientes (calorías, proteínas y grasas). En el Anuario se indican sólo los elementos nutrientes para atender las necesidades inmediatas de muchos usuarios respecto al análisis y la evaluación de la situación alimentaria y nutricional en los países indicados.

Es importante notar que las cantidades de alimentos disponibles se refieren simplemente a las de productos alimenticios que llegan al consumidor, pero no necesariamente a las de los alimentos realmente consumidos, que pueden ser inferiores a las cantidades indicadas, según el grado de las pérdidas de comestibles y nutrientes sufridas en el hogar familiar, por ejemplo, durante el almacenamiento, la preparación y cocinado, los desperdicios que se dejan en el plato o las cantidades suministradas como alimento a los animales domésticos y caseros, o las que se tiran.

Las cifras dadas no ofrecen una indicación de las diferencias que pueden existir en la alimentación de grupos distintos de población, por ejemplo diferentes grupos socioeconómicos, zonas ecológicas y áreas geográficas dentro de un país, ni proporcionan información alguna sobre variaciones estacionales en el abastecimiento alimentario total. Dichas cifras representan únicamente el promedio de suministros para el conjunto de la población, y no indican lo que en realidad consume cada individuo.

Se han preparado y enviado a los países, para que presenten sus observaciones sobre las mismas, hojas de balance de alimentos provisionales y detalladas para el promedio del período 1972-74, junto con series a largo plazo de las disponibilidades de alimentos por persona, para el promedio del período 1961-63 y para cada uno de los años 1964 a 1974, ordenadas por grupos de alimentos principales y expresadas en calorías, proteínas y grasas.

MEDIOS DE PRODUCCION

Maquinaria agrícola

Tractores. Los datos se refieren en general al número total de tractores de ruedas y de oruga (excluidos los tractores hortícolas) que se emplean en la agricultura.

Ordeñadoras mecánicas. En este cuadro figura el número de ordeñadoras mecánicas o, tratándose de determinados países, el de granjas que las usan. Los datos del Reino Unido excluyen las ordeñadoras mecánicas en Escocia e Irlanda del Norte. Los datos de Australia se refieren al número de unidades (por capacidad de vaca).

Plaguicidas

Los datos se refieren en general a las cantidades de plaguicidas empleados en la agricultura o vendidos a los agricultores. (Los datos de México se refieren a la producción; los datos de Nicaragua se refieren a las importaciones; los datos de los Estados Unidos se refieren a las ventas, a menos que se indique lo contrario.) Se consignan expresados en ingredientes activos, salvo en el caso de los países que figuran a continuación, cuyos datos se refieren al peso del preparado: Botswana, Colombia, Costa Rica, Cuba, Chile, Ecuador, Egipto, Francia, Guadalupe, Guatemala, Hungría, Islas Vírgenes (EE.UU.), Italia, Jordania, Kuwait, Marruecos, Martinica, Mauricio, Nicaragua, Nigeria, Puerto Rico, República Democrática Alemana, Reunión, Samoa, Sudáfrica, Suecia, Trinidad y Tabago, Uganda y Yugoslavia. El peso del preparado, por lo común, incluye el ingrediente activo más diluyentes y coadyuvantes; por esta razón y, asimismo, por lo incompleto de los datos, no ha sido posible calcular los totales.

Al examinar los cuadros de los productos que se mencionan a continuación, deberán tenerse presentes los puntos que se indican seguidamente:

DDT. Los datos de Argentina incluyen las exportaciones.

BHC. Los datos de Austria incluyen el lindano.

Lindano. Los datos de Hungría incluyen el BHC.

Aldrina y otros. Los datos de Austria incluyen el toxafeno. Hasta 1974, los datos de los Estados Unidos incluyen el toxafeno.

Otros hidrocarburos clorados. Los datos de Puerto Rico incluyen todos los hidrocarburos clorados, con excepción del DDT y el clordano. Los datos de Cuba, Ecuador, Hong Kong, Israel, Noruega, Nueva Zelandia y Sudáfrica se refieren a todos los hidrocarburos clorados.

Paratión. Los datos de Suecia se refieren a todos los compuestos orgánicos de fósforo.

Otros compuestos orgánicos de fósforo. Los datos de Cuba, Hong Kong, Islandia, Israel, Jordania, Noruega, Puerto Rico y Sudáfrica incluyen todos los compuestos orgánicos de fósforo.

Piretro. Los datos de los Estados Unidos se refieren a las importaciones.

Otros insecticidas botánicos. Los datos de los Estados Unidos se refieren a las importaciones. Los de Noruega comprenden todos los insecticidas botánicos.

Otros insecticidas. Los datos de Colombia, Dinamarca, Guadalupe, Guatemala, Martinica, Marruecos, Mauricio, Reunión, Trinidad y Tabago, Uganda y Yugoslavia incluyen todos los insecticidas. Los datos de Puerto Rico incluyen todos los insecticidas, a excepción de los que figuran por separado. Los datos de Bahrein incluyen todos los insecticidas, excepto el paratión.

Otros fungicidas. Los datos de Bahrein, Colombia, Cuba, Ecuador, Guadalupe, Guatemala, Hong Kong, Jordania, Marruecos, Mauricio, Puerto Rico, Reunión, Sudáfrica, Trinidad y Tabago, Uganda y Yugoslavia incluyen todos los fungicidas. Los datos de Martinica incluyen todos los fungicidas, pero no incluyen los compuestos de cobre, que se indican por separado.

2,4-D. Los datos de Canadá y Noruega incluyen mezclas de 2,4-D y 2,4,5-T. Los datos de la India incluyen otros herbicidas.

Otros herbicidas. Los datos de Bahrein, Colombia, Cuba, Dinamarca, Ecuador, Guadalupe, Guatemala, Hong Kong, Marruecos, Martinica, Puerto Rico, República Democrática Alemana, Reunión, Suiza, Uganda y Yugoslavia incluyen todos los herbicidas. Los datos de Nueva Zelandia se refieren a todos los herbicidas, salvo el 2,4,5-T. Los datos de Suecia incluyen todos los herbicidas excepto los que se indican por separado.

Bromuros. Los datos de Israel incluyen los fumigantes carbamatos.

Otros fumigantes. Los datos de Canadá, Guatemala, la República Federal de Alemania y Sudáfrica se refieren a todos los fumigantes.

Otros rodenticidas. Los datos de Dinamarca, Mauricio y Puerto Rico incluyen todos los rodenticidas. Los datos de los Estados Unidos se refieren a las importaciones.

Otros plaguicidas. Los datos de Costa Rica incluyen todos los insecticidas y fungicidas. Los de la República Democrática Alemana se refieren a todos los plaguicidas, exceptuados los herbicidas. Los datos para Nueva Zelandia se refieren a todos los insecticidas, salvo los hidrocarburos clorados, los fungicidas y fumigantes.

PRECIOS

Como medida encaminada al uso de una serie sencilla de coeficientes de conversión en diversas publicaciones estadísticas de la FAO, se ha utilizado el promedio ponderado de los coeficientes comerciales de conversión, según se ha publicado en *International Financial Statistics*, como tipo de cambio para convertir los datos relativos a las monedas nacionales en dólares EE.UU., armonizando así estas conversiones con las prácticas utilizadas para los valores comerciales.

Las series de precios se obtienen de muchas fuentes: cuestionarios especiales, grupos internacionales de productos básicos, literatura oficial y extraoficial y, en algunos casos, corresponsales especiales.

Puede todavía acudirse a la « Lista de las fuentes de los cuadros de precios », publicada en el *Anuario de Producción* de 1960 para las series de precios contenidas en ese número, como guía general, si bien no sigue la misma presentación de los cuadros adoptada en este volumen.

Las unidades cuantitativas nacionales a las que corresponden los precios originales se expresan en su equivalente en unidades métricas o británicas.

NOTAS SOBRE PAISES

Berlín

Los datos referentes a la República Federal de Alemania y a la República Democrática Alemana comprenden los correspondientes a Berlín, ciudad para la que no se dispone de datos por separado respecto a la mayoría de los productos.

China

Los datos correspondientes a China comprenden generalmente los de la provincia de Taiwán.

Las estimaciones de la producción agrícola y ganadera de China presentan una de las principales dificultades con que se tropieza en la preparación del *Anuario FAO de Producción*. China es el país más poblado del mundo y su producción agrícola, que representa al parecer casi el 12 por ciento del total mundial, sigue a la de los Estados Unidos y a la de la U.R.S.S.

Sin embargo, no se han publicado estadísticas oficiales apropiadas para China durante los últimos 19 años. Por consiguiente, las estimaciones hechas en este Anuario respecto a dicho país se basan en los datos oficiales fragmentarios comunicados por los medios chinos de información de masa y en evaluaciones de personas competentes que viven en China o en países y territorios vecinos, o que los visitan. También se han tenido en cuenta estadísticas comerciales procedentes de publicaciones de asociados comerciales de China, así como la información disponible sobre fenómenos meteorológicos, servicios de riego, mecanización de la agricultura, consumo de fertilizantes, etc.

Chipre

Debido a la situación actual de Chipre, es posible que los datos correspondientes a años recientes no abarquen a todo el país.

India y Pakistán

Los datos relativos a Cachemira-Jammu, cuya condición jurídica definitiva no se ha determinado aún, se incluyen generalmente en las cifras correspondientes a la India y no en las de Pakistán.

Los datos relativos a Sikkim se incluyen en los que corresponden a la India.

Jordania

Salvo en el Cuadro 1 (Aprovechamiento de tierras y riego), los datos a partir de 1967 se refieren sólo a la orilla oriental del Jordán.

Timor oriental

Los datos correspondientes a años recientes se incluyen por lo general en los datos de Indonesia.

CLASIFICACION DE LOS PAISES POR CLASES ECONOMICAS Y REGIONES

En todos los cuadros en que figuran totales continentales aparecen también los regionales, incluso cuando éstos coinciden exactamente con aquéllos o cuando, a base de los primeros, se hubieran podido calcular fácilmente para evitar toda posible ambigüedad.

Las clases económicas y regiones en las cuales se ha dividido el mundo, para los estudios analíticos de la FAO, son las siguientes:

CLASE I: ECONOMIAS DE MERCADO DESARROLLADAS

Región A - América del Norte: Canadá, Estados Unidos.

Región B - Europa occidental: Alemania (República Federal de) (inclusive Berlín occidental), Andorra, Austria, Bélgica-Luxemburgo, Dinamarca, España, Finlandia, Francia, Gibraltar, Grecia, Irlanda, Islandia, Islas Feroé, Italia, Liechtenstein, Malta, Mónaco, Noruega, Países Bajos, Portugal (inclusive Azores y Madera), Reino Unido (inclusive Islas Normandas, Isla de Man), San Marino, Santa Sede, Suecia, Suiza, Yugoslavia.

Región C - Oceanía: Australia, Nueva Zelandia.

Región D - Otras economías de mercado desarrolladas: Israel, Japón (inclusive islas Bonin y Ryukyu), Sudáfrica.

CLASE II: ECONOMIAS DE MERCADO EN DESARROLLO

Región A - Africa: Alto Volta, Angola, Argelia, Benin, Botswana, Burundi, Cabo Verde, Camerún, Comoras, Congo, Costa de Marfil, Chad, Djibouti, Etiopía, Gabón, Gambia, Ghana, Guinea, Guinea-Bissau, Guinea ecuatorial, Imperio Centroafricano, Kenya, Lesotho, Liberia, Madagascar, Malawi, Malí, Marruecos, Mauricio, Mauritania, Mozambique, Namibia, Níger, Nigeria, Norte de Africa española, Reunión, Rhodesia, Rwanda, Sáhara Occidental, Santa Elena, Santo Tomé y Príncipe, Senegal, Seychelles, Sierra Leona, Somalia, Swazilandia, Tanzania, Territorio británico del Océano Indico, Togo, Túnez, Uganda, Zaire, Zambia.

Región B - América Latina: Antigua, Antillas neerlandesas, Argentina, Bahamas, Barbados, Belize, Bolivia, Brasil, Colombia, Costa Rica, Cuba, Chile, Dominica, Ecuador (inclusive Archipiélago de Colón), El Salvador, Granada, Guadalupe, Guatemala, Guayana francesa, Guyana, Haití, Hondu-

ras, Islas Caimán, Islas Malvinas (Falkland), Islas del Turco y Caicos, Islas Vírgenes (EE.UU.), Islas Vírgenes (Reino Unido), Jamaica, Martinica, México, Montserrat, Nicaragua, Panamá, Panamá (Zona del Canal), Paraguay, Perú, Puerto Rico, República Dominicana, San Cristóbal y Nieves-Anguilla, Santa Lucía, San Vicente, Surinam, Trinidad y Tabago, Uruguay, Venezuela.

Región C - Cercano Oriente: Africa: Egipto, Libia, Sudán. *Asia:* Afganistán, Arabia Saudita, Bahrein, Chipre, Emiratos Arabes Unidos, Gaza, faja de (Palestina), Irak, Irán, Jordania, Kuwait, Líbano, Omán, Qatar, República Arabe del Yemen, Siria, Turquía, Yemen Democrático.

Región D - Lejano Oriente: Bangladesh, Bhutan, Birmania, Brunei, Corea (República de), Filipinas, Hong Kong, India, Indonesia, Lao, Macao, Malasia (Malasia Peninsular, Sabah, Sarawak), Maldivas, Nepal, Pakistán, Singapur, Sri Lanka, Tailandia, Timor oriental.

Región E - Otras economías de mercado en desarrollo: América: Bermudas, Groenlandia, San Pedro y Miquelón. *Oceanía:* Fiji, Guam, Isla Christmas (Australia), Isla Niué, Isla Norfolk, Isla Pitcairn, Isla Wake, Islas Canton y Enderbury, Islas Cocos, Islas Cook, Islas del Pacífico (Territorios en fideicomiso), Islas Gilbert, Islas Johnston, Islas Midway, Islas Salomón, Islas Wallis y Futuna, Nauru, Nueva Caledonia, Nuevas Hébridas, Papua Nueva Guinea, Polinesia francesa, Samoa, Samoa americana, Tokelau, Tonga, Tuvalu.

CLASE III: ECONOMIAS DE PLANIFICACION CENTRALIZADA

Región A - Asia: Corea (República Democrática Popular de), China, Kampuchea Democrática, Mongolia, Viet Nam.

Región B - Europa oriental y U.R.S.S.: Albania, Bulgaria, Checoslovaquia, Hungría, Polonia, República Democrática Alemana (inclusive Berlín oriental), Rumania, U.R.S.S.

TODOS LOS PAISES DESARROLLADOS

Incluye las economías de mercado desarrolladas y la Región B de las Economías de planificación centralizada.

TODOS LOS PAISES EN DESARROLLO

Incluye las economías de mercado en desarrollo y la Región A de las Economías de planificación centralizada.

Notice

The designations employed and the presentation of material in this publication do not imply the expression of any opinion whatsoever on the part of the Food and Agriculture Organization of the United Nations concerning the legal status of any country, territory, city or area or of its authorities, or concerning the delimitation of its frontiers or boundaries. The designations " developed " and " developing " economies are intended for statistical convenience and do not necessarily express a judgement about the stage reached by a particular country or area in the development process.

Avertissement

Les appellations employées dans cette publication et la présentation des données qui y figurent n'impliquent, de la part de l'Organisation des Nations Unies pour l'alimentation et l'agriculture, aucune prise de position quant au statut juridique des pays, territoires, villes ou zones, ou de leurs autorités, ni quant au tracé de leurs frontières ou limites. Les expressions « pays développé » et « pays en développement » sont utilisées pour des raisons de commodité statistique et n'expriment pas nécessairement un jugement quant au niveau de développement atteint par tel ou tel pays ou région.

Aviso

Las denominaciones empleadas en esta publicación y la forma en que aparecen presentados los datos que contiene no implican, de parte de la Organización de las Naciones Unidas para la Agricultura y la Alimentación, juicio alguno sobre la condición jurídica de países, territorios, ciudades o zonas, o de sus autoridades, ni respecto de la delimitación de sus fronteras o límites. Las definiciones de economías « desarrolladas » y « en desarrollo » se usan para fines estadísticos y no representan un juicio acerca del nivel alcanzado en el proceso de desarrollo por un país o área determinados.

LIST OF COUNTRIES, CONTINENTS, ECONOMIC CLASSES AND REGIONS
shown in the order in which they appear in the tables

LISTE DES PAYS, CONTINENTS, CATÉGORIES ÉCONOMIQUES ET RÉGIONS
suivant le même ordre que dans les tableaux

LISTA DE PAISES, CONTINENTES, CLASES ECONOMICAS Y REGIONES
en el orden en que aparecen en los cuadros

Abbreviated name in English	English	Français	Español
WORLD	WORLD	MONDE	MUNDO
AFRICA	AFRICA	AFRIQUE	AFRICA
ALGERIA	Algeria	Algérie	Argelia
ANGOLA	Angola	Angola	Angola
BENIN	Benin	Bénin	Benin
BOTSWANA	Botswana	Botswana	Botswana
BR IND OC TR	British Indian Ocean Territory	Territoire britannique de l'océan Indien	Territorio británico del Océano Indico
BURUNDI	Burundi	Burundi	Burundi
CAMEROON	Cameroon	Cameroun	Camerún
CAP VERDE	Cape Verde	Cap-Vert	Cabo Verde
CENT AFR EMP	Central African Empire	Empire centrafricain	Imperio Centroafricano
CHAD	Chad	Tchad	Chad
COMOROS	Comoros	Comores	Comoras
CONGO	Congo	Congo	Congo
DJIBOUTI	Djibouti	Djibouti	Djibouti
EGYPT	Egypt	Egypte	Egipto
EQ GUINEA	Equatorial Guinea	Guinée équatoriale	Guinea ecuatorial
ETHIOPIA	Ethiopia	Ethiopie	Etiopía
GABON	Gabon	Gabon	Gabón
GAMBIA	Gambia	Gambie	Gambia
GHANA	Ghana	Ghana	Ghana
GUINEA	Guinea	Guinée	Guinea
GUIN BISSAU	Guinea-Bissau	Guinée-Bissau	Guinea-Bissau
IVORY COAST	Ivory Coast	Côte-d'Ivoire	Costa de Marfil
KENYA	Kenya	Kenya	Kenya
LESOTHO	Lesotho	Lesotho	Lesotho
LIBERIA	Liberia	Libéria	Liberia
LIBYA	Libya	Libye	Libia
MADAGASCAR	Madagascar	Madagascar	Madagascar
MALAWI	Malawi	Malawi	Malawi
MALI	Mali	Mali	Malí
MAURITANIA	Mauritania	Mauritanie	Mauritania
MAURITIUS	Mauritius	Maurice	Mauricio
MOROCCO	Morocco	Maroc	Marruecos
MOZAMBIQUE	Mozambique	Mozambique	Mozambique
NAMIBIA	Namibia	Namibie	Namibia
NIGER	Niger	Niger	Níger
NIGERIA	Nigeria	Nigéria	Nigeria
REUNION	Reunion	Réunion	Reunión
RHODESIA	Rhodesia	Rhodésie	Rhodesia
RWANDA	Rwanda	Rwanda	Rwanda
ST HELENA	St. Helena	Ste-Hélène	Santa Elena
SAO TOME ETC	São Tomé and Principe	St-Thomas et Prince	Santo Tomé y Príncipe
SENEGAL	Senegal	Sénégal	Senegal
SEYCHELLES	Seychelles	Seychelles	Seychelles
SIERRA LEONE	Sierra Leone	Sierra Leone	Sierra Leona

Abbreviated name in English	*English*	*Français*	*Español*
SOMALIA	Somalia	Somalie	Somalia
SOUTH AFRICA	South Africa	Afrique du Sud	Sudáfrica
SP NO AFRICA	Spanish North Africa	Afrique du Nord espagnole	Africa del Norte española
SUDAN	Sudan	Soudan	Sudán
SWAZILAND	Swaziland	Swaziland	Swazilandia
TANZANIA	Tanzania	Tanzanie	Tanzania
TOGO	Togo	Togo	Togo
TUNISIA	Tunisia	Tunisie	Túnez
UGANDA	Uganda	Ouganda	Uganda
UPPER VOLTA	Upper Volta	Haute-Volta	Alto Volta
WESTN SAHARA	Western Sahara	Sahara occidental	Sáhara occidental
ZAIRE	Zaire	Zaïre	Zaire
ZAMBIA	Zambia	Zambie	Zambia

N C AMERICA	NORTH AND CENTRAL AMERICA	AMÉRIQUE DU NORD ET CENTRALE	AMERICA DEL NORTE Y CENTRAL
ANTIGUA	Antigua	Antigoa	Antigua
BAHAMAS	Bahamas	Bahama	Bahamas
BARBADOS	Barbados	Barbade	Barbados
BELIZE	Belize	Belize	Belize
BERMUDA	Bermuda	Bermudes	Bermudas
CANADA	Canada	Canada	Canadá
CAYMAN IS	Cayman Islands	Iles Caïmanes	Islas Caimán
COSTA RICA	Costa Rica	Costa Rica	Costa Rica
CUBA	Cuba	Cuba	Cuba
DOMINICA	Dominica	Dominique	Dominica
DOMINICAN RP	Dominican Republic	République Dominicaine	República Dominicana
EL SALVADOR	El Salvador	El Salvador	El Salvador
GREENLAND	Greenland	Groenland	Groenlandia
GRENADA	Grenada	Grenade	Granada
GUADELOUPE	Guadeloupe	Guadeloupe	Guadalupe
GUATEMALA	Guatemala	Guatemala	Guatemala
HAITI	Haiti	Haïti	Haití
HONDURAS	Honduras	Honduras	Honduras
JAMAICA	Jamaica	Jamaïque	Jamaica
MARTINIQUE	Martinique	Martinique	Martinica
MEXICO	Mexico	Mexique	México
MONTSERRAT	Montserrat	Montserrat	Montserrat
NETH ANTILLE	Netherlands Antilles	Antilles néerlandaises	Antillas neerlandesas
NICARAGUA	Nicaragua	Nicaragua	Nicaragua
PANAMA	Panama	Panama	Panamá
PANAMA CA ZN	Panama Canal Zone	Zone du canal de Panama	Panamá (Zona del Canal)
PUERTO RICO	Puerto Rico	Porto Rico	Puerto Rico
ST KITTS ETC	St. Kitts-Nevis-Anguilla	St-Christophe-Nevis-Anguilla	San Cristóbal y Nieves-Anguilla
ST LUCIA	St. Lucia	Ste-Lucie	Santa Lucía
ST PIER ETC	St. Pierre and Miquelon	St-Pierre-et-Miquelon	San Pedro y Miquelón
ST VINCENT	St. Vincent	St-Vincent	San Vicente
TRINIDAD ETC	Trinidad and Tobago	Trinité-et-Tobago	Trinidad y Tabago
TURKS CAICOS	Turks and Caicos Islands	Iles Turques et Caïques	Islas del Turco y Caicos
USA	United States	Etats-Unis	Estados Unidos
VIRGIN IS UK	Virgin Islands (U.K.)	Iles Vierges (R.-U.)	Islas Vírgenes (Reino Unido)
VIRGIN IS US	Virgin Islands (U.S.)	Iles Vierges (E.-U.)	Islas Vírgenes (EE.UU.)

SOUTH AMERIC	SOUTH AMERICA	AMÉRIQUE DU SUD	AMERICA DEL SUR
ARGENTINA	Argentina	Argentine	Argentina
BOLIVIA	Bolivia	Bolivie	Bolivia
BRAZIL	Brazil	Brésil	Brasil
CHILE	Chile	Chili	Chile
COLOMBIA	Colombia	Colombie	Colombia
ECUADOR	Ecuador	Equateur	Ecuador
FALKLAND IS	Falkland Islands (Malvinas)	Iles Falkland (Malvinas)	Islas Malvinas (Falkland)
FR GUIANA	French Guiana	Guyane française	Guayana francesa
GUYANA	Guyana	Guyane	Guyana
PARAGUAY	Paraguay	Paraguay	Paraguay

Abbreviated name in English	English	Français	Español
PERU	Peru	Pérou	Perú
SURINAM	Surinam	Surinam	Surinam
URUGUAY	Uruguay	Uruguay	Uruguay
VENEZUELA	Venezuela	Venezuela	Venezuela
ASIA	ASIA	ASIE	ASIA
AFGHANISTAN	Afghanistan	Afghanistan	Afganistán
BAHRAIN	Bahrain	Bahreïn	Bahrein
BANGLADESH	Bangladesh	Bangladesh	Bangladesh
BHUTAN	Bhutan	Bhoutan	Bhután
BRUNEI	Brunei	Brunéi	Brunei
BURMA	Burma	Birmanie	Birmania
CHINA	China	Chine	China
CYPRUS	Cyprus	Chypre	Chipre
EAST TIMOR	East Timor	Timor oriental	Timor oriental
GAZA STRIP	Gaza Strip (Palestine)	Zone de Gaza (Palestine)	Zona de Gaza (Palestina)
HONG KONG	Hong Kong	Hong-kong	Hong Kong
INDIA	India	Inde	India
INDONESIA	Indonesia	Indonésie	Indonesia
IRAN	Iran	Iran	Irán
IRAQ	Iraq	Irak	Irak
ISRAEL	Israel	Israël	Israel
JAPAN	Japan	Japon	Japón
JORDAN	Jordan	Jordanie	Jordania
KAMPUCHEA DM	Kampuchea, Democratic	Kampuchea démocratique	Kampuchea Democrática
KOREA DPR	Korea, Democratic People's Republic of	Corée, République populaire démocratique de	Corea, República Popular Democrática de
KOREA REP	Korea, Republic of	Corée, République de	Corea, República de
KUWAIT	Kuwait	Koweït	Kuwait
LAO	Lao	Lao	Lao
LEBANON	Lebanon	Liban	Líbano
MACAU	Macau	Macao	Macao
MAL: PENINSUL	Malaysia: Peninsular Malaysia	Malaisie: Malaisie péninsulaire	Malasia: Malasia Peninsular
MAL: SABAH	Malaysia: Sabah	Malaisie: Sabah	Malasia: Sabah
MAL: SARAWAK	Malaysia: Sarawak	Malaisie: Sarawak	Malasia: Sarawak
MALDIVES	Maldives	Maldives	Maldivas
MONGOLIA	Mongolia	Mongolie	Mongolia
NEPAL	Nepal	Népal	Nepal
OMAN	Oman	Oman	Omán
PAKISTAN	Pakistan	Pakistan	Pakistán
PHILIPPINES	Philippines	Philippines	Filipinas
QATAR	Qatar	Katar	Qatar
SAUDI ARABIA	Saudi Arabia	Arabie saoudite	Arabia Saudita
SINGAPORE	Singapore	Singapour	Singapur
SRI LANKA	Sri Lanka	Sri Lanka	Sri Lanka
SYRIA	Syria	Syrie	Siria
THAILAND	Thailand	Thaïlande	Tailandia
TURKEY	Turkey	Turquie	Turquía
U A EMIRATES	United Arab Emirates	Emirats arabes unis	Emiratos Arabes Unidos
VIET NAM	Viet Nam	Viet Nam	Viet Nam
YEMEN AR	Yemen Arab Republic	République arabe du Yémen	República Arabe del Yemen
YEMEN DEM	Yemen, Democratic	Yémen démocratique	Yemen Democrático
EUROPE	EUROPE	EUROPE	EUROPA
ALBANIA	Albania	Albanie	Albania
ANDORRA	Andorra	Andorre	Andorra
AUSTRIA	Austria	Autriche	Austria
BELGIUM-LUX	Belgium-Luxembourg	Belgique-Luxembourg	Bélgica-Luxemburgo
BULGARIA	Bulgaria	Bulgarie	Bulgaria
CZECHOSLOVAK	Czechoslovakia	Tchécoslovaquie	Checoslovaquia
DENMARK	Denmark	Danemark	Dinamarca
FAEROE IS	Faeroe Islands	Iles Féroé	Islas Feroé
FINLAND	Finland	Finlande	Finlandia
FRANCE	France	France	Francia

Abbreviated name in English	English	Français	Español
GERMAN DR	German Democratic Republic	République démocratique allemande	República Democrática Alemana
GERMANY FED	Germany, Federal Republic of	Allemagne, République fédérale d'	Alemania, República Federal de
GIBRALTAR	Gibraltar	Gibraltar	Gibraltar
GREECE	Greece	Grèce	Grecia
HOLY SEE	Holy See	Saint-Siège	Santa Sede
HUNGARY	Hungary	Hongrie	Hungría
ICELAND	Iceland	Islande	Islandia
IRELAND	Ireland	Irlande	Irlanda
ITALY	Italy	Italie	Italia
LIECHTENSTEN	Liechtenstein	Liechtenstein	Liechtenstein
MALTA	Malta	Malte	Malta
MONACO	Monaco	Monaco	Mónaco
NETHERLANDS	Netherlands	Pays-Bas	Países Bajos
NORWAY	Norway	Norvège	Noruega
POLAND	Poland	Pologne	Polonia
PORTUGAL	Portugal	Portugal	Portugal
ROMANIA	Romania	Roumanie	Rumania
SAN MARINO	San Marino	Saint-Marin	San Marino
SPAIN	Spain	Espagne	España
SWEDEN	Sweden	Suède	Suecia
SWITZERLAND	Switzerland	Suisse	Suiza
UK	United Kingdom	Royaume-Uni	Reino Unido
YUGOSLAVIA	Yugoslavia	Yougoslavie	Yugoslavia
OCEANIA	OCEANIA	OCÉANIE	OCEANIA
AMER SAMOA	American Samoa	Samoa américaines	Samoa americana
AUSTRALIA	Australia	Australie	Australia
CANTON IS	Canton and Enderbury Islands	Iles Canton et Enderbury	Islas Cantón y Enderbury
CHRISTMAS IS	Christmas Island (Aust.)	Ile Christmas (aust.)	Isla Christmas (Aust.)
COCOS IS	Cocos (Keeling) Islands	Iles Cocos (Keeling)	Islas Cocos (Keeling)
COOK ISLANDS	Cook Islands	Iles Cook	Islas Cook
FIJI	Fiji	Fidji	Fiji
FR POLYNESIA	French Polynesia	Polynésie française	Polinesia francesa
GILBERT IS	Gilbert Islands	Iles Gilbert	Islas Gilbert
GUAM	Guam	Guam	Guam
JOHNSTON IS	Johnston Island	Ile Johnston	Isla Johnston
MIDWAY IS	Midway Islands	Iles Midway	Islas Midway
NAURU	Nauru	Nauru	Nauru
NEWCALEDONIA	New Caledonia	Nouvelle-Calédonie	Nueva Caledonia
NEW HEBRIDES	New Hebrides	Nouvelles-Hébrides	Nuevas Hébridas
NEW ZEALAND	New Zealand	Nouvelle-Zélande	Nueva Zelandia
NIUE ISLAND	Niue Island	Ile Nioué	Isla Niué
NORFOLK IS	Norfolk Island	Ile Norfolk	Isla Norfolk
PACIFIC IS	Pacific Islands (Trust Territory)	Iles du Pacifique (Territoire sous tutelle)	Islas del Pacífico (Territorios en fideicomiso)
PAPUA N GUIN	Papua New Guinea	Papouasie Nouvelle-Guinée	Papua Nueva Guinea
PITCAIRN IS	Pitcairn Island	Ile Pitcairn	Isla Pitcairn
SAMOA	Samoa	Samoa	Samoa
SOLOMON IS	Solomon Islands	Iles Salomon	Islas Salomón
TOKELAU	Tokelau	Tokélaou	Tokelau
TONGA	Tonga	Tonga	Tonga
TUVALU	Tuvalu	Tuvalu	Tuvalu
WAKE ISLAND	Wake Island	Ile de Wake	Isla Wake
WALLIS ETC	Wallis and Futuna Islands	Iles Wallis et Futuna	Islas Wallis y Futuna
USSR	Union of Soviet Socialist Republics	Union des Républiques socialistes soviétiques	Unión de Repúblicas Socialistas Soviéticas

Abbreviated name in English	English	Français	Español
DEV.PED M E	Developed market economies	Pays développés à économies de marché	Economías de mercado desarrolladas
N AMERICA	North America	Amérique du Nord	América del Norte
W EUROPE	Western Europe	Europe occidentale	Europa occidental
OCEANIA	Oceania	Océanie	Oceanía
OTH DEV.PED	Other developed market economies	Autres pays développés à économies de marché	Otras economías de mercado desarrolladas
DEV.PING M E	Developing market economies	Pays en développement à économies de marché	Economías de mercado en desarrollo
AFRICA	Africa	Afrique	Africa
LAT AMERICA	Latin America	Amérique latine	América Latina
NEAR EAST	Near East	Proche-Orient	Cercano Oriente
FAR EAST	Far East	Extrême-Orient	Lejano Oriente
OTH DV.PING	Other developing market economies	Autres pays en développement à économies de marché	Otras economías de mercado en desarrollo
CENTR PLANND	Centrally planned economies	Pays à économies centralement planifiées	Economías de planificación centralizada
ASIAN C P E	Asian centrally planned economies	Pays d'Asie à économies centralement planifiées	Economías de planificación centralizada de Asia
E EUR + USSR	Eastern Europe and U.S.S.R.	Europe orientale et U.R.S.S.	Europa oriental y U.R.S.S.
DEV.PED ALL	All developed countries	Tous les pays développés	Todos los países desarrollados
DEV.PING ALL	All developing countries	Tous les pays en développement	Todos los países en desarrollo

NOTE: The geographical coverage of the Economic Classes and Regions is given on page 11.

NOTE: Les pays inclus dans les Catégories économiques et les régions sont indiqués à la page 23.

NOTA: Los países que se incluyen en las Clases Económicas y las Regiones se indican en la página 35.

LAND
TERRES
TIERRAS

TABLE
TABLEAU **1**
CUADRO

LAND USE   UTILISATION DES TERRES   USO DE TIERRAS

1000 HA

	1961-65	1966	1971	1976
WORLD				
TOTAL AREA	13395159	13395161	13395217	13395241
LAND AREA	13078565	13078540	13078547	13078328
ARAB&PERM CR	1393100	1416510	1462710	1488050
ARABLE LAND	1314381	1334876	1376034	1397734
PERM CROPS	78717	81634	86676	90316
PERM PASTURE	3047932	3056365	3066849	3058083
FOREST&WOODL	4158547	4149520	4146326	4145215
OTHER LAND	4478986	4456145	4402662	4386980
AFRICA				
TOTAL AREA	3031168	3031168	3031168	3031168
LAND AREA	2964697	2964697	2964697	2964613
ARAB&PERM CR	189370	196134	204021	209375
ARABLE LAND	177269	182833	190228	194910
PERM CROPS	12098	13301	13793	14465
PERM PASTURE	807369	803163	801481	800437
FOREST&WOODL	666358	655226	645429	639602
OTHER LAND	1301600	1310174	1313766	1315199
ALGERIA				
TOTAL AREA	238174	238174	238174	238174
LAND AREA	238174	238174	238174	238174
ARAB&PERM CR	6863	6784	6981	7110F
ARABLE LAND	6261	6221	6380	6500F
PERM CROPS	601	563	601	610F
PERM PASTURE	37780	37383	38452	38452*
FOREST&WOODL	2549	2443	2424	2424*
OTHER LAND	190982	191564	190317	190188
ANGOLA				
TOTAL AREA	124670	124670	124670	124670
LAND AREA	124670	124670	124670	124670
ARAB&PERM CR	1570F	1700F	1820F	1830F
ARABLE LAND	1050F	1150F	1270F	1280F
PERM CROPS	520F	550*	550*	550*
PERM PASTURE	29000*	29000*	29000*	29000*
FOREST&WOODL	72660*	72660*	72660*	72660*
OTHER LAND	21440	21310	21190	21180
BENIN				
TOTAL AREA	11262	11262	11262	11262
LAND AREA	11062	11062	11062	11062
ARAB&PERM CR	2940	2950F	2950F	2950F
ARABLE LAND	2930	2940F	2940F	2940F
PERM CROPS	10F	10F	10F	10F
PERM PASTURE	442	442*	442*	442*
FOREST&WOODL	2157	2157*	2144*	2144*
OTHER LAND	5523	5513	5526	5526
BOTSWANA				
TOTAL AREA	60037	60037	60037	60037
LAND AREA	58537	58537	58537	58537
ARAB&PERM CR	1000F	1040F	1200F	1360
ARABLE LAND	1000F	1039F	1199F	1360
PERM CROPS		1	1	
PERM PASTURE	41800F	42000F	43000F	43983
FOREST&WOODL	958	958	962	962*
OTHER LAND	14779	14539	13375	12232
BR IND OC TR				
TOTAL AREA	8	8	8	8
LAND AREA	8	8	8	8
OTHER LAND	8	8	8	8
BURUNDI				
TOTAL AREA	2783	2783	2783	2783
LAND AREA	2565	2565	2565	2565
ARAB&PERM CR	952	993	1189	1260F
ARABLE LAND	801	837	1018	1075F
PERM CROPS	151	156	171	185F
PERM PASTURE	648	628	435	435*
FOREST&WOODL	67	69	70	80*
OTHER LAND	898	875	871	790
CAMEROON				
TOTAL AREA	47544	47544	47544	47544
LAND AREA	46944	46944	46944	46944
ARAB&PERM CR	7140F	7200F	7300	7345F
ARABLE LAND	6616F	6650F	6722	6760F
PERM CROPS	524	550F	578	585F
PERM PASTURE	8866*	8866*	8300	8300*
FOREST&WOODL	30000*	30000*	30000	30000*
OTHER LAND	938	878	1344	1299
CAPE VERDE				
TOTAL AREA	403	403	403	403
LAND AREA	403	403	403	403
ARAB&PERM CR	40*	40F	40*	40*
ARABLE LAND	38*	38*	38*	38*
PERM CROPS	2*	2*	2*	2*
PERM PASTURE	25*	25*	25*	25*
FOREST&WOODL	1	1	1	1
OTHER LAND	337	337	337	337

	1961-65	1966	1971	1976
CENT AFR EMP				
TOTAL AREA	62298	62298	62298	62298
LAND AREA	62298	62298	62298	62298
ARAB&PERM CR	5901	5900	5500*	5910*
ARABLE LAND	5842	5840	5840*	5840*
PERM CROPS	58	60	60*	70F
PERM PASTURE	100	100	100*	100*
FOREST&WOODL	7400	7400	7400*	7400*
OTHER LAND	48897	48898	48898	48888
CHAD				
TOTAL AREA	128400	128400	128400	128400
LAND AREA	126000	126000	126000	125920
ARAB&PERM CR	7000	7000	7000	7000*
ARABLE LAND	6997	6997	6997	6997*
PERM CROPS	3*	3*	3*	3*
PERM PASTURE	45000	45000	45000	45000
FOREST&WOODL	16540	16500	16500	16500
OTHER LAND	57460	57500	57500	57420
COMOROS				
TOTAL AREA	217	217	217	217
LAND AREA	217	217	217	217
ARAB&PERM CR	84	90*	90*	90*
ARABLE LAND	69F	75F	75F	75F
PERM CROPS	15F	15F	15F	15F
PERM PASTURE	15	15*	15*	15*
FOREST&WOODL	35	35*	35*	35*
OTHER LAND	83	77	77	77
CONGO				
TOTAL AREA	34200	34200	34200	34200
LAND AREA	34150	34150	34150	34150
ARAB&PERM CR	628	622F	632F	662F
ARABLE LAND	616	610F	620F	650F
PERM CROPS	12	12*	12*	12*
PERM PASTURE	14300*	14300*	14300*	14300*
FOREST&WOODL	17300	16500F	16000F	15000F
OTHER LAND	1922	2728	3218	4188
DJIBOUTI				
TOTAL AREA	2200	2200	2200	2200
LAND AREA	2198	2198	2198	2198
ARAB&PERM CR	1*	1*	1*	1*
ARABLE LAND	1*	1*	1*	1*
PERM PASTURE	244*	244*	244*	244*
FOREST&WOODL	8	7F	6*	6*
OTHER LAND	1945	1946	1947	1947
EGYPT				
TOTAL AREA	100145	100145	100145	100145
LAND AREA	99545	99545	99545	99545
ARAB&PERM CR	2548	2780	2852	2826F
ARABLE LAND	2470	2683	2735	2690F
PERM CROPS	78	97	117	136F
FOREST&WOODL	2	2	2	2
OTHER LAND	96995	96763	96691	96717
EQ GUINEA				
TOTAL AREA	2805	2805	2805	2805
LAND AREA	2805	2805	2805	2805
ARAB&PERM CR	217	225F	230F	232F
ARABLE LAND	117	125F	130F	132F
PERM CROPS	100F	100F	100F	100F
PERM PASTURE	104	104*	104*	104*
FOREST&WOODL	1700	1700*	1700*	1700*
OTHER LAND	784	776	771	769
ETHIOPIA				
TOTAL AREA	122190	122190	122190	122190
LAND AREA	110100	110100	110100	110100
ARAB&PERM CR	12044	12670F	13500	13730*
ARABLE LAND	11543	12070F	12770	13000*
PERM CROPS	501	600F	730	730*
PERM PASTURE	66796	66480F	65400	64700F
FOREST&WOODL	8652	8750F	8820	8860*
OTHER LAND	22608	22200	22380	22810
GABON				
TOTAL AREA	26767	26767	26767	26767
LAND AREA	25767	25767	25767	25767
ARAB&PERM CR	134	150F	155F	155F
ARABLE LAND	108F	124F	126F	126F
PERM CROPS	26F	26F	29F	29F
PERM PASTURE	5050F	5050F	4900F	4800F
FOREST&WOODL	20000	20000	20000	20000
OTHER LAND	583	567	712	812
GAMBIA				
TOTAL AREA	1130	1130	1130	1130
LAND AREA	1000	1000	1000	1000
ARAB&PERM CR	197	210F	235F	260F
ARABLE LAND	197	210F	235F	260*
PERM PASTURE	400	400	360F	340F

TABLE
TABLEAU 1
CUADRO

LAND USE   UTILISATION DES TERRES   USO DE TIERRAS

1000 HA

	1961-65	1966	1971	1976
GAMBIA CONT.				
FOREST&WOODL	70F	70F	65	65F
OTHER LAND	333	320	340	335
GHANA				
TOTAL AREA	23854	23854	23854	23854
LAND AREA	23002	23002	23002	23002
ARAB&PERM CR	2541	2727	2600F	2700F
ARABLE LAND	830	724	1000F	1050F
PERM CROPS	1711	2003	1600F	1650F
PERM PASTURE	12210	11237*	10800F	10700F
FOREST&WOODL	2484	2447*	2447*	2447*
OTHER LAND	5767	6591	7155	7155
GUINEA				
TOTAL AREA	24586	24586	24586	24586
LAND AREA	24586	24586	24586	24586
ARAB&PERM CR	3950F	4030F	4130F	4170F
ARABLE LAND	3880F	3960F	4060F	4100F
PERM CROPS	70F	70F	70F	70F
PERM PASTURE	3000F	3000F	3000F	3000F
FOREST&WOODL	1046*	1046*	1100F	1100F
OTHER LAND	16590	16510	16356	16316
GUIN BISSAU				
TOTAL AREA	3612	3612	3612	3612
LAND AREA	2800	2800	2800	2800
ARAB&PERM CR	263	263*	275*	285F
ARABLE LAND	235	235*	245*	255F
PERM CROPS	28	28*	30*	30*
PERM PASTURE	1280*	1280*	1280*	1280*
FOREST&WOODL	1100F	1100F	1070F	1070F
OTHER LAND	157	157	175	165
IVORY COAST				
TOTAL AREA	32246	32246	32246	32246
LAND AREA	31800	31800	31800	31800
ARAB&PERM CR	8650F	8802	8887	9160F
ARABLE LAND	7680F	7802	7807	8050F
PERM CROPS	970F	1000	1080	1110F
PERM PASTURE	8000*	8000*	8000*	8000*
FOREST&WOODL	12260	11000F	9000*	9000*
OTHER LAND	2890	3998	5913	5640
KENYA				
TOTAL AREA	58264	58264	58264	58264
LAND AREA	56925	56925	56925	56925
ARAB&PERM CR	1723	1885F	2090F	2160F
ARABLE LAND	1447	1600F	1780F	1815F
PERM CROPS	276	285F	310F	345F
PERM PASTURE	3918	3880F	3830F	3780F
FOREST&WOODL	1935*	1934	1848	1874*
OTHER LAND	49349	49226	49157	49111
LESOTHO				
TOTAL AREA	3035	3035	3035	3035
LAND AREA	3035	3035	3035	3035
ARAB&PERM CR	355	360F	360F	355F
ARABLE LAND	355	360F	360F	355F
PERM PASTURE	2494	2500*	2500*	2500*
OTHER LAND	186	175	175	180
LIBERIA				
TOTAL AREA	11137	11137	11137	11137
LAND AREA	9632	9632	9632	9632
ARAB&PERM CR	375F	370F	366	366*
ARABLE LAND	135F	130F	126	126
PERM CROPS	240*	240*	240	240*
PERM PASTURE	240*	240*	240*	240*
FOREST&WOODL	3869	3300F	2500*	2500*
OTHER LAND	5148	5722	6526	6526
LIBYA				
TOTAL AREA	175954	175954	175954	175954
LAND AREA	175954	175954	175954	175954
ARAB&PERM CR	2509	2509	2521	2544F
ARABLE LAND	2375	2375	2377	2400F
PERM CROPS	134	134	144	144*
PERM PASTURE	7400F	7250F	7000	6780F
FOREST&WOODL	491*	501*	534	534*
OTHER LAND	165554	165694	165899	166096
MADAGASCAR				
TOTAL AREA	58704	58704	58704	58704
LAND AREA	58154	58154	58154	58154
ARAB&PERM CR	2770	2830F	2856	2862F
ARABLE LAND	2530F	2560F	2576F	2580F
PERM CROPS	240F	270F	280F	282F
PERM PASTURE	34000	34000	34000	34000
FOREST&WOODL	12470	12470	12470	12472*
OTHER LAND	8914	8854	8828	8820
MALAWI				
TOTAL AREA	11848	11848	11848	11848
LAND AREA	9408	9408	9408	9408

	1961-65	1966	1971	1976
MALAWI CONT.				
ARAB&PERM CR	1971	1996	2158F	2288F
ARABLE LAND	1953	1978	2140F	2270F
PERM CROPS	18	18	18	18*
PERM PASTURE	1840*	1840*	1840*	1840*
FOREST&WOODL	2314	2314	2314	2314
OTHER LAND	3283	3258	3096	2966
MALI				
TOTAL AREA	124000	124000	124000	124000
LAND AREA	122000	122000	122000	122000
ARAB&PERM CR	7820	9400F	9750F	9800F
ARABLE LAND	7817	9397F	9747F	9797F
PERM CROPS	3	3*	3*	3*
PERM PASTURE	30000*	30000*	30000*	30000*
FOREST&WOODL	4470	4457	4457*	4457*
OTHER LAND	79710	78143	77793	77743
MAURITANIA				
TOTAL AREA	103070	103070	103070	103070
LAND AREA	103040	103040	103040	103040
ARAB&PERM CR	994F	1005F	1005F	1005F
ARABLE LAND	990F	1000F	1000F	1000F
PERM CROPS	4	5F	5F	5F
PERM PASTURE	39250	39250*	39250*	39250*
FOREST&WOODL	15134	15134*	15134*	15134*
OTHER LAND	47662	47651	47651	47651
MAURITIUS				
TOTAL AREA	186	186	186	186
LAND AREA	185	185	185	185
ARAB&PERM CR	93	92	105	107F
ARABLE LAND	91	89	100	100F
PERM CROPS	2F	3F	5F	7*
PERM PASTURE	7*	7*	7	7*
FOREST&WOODL	64F	60F	58	58*
OTHER LAND	21	26	15	13
MOROCCO				
TOTAL AREA	44655	44655	44655	44655
LAND AREA	44634	44634	44634	44630
ARAB&PERM CR	7066	7338F	7504	7830F
ARABLE LAND	6666	6900F	7075	7400F
PERM CROPS	400	438	429	430F
PERM PASTURE	12500*	12500*	12500	12500*
FOREST&WOODL	5302	5160	5164	5190F
OTHER LAND	19766	19636	19466	19110
MOZAMBIQUE				
TOTAL AREA	78303	78303	78303	78303
LAND AREA	76553	76553	76553	76553
ARAB&PERM CR	2669	2694	3009*	3080F
ARABLE LAND	2456	2474*	2785*	2850F
PERM CROPS	213	220*	224*	230F
PERM PASTURE	44000*	44000*	44000*	44000*
FOREST&WOODL	19400	19400	19400	19400
OTHER LAND	10484	10459	10144	10073
NAMIBIA				
TOTAL AREA	82429	82429	82429	8242?
LAND AREA	82329	82329	82329	82329
ARAB&PERM CR	644	646F	653F	655
ARABLE LAND	643	645F	652F	654F
PERM CROPS	1*	1*	1*	1*
PERM PASTURE	52906*	52906*	52906*	52906*
FOREST&WOODL	10427*	10427*	10427*	10427*
OTHER LAND	18352	18350	18343	18341
NIGER				
TOTAL AREA	126700	126700	126700	126700
LAND AREA	126670	126670	126670	126670
ARAB&PERM CR	15000	15000	15000	15000*
ARABLE LAND	15000	15000	15000	15000*
PERM PASTURE	2900	2900	3000	3000*
FOREST&WOODL	14222	13560F	12450F	11350F
OTHER LAND	94548	95210	96220	97320
NIGERIA				
TOTAL AREA	92377	92377	92377	92377
LAND AREA	91077	91077	91077	91077
ARAB&PERM CR	22199	22800F	23650F	23840F
ARABLE LAND	21397F	21920F	22690F	22850F
PERM CROPS	802F	880F	960F	990F
PERM PASTURE	19100F	19400F	19843	20800F
FOREST&WOODL	31069	31069	31069	31069
OTHER LAND	18709	17808	16515	15368
REUNION				
TOTAL AREA	252	252	252	252
LAND AREA	251	251	251	251
ARAB&PERM CR	61	62	62	56
ARABLE LAND	61	62	61	55
PERM CROPS			1	1
PERM PASTURE	8*	8*	8*	8
FOREST&WOODL	96	98	98	98

TABLE
TABLEAU 1
CUADRO

LAND USE UTILISATION DES TERRES USO DE TIERRAS

1000 HA

	1961-65	1966	1971	1976
REUNION CONT.				
OTHER LAND	86	83	83	89
RHODESIA				
TOTAL AREA	39058	39058	39058	39058
LAND AREA	38767	38767	38767	38767
ARAB&PERM CR	1996F	2150F	2400F	2480F
ARABLE LAND	1981F	2135F	2385F	2465F
PERM CROPS	15*	15*	15*	15*
PERM PASTURE	4856*	4856*	4856*	4856*
FOREST&WOODL	23810	23810	23810	23810
OTHER LAND	8105	7951	7701	7621
RWANDA				
TOTAL AREA	2634	2634	2634	2634
LAND AREA	2495	2495	2495	2495
ARAB&PERM CR	635F	665F	748	939F
ARABLE LAND	495F	510F	560	700F
PERM CROPS	140F	155F	188	239
PERM PASTURE	872	850F	817	565F
FOREST&WOODL	328	328	329	290F
OTHER LAND	660	652	601	701
ST HELENA				
TOTAL AREA	31	31	31	31
LAND AREA	31	31	31	31
ARAB&PERM CR	2	2*	2*	2*
ARABLE LAND	2	2*	2*	2*
PERM PASTURE	2	2	2	2
FOREST&WOODL	1	1	1	1
OTHER LAND	26	26	26	26
SAO TOME ETC				
TOTAL AREA	96	96	96	96
LAND AREA	96	96	96	96
ARAB&PERM CR	34*	35*	35*	36*
ARABLE LAND	1F	1F	1F	1F
PERM CROPS	33*	34*	34*	35*
PERM PASTURE	1F	1F	1F	1F
OTHER LAND	61	60	60	59
SENEGAL				
TOTAL AREA	19619	19619	19619	19619
LAND AREA	19200	19200	19200	19200
ARAB&PERM CR	2215	2422	2307	2404F
ARABLE LAND	2212	2419	2303	2400F
PERM CROPS	3F	3F	4*	4*
PERM PASTURE	5700	5700	5700	5700*
FOREST&WOODL	5318	5318	5318	5318*
OTHER LAND	5967	5760	5875	5778
SEYCHELLES				
TOTAL AREA	28	28	28	28
LAND AREA	27	27	27	27
ARAB&PERM CR	5	5	5	5
ARABLE LAND	1	1	1	1
PERM CROPS	4	4	4	4
FOREST&WOODL	5	5	5	5
OTHER LAND	17	17	17	17
SIERRA LEONE				
TOTAL AREA	7174	7174	7174	7174
LAND AREA	7162	7162	7162	7162
ARAB&PERM CR	4002F	4024F	4060F	4094F
ARABLE LAND	3876F	3892F	3920F	3948F
PERM CROPS	126F	132F	140*	146F
PERM PASTURE	2203	2204*	2204*	2204*
FOREST&WOODL	301	301*	290*	290*
OTHER LAND	656	633	608	574
SOMALIA				
TOTAL AREA	63766	63766	63766	63766
LAND AREA	62734	62734	62734	62734
ARAB&PERM CR	969F	989F	1022F	1065F
ARABLE LAND	955F	975F	1010F	1050F
PERM CROPS	14*	14*	12F	15F
PERM PASTURE	28850*	28850*	28850*	28850*
FOREST&WOODL	10348	10120*	9500F	8700F
OTHER LAND	22567	22775	23362	24119
SOUTH AFRICA				
TOTAL AREA	122104	122104	122104	122104
LAND AREA	122104	122104	122104	122104
ARAB&PERM CR	12610	12920*	14620*	14520F
ARABLE LAND	11910	12000*	13600*	13420F
PERM CROPS	700	920F	1020F	1100F
PERM PASTURE	86670	84000F	82700F	81600F
FOREST&WOODL	4400	4600*	4600*	4600*
OTHER LAND	18424	20584	20184	21384
SUDAN				
TOTAL AREA	250581	250581	250581	250581
LAND AREA	237600	237600	237600	237600
ARAB&PERM CR	6180F	6250F	6392	7495F
ARABLE LAND	6154F	6220F	6355	7450F

	1961-65	1966	1971	1976
SUDAN CONT.				
PERM CROPS	26F	30F	37	45F
PERM PASTURE	24000	24000	24000	24000
FOREST&WOODL	91500	91500	91500	91500
OTHER LAND	115920	115850	115708	114605
SWAZILAND				
TOTAL AREA	1736	1736	1736	1736
LAND AREA	1720	1720	1720	1720
ARAB&PERM CR	134	151*	152	173F
ARABLE LAND	132	149*	149	170F
PERM CROPS	2	2	3	3*
PERM PASTURE	1342*	1342*	1342*	1100F
FOREST&WOODL	129	129	116*	92F
OTHER LAND	115	98	110	355
TANZANIA				
TOTAL AREA	94509	94509	94509	94509
LAND AREA	88604	88604	88604	88604
ARAB&PERM CR	3349	4247	5280F	6290F
ARABLE LAND	2613	3388	4300F	5200F
PERM CROPS	736	859	980F	1090F
PERM PASTURE	45240	45120F	44920F	44720F
FOREST&WOODL	33942	31074	31074	31074
OTHER LAND	6073	8163	7330	6520
TOGO				
TOTAL AREA	5600	5600	5600	5600
LAND AREA	5360	5360	5360	5360
ARAB&PERM CR	2135	2201F	2278F	2285F
ARABLE LAND	2089	2140F	2200F	2220F
PERM CROPS	46	61	78*	65F
PERM PASTURE	200	200	200*	200*
FOREST&WOODL	506	500F	380	350F
OTHER LAND	2519	2459	2502	2525
TUNISIA				
TOTAL AREA	16361	16361	16361	16361
LAND AREA	15536	15536	15536	15536
ARAB&PERM CR	4406	4334	4330F	4410F
ARABLE LAND	3254	3180	3170F	3250F
PERM CROPS	1152	1154	1160F	1160F
PERM PASTURE	3250	3250*	3250*	3250*
FOREST&WOODL	674	650F	576	530F
OTHER LAND	7206	7302	7380	7346
UGANDA				
TOTAL AREA	23604	23604	23604	23604
LAND AREA	19971	19971	19971	19971
ARAB&PERM CR	4427	4885	5030F	5380F
ARABLE LAND	3434	3776	3800F	4000F
PERM CROPS	992	1109	1230F	1380F
PERM PASTURE	5000*	5000*	5000*	5000*
FOREST&WOODL	7560F	6000F	3500F	2759F
OTHER LAND	2984	4086	6441	6832
UPPER VOLTA				
TOTAL AREA	27420	27420	27420	27420
LAND AREA	27380	27380	27380	27380
ARAB&PERM CR	4989	5110F	5372F	5613*
ARABLE LAND	4980	5100F	5360F	5600*
PERM CROPS	9F	10F	12*	13*
PERM PASTURE	13757	13750*	13755*	13755*
FOREST&WOODL	4234	4200F	4101	3600F
OTHER LAND	4400	4320	4152	4412
WESTN SAHARA				
TOTAL AREA	26600	26600	26600	26600
LAND AREA	26600	26600	26600	26600
ARAB&PERM CR			2	2*
ARABLE LAND			2	2*
PERM PASTURE	5000	5000	5000	5000
OTHER LAND	21600	21600	21598	21598
ZAIRE				
TOTAL AREA	234541	234541	234541	234541
LAND AREA	226760	226760	226760	226760
ARAB&PERM CR	5550F	5700F	5950F	6150F
ARABLE LAND	5168F	5250F	5460F	5600F
PERM CROPS	382F	450F	490F	550F
PERM PASTURE	24803*	24803*	24803*	24803*
FOREST&WOODL	127419	124330F	122670F	121050F
OTHER LAND	68988	71927	73337	74757
ZAMBIA				
TOTAL AREA	75261	75261	75261	75261
LAND AREA	74072	74072	74072	74072
ARAB&PERM CR	4820	4880F	4980F	5008F
ARABLE LAND	4815	4874F	4973F	5000F
PERM CROPS	5F	6F	7F	8F
PERM PASTURE	30000F	30000F	30000F	30000F
FOREST&WOODL	37631	37631*	37330*	37300F
OTHER LAND	1621	1561	1762	1764
N C AMERICA				
TOTAL AREA	2246443	2246443	2246443	2246443

TABLE TABLEAU CUADRO **1**

LAND USE UTILISATION DES TERRES USO DE TIERRAS

1000 HA

	1961-65	1966	1971	1976
N C AMERICA CONT.				
LAND AREA	2140536	2140536	2140536	2140488
ARAB&PERM CR	256978	256695	271898	271542
ARABLE LAND	251446	251032	265760	264974
PERM CROPS	5533	5663	6138	6568
PERM PASTURE	369583	365416	352257	346735
FOREST&WOODL	728975	725164	721159	718311
OTHER LAND	785000	793261	795222	803900
ANTIGUA				
TOTAL AREA	44	44	44	44
LAND AREA	44	44	44	44
ARAB&PERM CR	8	8	8*	8*
ARABLE LAND	8	8	8*	8*
PERM PASTURE	2	3	3*	3*
FOREST&WOODL	6	6	7*	7*
OTHER LAND	28	27	26	26
BAHAMAS				
TOTAL AREA	1394	1394	1394	1394
LAND AREA	1007	1007	1007	1007
ARAB&PERM CR	14	14	16*	16*
ARABLE LAND	1*	1	2*	2*
PERM CROPS	13	13	14*	14*
PERM PASTURE	1	1	1	1*
FOREST&WOODL	324	324	324	324*
OTHER LAND	668	668	666	666
BARBADOS				
TOTAL AREA	43	43	43	43
LAND AREA	43	43	43	43
ARAB&PERM CR	30	33*	33F	33F
ARABLE LAND	30	33*	33F	33F
PERM PASTURE	4*	4*	4*	4*
OTHER LAND	9	6	6	6
BELIZE				
TOTAL AREA	2296	2296	2296	2296
LAND AREA	2280	2280	2280	2280
ARAB&PERM CR	31	38	48F	49F
ARABLE LAND	21	27	28F	30F
PERM CROPS	10	11	20F	19*
PERM PASTURE	16	17	18*	21*
FOREST&WOODL	1054	1049	1047	1012
OTHER LAND	1179	1176	1167	1198
BERMUDA				
TOTAL AREA	5	5	5	5
LAND AREA	5	5	5	5
FOREST&WOODL	1	1	1	1
OTHER LAND	4	4	4	4
CANADA				
TOTAL AREA	997614	997614	997614	997614
LAND AREA	922107	922107	922107	922107
ARAB&PERM CR	42159	43404	43767	43709
ARABLE LAND	42079	43324	43687	43629
PERM CROPS	80*	80*	80*	80*
PERM PASTURE	20994	20957	24896	23461
FOREST&WOODL	322271*	322271*	324600F	326129
OTHER LAND	536683	535475	528844	528808
CAYMAN IS				
TOTAL AREA	26	26	26	26
LAND AREA	26	26	26	26
PERM PASTURE	2*	2*	2*	2*
FOREST&WOODL	6*	6*	6*	6*
OTHER LAND	18	18	18	18
COSTA RICA				
TOTAL AREA	5070	5070	507C	5070
LAND AREA	5066	5066	5066	5066
ARAB&PERM CR	484	485	497	490*
ARABLE LAND	285*	285*	285*	283*
PERM CROPS	199	200	212	207*
PERM PASTURE	969	1150F	1390	1558F
FOREST&WOODL	2848F	2740F	2518	2518*
OTHER LAND	765	691	661	500
CUBA				
TOTAL AREA	11452	11452	11452	11452
LAND AREA	11452	11452	11452	11452
ARAB&PERM CR	1790*	1744*	2673	3110F
ARABLE LAND	1616*	1551*	2277	2450F
PERM CROPS	174	193	396	660F
PERM PASTURE	2349	2766	2400	2750F
FOREST&WOODL	1616	1570F	1380F	1230F
OTHER LAND	5697	5372	4999	4362
DOMINICA				
TOTAL AREA	75	75	75	75
LAND AREA	75	75	75	75
ARAB&PERM CR	16	17	17*	17*
ARABLE LAND	8	7	7*	7*
PERM CROPS	8	10	10*	10*

	1961-65	1966	1971	1976
DOMINICA CONT.				
PERM PASTURE	2	2	2*	2*
FOREST&WOODL	50	50	35*	35*
OTHER LAND	7	6	21	21
DOMINICAN RP				
TOTAL AREA	4873	4873	4873	4873
LAND AREA	4838	4838	4838	4838
ARAB&PERM CR	860F	895F	972	995F
ARABLE LAND	580F	600F	635	645F
PERM CROPS	280F	295F	337	350F
PERM PASTURE	1020F	1170F	1436	1470F
FOREST&WOODL	1100	1100	1100	1104*
OTHER LAND	1858	1673	1330	1269
EL SALVADOR				
TOTAL AREA	2104	2104	2104	2104
LAND AREA	2121	2121	2121	2072
ARAB&PERM CR	655	645F	651	669
ARABLE LAND	489	470F	488	475
PERM CROPS	166	175F	163	194
PERM PASTURE	606	610F	665	670F
FOREST&WOODL	230	238F	250	263
OTHER LAND	630	628	555	470
GREENLAND				
TOTAL AREA	34170	34170	3417C	34170
LAND AREA	34170	34170	34170	34170
PERM PASTURE	5*	5*	5*	5*
FOREST&WOODL	10	10	10	10
OTHER LAND	34155	34155	34155	34155
GRENADA				
TOTAL AREA	34	34	34	34
LAND AREA	34	34	34	34
ARAB&PERM CR	16	16	16	16
ARABLE LAND	2	2	2	2
PERM CROPS	14	14	14	14
PERM PASTURE	1	1	1	1
FOREST&WOODL	4	4	4	4
OTHER LAND	13	13	13	13
GUADELOUPE				
TOTAL AREA	178	178	178	178
LAND AREA	176	176	176	176
ARAB&PERM CR	51	49	57	56
ARABLE LAND	39	44	46	45
PERM CROPS	11	5	11	11*
PERM PASTURE	15	18	17	17
FOREST&WOODL	60	56	62	70
OTHER LAND	50	53	40	33
GUATEMALA				
TOTAL AREA	10889	10889	10889	10889
LAND AREA	10789	10789	10789	10789
ARAB&PERM CR	1442	1505F	1545F	1735F
ARABLE LAND	1125	1185F	1220F	1400F
PERM CROPS	317	320F	325F	335F
PERM PASTURE	1039	990F	925F	890F
FOREST&WOODL	6400	6000F	5800*	5800*
OTHER LAND	1908	2294	2519	2364
HAITI				
TOTAL AREA	2775	2775	2775	2775
LAND AREA	2756	2756	2756	2756
ARAB&PERM CR	702F	745F	820F	870
ARABLE LAND	430F	460F	510F	535F
PERM CROPS	270F	285F	310F	335F
PERM PASTURE	596F	615F	640F	530
FOREST&WOODL	340F	310F	260F	200
OTHER LAND	1118	1086	1036	1156
HONDURAS				
TOTAL AREA	11209	11209	11209	11209
LAND AREA	11189	11189	11189	11189
ARAB&PERM CR	821	830F	845F	890F
ARABLE LAND	682F	685F	695F	715F
PERM CROPS	139F	145F	150F	175F
PERM PASTURE	2000F	2000F	2000F	2000F
FOREST&WOODL	7100*	7100*	7100*	7100*
OTHER LAND	1268	1259	1244	1199
JAMAICA				
TOTAL AREA	1099	1099	1099	1099
LAND AREA	1083	1083	1083	1083
ARAB&PERM CR	233	245F	250F	265F
ARABLE LAND	187F	195F	195F	205F
PERM CROPS	46F	50F	55F	60F
PERM PASTURE	256	245F	235F	215F
FOREST&WOODL	508F	505F	500F	492*
OTHER LAND	86	88	98	111
MARTINIQUE				
TOTAL AREA	110	110	110	110
LAND AREA	106	106	106	106

TABLE
TABLEAU 1
CUADRO

LAND USE UTILISATION DES TERRES USO DE TIERRAS

1000 HA

	1961-65	1966	1971	1976
MARTINIQUE CONT.				
ARAB&PERM CR	32	32	29	26
ARABLE LAND	5	6	3	7
PERM CROPS	27	26	26	19
PERM PASTURE	20	20	20	25
FOREST&WOODL	27	26	32	32
OTHER LAND	27	28	25	23
MEXICO				
TOTAL AREA	202206	202206	202206	202206
LAND AREA	197255	197255	197255	197255
ARAB&PERM CR	24908F	26000F	27050*	27790F
ARABLE LAND	23486F	24470F	25290*	26000F
PERM CROPS	1422F	1530F	1760*	1790F
PERM PASTURE	73820F	72100F	69200F	66700F
FOREST&WOODL	80620F	78000F	74000F	71100F
OTHER LAND	17907	21155	27005	31665
MONTSERRAT				
TOTAL AREA	10	10	10	10
LAND AREA	10	10	10	10
ARAB&PERM CR	3	2	2	1*
ARABLE LAND	3	2	2	1*
PERM PASTURE	1	1	1	1*
FOREST&WOODL	2	1	1	4*
OTHER LAND	4	6	6	4
NETH ANTILLE				
TOTAL AREA	96	96	96	96
LAND AREA	96	96	96	96
ARAB&PERM CR	6F	7F	8F	8F
ARABLE LAND	6F	7F	8F	8F
OTHER LAND	90	89	88	88
NICARAGUA				
TOTAL AREA	13000	13000	13000	13000
LAND AREA	11875	11875	11875	11875
ARAB&PERM CR	1335F	1400F	1435F	1505
ARABLE LAND	1180F	1240F	1270F	1329
PERM CROPS	155	160F	165*	176
PERM PASTURE	3384*	3384*	3384*	3384
FOREST&WOODL	5853*	5853*	5853*	6282
OTHER LAND	1303	1238	1203	704
PANAMA				
TOTAL AREA	7565	7565	7565	7565
LAND AREA	7505	7505	7505	7505
ARAB&PERM CR	560	555F	542	565F
ARABLE LAND	437	435F	431	450F
PERM CROPS	124	120F	111	115F
PERM PASTURE	899	1000F	1141	1150F
FOREST&WOODL	4100*	4100*	4100*	4100*
OTHER LAND	1946	1850	1722	1690
PANAMA CA ZN				
TOTAL AREA	143	143	143	143
LAND AREA	94	94	94	94
ARAB&PERM CR	1*	1*	1*	1*
ARABLE LAND	1*	1*	1*	1*
PERM PASTURE	11*	11*	11*	11*
FOREST&WOODL	56*	56*	56*	56*
OTHER LAND	26	26	26	26
PUERTO RICO				
TOTAL AREA	890	890	890	890
LAND AREA	886	886	886	886
ARAB&PERM CR	287	267	193F	159
ARABLE LAND	192	185F	130F	97*
PERM CROPS	96	82F	63F	62*
PERM PASTURE	310	314	330*	334
FOREST&WOODL	118	118	150F	171
OTHER LAND	171	187	213	222
ST KITTS ETC				
TOTAL AREA	36	36	36	36
LAND AREA	36	36	36	36
ARAB&PERM CR	16	16	14	14*
ARABLE LAND	10*	10*	8	8*
PERM CROPS	6*	6*	6	6*
PERM PASTURE	4	4	1	1*
FOREST&WOODL	7	7	7	6*
OTHER LAND	9	9	14	15
ST LUCIA				
TOTAL AREA	62	62	62	62
LAND AREA	61	61	61	62
ARAB&PERM CR	20	21	27*	27*
ARABLE LAND	11	11	15*	15*
PERM CROPS	10	10	12*	12*
PERM PASTURE	3	3	3	3*
FOREST&WOODL	13	13	12F	11*
OTHER LAND	25	24	19	21
ST PIER ETC				
TOTAL AREA	24	24	24	24

	1961-65	1966	1971	1976
ST PIER ETC CONT.				
LAND AREA	23	23	23	23
ARAB&PERM CR	3*	3*	3*	3*
ARABLE LAND	3*	3*	3*	3*
FOREST&WOODL	1*	1*	1*	1*
OTHER LAND	19	19	19	19
ST VINCENT				
TOTAL AREA	34	34	34	34
LAND AREA	34	34	34	34
ARAB&PERM CR	16	18	18	18*
ARABLE LAND	13	14	13	13*
PERM CROPS	4	4	5	5*
PERM PASTURE	1	1	1	1*
FOREST&WOODL	15	14	14	14*
OTHER LAND	2	1	1	1
TRINIDAD ETC				
TOTAL AREA	513	513	513	513
LAND AREA	513	513	513	513
ARAB&PERM CR	139	139*	140*	157*
ARABLE LAND	57	57*	60F	70F
PERM CROPS	82	82*	80*	87*
PERM PASTURE	6	8F	11*	11*
FOREST&WOODL	232	232	226*	226*
OTHER LAND	136	134	136	119
TURKS CAICOS				
TOTAL AREA	43	43	43	43
LAND AREA	43	43	43	43
ARAB&PERM CR	1*	1*	1*	1*
ARABLE LAND	1*	1*	1*	1*
OTHER LAND	42	42	42	42
USA				
TOTAL AREA	936312	936312	936312	936312
LAND AREA	912689	912689	912689	912689
ARAB&PERM CR	180331	177550	190211F	188330F
ARABLE LAND	178453	175705	188400F	186500F
PERM CROPS	1878	1845*	1811*	1830F
PERM PASTURE	261235	258000*	243500F	241500F
FOREST&WOODL	294000	293400F	291700F	290000F
OTHER LAND	177123	183739	187278	192859
VIRGIN IS UK				
TOTAL AREA	15	15	15	15
LAND AREA	15	15	15	15
ARAB&PERM CR	2F	3F	3F	3F
ARABLE LAND	1F	2F	2F	2F
PERM CROPS	1F	1F	1F	1F
PERM PASTURE	4F	5F	5F	5F
FOREST&WOODL	1*	1*	1*	1*
OTHER LAND	8	6	6	6
VIRGIN IS US				
TOTAL AREA	34	34	34	34
LAND AREA	34	34	34	34
ARAB&PERM CR	6	7*	6	6
ARABLE LAND	5	6*	5	5
PERM CROPS	1	1	1	1
PERM PASTURE	8	9	9	9
FOREST&WOODL	2	2	2	2
OTHER LAND	18	16	17	17
SOUTH AMERIC				
TOTAL AREA	1781980	1781980	1781980	1781980
LAND AREA	1753677	1753677	1753677	1753691
ARAB&PERM CR	82367	87820	96220	104068
ARABLE LAND	62607	67652	74984	81892
PERM CROPS	19759	20168	21236	22176
PERM PASTURE	407332	416574	431912	441834
FOREST&WOODL	949349	943120	933733	924263
OTHER LAND	314629	306163	291812	283526
ARGENTINA				
TOTAL AREA	276689	276689	276689	276689
LAND AREA	273669	273669	273669	273669
ARAB&PERM CR	28098	30248	33850F	35000F
ARABLE LAND	19598	21608	24400F	25000F
PERM CROPS	8500F	8640	9450F	10000F
PERM PASTURE	146500F	145802	144300F	143600F
FOREST&WOODL	61000	60850F	60600F	60270
OTHER LAND	38071	36769	34919	34799
BOLIVIA				
TOTAL AREA	109858	109858	109858	109858
LAND AREA	108547	108547	108547	108547
ARAB&PERM CR	1503	1730F	2335	3336F
ARABLE LAND	1348	1600F	2260	3250F
PERM CROPS	155	130F	75	86F
PERM PASTURE	28353	28000F	27500F	27100F
FOREST&WOODL	59950	59200F	57950F	56900F
OTHER LAND	18741	19617	20762	21211
BRAZIL				
TOTAL AREA	851197	851197	851197	851197

TABLE
TABLEAU 1
CUADRO

LAND USE   UTILISATION DES TERRES   USO DE TIERRAS

1000 HA

	1961-65	1966	1971	1976
BRAZIL CONT.				
LAND AREA	845651	845651	845651	845651
ARAB&PERM CR	30254F	31910F	34500F	37630F
ARABLE LAND	22400F	24000F	26500F	29500F
PERM CROPS	7854F	7910F	8000F	8130F
PERM PASTURE	131880F	141400F	156000F	165000F
FOREST&WOODL	526800	522600F	515600F	508000F
OTHER LAND	156717	149741	139551	135021
CHILE				
TOTAL AREA	75695	75695	75695	75695
LAND AREA	74880	74880	74880	74880
ARAB&PERM CR	4206	4598F	5348F	5828F
ARABLE LAND	4007	4400F	5150F	5630F
PERM CROPS	199	198	198	198
PERM PASTURE	9850	10300F	11200F	11750F
FOREST&WOODL	20686	20686	20686	20686
OTHER LAND	40138	39296	37646	36616
COLOMBIA				
TOTAL AREA	113891	113891	113891	113891
LAND AREA	103870	103870	103870	103870
ARAB&PERM CR	5051	5050F	5054	5160F
ARABLE LAND	3545	3570F	3573	3630F
PERM CROPS	1506	1480F	1481	1530F
PERM PASTURE	16682F	16850F	17150F	17400F
FOREST&WOODL	77190*	77190*	77190*	77190*
OTHER LAND	4947	4780	4476	4120
ECUADOR				
TOTAL AREA	28356	28356	28356	28356
LAND AREA	27684	27684	27684	27684
ARAB&PERM CR	2655	3300F	3820F	5096
ARABLE LAND	2112	2500F	2840*	3996*
PERM CROPS	543	800F	980*	1100*
PERM PASTURE	2200	2200	2200*	2200*
FOREST&WOODL	18800F	18200F	18030F	17886
OTHER LAND	4029	3984	3634	2502
FALKLAND IS				
TOTAL AREA	1217	1217	1217	1217
LAND AREA	1217	1217	1217	1217
PERM PASTURE	1083	1196	1200*	1200*
OTHER LAND	134	21	17	17
FR GUIANA				
TOTAL AREA	9100	9100	9100	9100
LAND AREA	8915	8915	8915	8915
ARAB&PERM CR	3	2*	2	2*
ARABLE LAND	3	2*	2	2*
PERM PASTURE	3*	3*	5	5*
FOREST&WOODL	8617	8400F	8001	8001
OTHER LAND	292	510	907	907
GUYANA				
TOTAL AREA	21497	21497	21497	21497
LAND AREA	19671	19671	19671	19685
ARAB&PERM CR	360F	365F	372F	379F
ARABLE LAND	350F	355F	360F	364F
PERM CROPS	10F	10F	12F	15F
PERM PASTURE	999	999	999	999*
FOREST&WOODL	18190	18190	18190	18190*
OTHER LAND	122	117	110	117
PARAGUAY				
TOTAL AREA	40675	40675	40675	40675
LAND AREA	39730	39730	39730	39730
ARAB&PERM CR	852	928	950F	1035F
ARABLE LAND	737	809	805F	870F
PERM CROPS	115	119	145F	165F
PERM PASTURE	13800F	14100F	14600F	15100F
FOREST&WOODL	20828	20634	20502*	20420F
OTHER LAND	4250	4068	3678	3175
PERU				
TOTAL AREA	128522	128522	128522	128522
LAND AREA	128000	128000	128000	128000
ARAB&PERM CR	2351	2625	2822	3330F
ARABLE LAND	2171	2417	2558	3000F
PERM CROPS	180	208	264	330F
PERM PASTURE	27977	27120	27120	27120F
FOREST&WOODL	73800*	73800*	73800*	73800*
OTHER LAND	23872	24455	24258	23750
SURINAM				
TOTAL AREA	16327	16327	16327	16327
LAND AREA	16147	16147	16147	16147
ARAB&PERM CR	37	37	41F	45F
ARABLE LAND	30	31	33F	35F
PERM CROPS	7	6	8*	10F
PERM PASTURE	7	7	9*	10F
FOREST&WOODL	14930F	14832	14600F	14350F
OTHER LAND	1173	1271	1497	1742
URUGUAY				
TOTAL AREA	17751	17751	17751	17751

	1961-65	1966	1971	1976
URUGUAY CONT.				
LAND AREA	17491	17491	17491	17491
ARAB&PERM CR	1779F	1812F	1851*	1905F
ARABLE LAND	1726F	1760F	1803*	1850F
PERM CROPS	53	52	48*	55F
PERM PASTURE	13769	13697	13629*	13550F
FOREST&WOODL	588	568	614	600F
OTHER LAND	1355	1414	1397	1436
VENEZUELA				
TOTAL AREA	91205	91205	91205	91205
LAND AREA	88205	88205	88205	88205
ARAB&PERM CR	5218	5215F	5275F	5322F
ARABLE LAND	4580	4600F	4700F	4765F
PERM CROPS	637	615F	575F	557*
PERM PASTURE	14229	14900F	16000F	16800F
FOREST&WOODL	47970	47970	47970	47970
OTHER LAND	20788	20120	18960	18113
ASIA				
TOTAL AREA	2757442	2757442	2757442	2757442
LAND AREA	2676669	2676669	2676669	2676621
ARAB&PERM CR	447256	455508	467087	481255
ARABLE LAND	425433	432975	442226	454908
PERM CROPS	21824	22533	24861	26347
PERM PASTURE	543026	547293	549115	538310
FOREST&WOODL	564166	573356	589276	603645
OTHER LAND	1122221	1100512	1071191	1053411
AFGHANISTAN				
TOTAL AREA	64750	64750	64750	64750
LAND AREA	64750	64750	64750	64750
ARAB&PERM CR	7804	7913	8447F	8535F
ARABLE LAND	7740	7835	7900F	7980F
PERM CROPS	64	78	547	555F
PERM PASTURE	6117	6087	5990F	5940F
FOREST&WOODL	2000	2000	1900	1900*
OTHER LAND	48829	48750	48413	48375
BAHRAIN				
TOTAL AREA	62	62	62	62
LAND AREA	62	62	62	62
ARAB&PERM CR	2	2	2	2*
ARABLE LAND	1	1	1	1*
PERM CROPS	1	1	1	1*
PERM PASTURE	4	4	4	4*
OTHER LAND	56	56	56	56
BANGLADESH				
TOTAL AREA	14400	14400	14400	14400
LAND AREA	13391	13391	13391	13391
ARAB&PERM CR	8919	9050F	9095	9392F
ARABLE LAND	8719*	8850F	8883	9180F
PERM CROPS	200*	200*	212	212*
PERM PASTURE	600F	600F	600F	600F
FOREST&WOODL	2221	2242*	2229	2201*
OTHER LAND	1651	1499	1467	1198
BHUTAN				
TOTAL AREA	4700	4700	4700	4700
LAND AREA	4700	4700	4700	4700
ARAB&PERM CR	132F	162F	212F	252F
ARABLE LAND	130F	160F	210F	250F
PERM CROPS	2F	2F	2F	2F
PERM PASTURE	20F	20F	20F	20F
FOREST&WOODL	3000	3000	3000	3000
OTHER LAND	1548	1518	1468	1428
BRUNEI				
TOTAL AREA	577	577	577	577
LAND AREA	527	527	527	527
ARAB&PERM CR	16	17*	13	13*
ARABLE LAND	4	6	4	4*
PERM CROPS	12	11*	9	9*
PERM PASTURE	6*	6*	6	6*
FOREST&WOODL	506	490*	435	425F
OTHER LAND		14	73	83
BURMA				
TOTAL AREA	67655	67655	67655	67655
LAND AREA	65888	65888	65888	65888
ARAB&PERM CR	10168F	10366F	10435F	9996
ARABLE LAND	9900F	9920F	9970F	9514
PERM CROPS	268	446	465F	482
PERM PASTURE	336	354	370F	362
FOREST&WOODL	45274	45274	45274	45274
OTHER LAND	10110	9894	9809	10256
CHINA				
TOTAL AREA	959696F	959696F	959696F	959696F
LAND AREA	930496F	930496F	930496F	930496F
ARAB&PERM CR	118940F	121800F	127000F	129500F
ARABLE LAND	118320F	121160F	126220F	128570F
PERM CROPS	620F	640F	780F	930F
PERM PASTURE	204000F	206400F	210000F	214000F

TABLE / TABLEAU / CUADRO **1**

LAND USE   UTILISATION DES TERRES   USO DE TIERRAS

1000 HA

	1961-65	1966	1971	1976
CHINA CONT.				
FOREST&WOODL	109180F	118500F	136000F	155500F
OTHER LAND	498376F	483796F	457496F	431496F
CYPRUS				
TOTAL AREA	925	925	925	925
LAND AREA	924	924	924	924
ARAB&PERM CR	433	432	432	432*
ARABLE LAND	367	365	365	365*
PERM CROPS	66	67	67	67*
PERM PASTURE	93	93	93	93*
FOREST&WOODL	171	171	171	171*
OTHER LAND	227	228	228	228
EAST TIMOR				
TOTAL AREA	1493	1493	1493	1493
LAND AREA	1493	1493	1493	1493
ARAB&PERM CR	80F	80F	80F	80F
ARABLE LAND	70F	70F	70F	70F
PERM CROPS	10F	10F	10F	10F
PERM PASTURE	150F	150F	150F	150F
FOREST&WOODL	1000*	1000*	1000*	1100*
OTHER LAND	263	263	263	163
HONG KONG				
TOTAL AREA	104	104	104	104
LAND AREA	102	102	102	102
ARAB&PERM CR	14	13	12	11
ARABLE LAND	13	12	11	10
PERM CROPS	1	1	1	1
FOREST&WOODL	10	10	11	12
OTHER LAND	78	79	79	79
INDIA				
TOTAL AREA	328759	328759	328759	328759
LAND AREA	297319	297319	297319	297319
ARAB&PERM CR	161998	162720	164440	168880F
ARABLE LAND	157185	158590	160140	164800F
PERM CROPS	4813	4130	4300	4080F
PERM PASTURE	14293	14910	13260	12950F
FOREST&WOODL	58207	60280	63920	65550*
OTHER LAND	62821	59409	55699	49939
INDONESIA				
TOTAL AREA	190435	190435	190435	190435
LAND AREA	181135	181135	181135	181135
ARAB&PERM CR	16740F	17400F	18100	19418
ARABLE LAND	12240F	12600F	13000	14168
PERM CROPS	4500F	4800F	5100	5250F
PERM PASTURE	11400F	11000F	9875	9875*
FOREST&WOODL	123800	123600F	122600F	121800
OTHER LAND	29195	29135	30560	30042
IRAN				
TOTAL AREA	164800	164800	164800	164800
LAND AREA	163600	163600	163600	163600
ARAB&PERM CR	15358F	15480F	15753F	15950
ARABLE LAND	15000F	15030F	15180F	15330
PERM CROPS	358	450F	573	620
PERM PASTURE	7496	10000F	11000	11000*
FOREST&WOODL	18000	18000	18000	18000
OTHER LAND	122746	120120	118847	118650
IRAQ				
TOTAL AREA	43492	43492	43492	43492
LAND AREA	43397	43397	43397	43397
ARAB&PERM CR	4810F	4925F	4999	5250F
ARABLE LAND	4700F	4800F	4848	5100F
PERM CROPS	110F	125F	151	190F
PERM PASTURE	4060	4000F	4000F	4000*
FOREST&WOODL	1953	1900F	1851	1500*
OTHER LAND	32574	32572	32547	32607
ISRAEL				
TOTAL AREA	2077	2077	2077	2077
LAND AREA	2033	2033	2033	2033
ARAB&PERM CR	401	411	411	433
ARABLE LAND	316	322	326	342
PERM CROPS	85	89	85	91
PERM PASTURE	751	822	818	818
FOREST&WOODL	94	101	111	116
OTHER LAND	787	699	693	666
JAPAN				
TOTAL AREA	37231	37231	37231	37231
LAND AREA	37103	37103	37103	37103
ARAB&PERM CR	5893	5839	5389	5030
ARABLE LAND	5448	5296	4773	4415
PERM CROPS	445	543	616	615
PERM PASTURE	124	157	352	506
FOREST&WOODL	25558	25400F	25043	24867*
OTHER LAND	5528	5707	6319	6700
JORDAN				
TOTAL AREA	9774	9774	9774	9774

	1961-65	1966	1971	1976
JORDAN CONT.				
LAND AREA	9718	9718	9718	9718
ARAB&PERM CR	1177	1300	1315F	1365F
ARABLE LAND	1056	1159	1140F	1175F
PERM CROPS	121	141	175F	190F
PERM PASTURE	100	100	100*	100*
FOREST&WOODL	125	125	125*	125*
OTHER LAND	8316	8193	8178	8128
KAMPUCHEA DM				
TOTAL AREA	18104	18104	18104	18104
LAND AREA	17652	17652	17652	17652
ARAB&PERM CR	2691	2983	3046F	3046F
ARABLE LAND	2576	2831	2900F	2900F
PERM CROPS	115	152	146F	146F
PERM PASTURE	580	580	580*	580F
FOREST&WOODL	13372	13372	13372	13372
OTHER LAND	1009	717	654	654
KOREA DPR				
TOTAL AREA	12054	12054	12054	12054
LAND AREA	12041	12041	12041	12041
ARAB&PERM CR	1905	1950F	2050F	2170F
ARABLE LAND	1863F	1900F	1980F	2080F
PERM CROPS	42F	50F	70F	90F
PERM PASTURE	50F	50F	50F	50F
FOREST&WOODL	8970	8970	8970	8970
OTHER LAND	1116	1071	971	851
KOREA REP				
TOTAL AREA	9848	9848	9848	9848
LAND AREA	9819	9819	9819	9819
ARAB&PERM CR	2133	2293	2271	2238
ARABLE LAND	2059	2187	2134	2060
PERM CROPS	74	106	137	178
PERM PASTURE	7	18	18	18
FOREST&WOODL	6663	6656	6628	6628
OTHER LAND	1016	852	902	935
KUWAIT				
TOTAL AREA	1782	1782	1782	1782
LAND AREA	1782	1782	1782	1782
ARAB&PERM CR	1	1	1	1
ARABLE LAND	1	1	1	1
PERM PASTURE	134	134	134	134
FOREST&WOODL	2	2	2	2
OTHER LAND	1645	1645	1645	1645
LAO				
TOTAL AREA	23680	23680	23680	23680
LAND AREA	23080	23080	23080	23080
ARAB&PERM CR	900	850F	950	961F
ARABLE LAND	890F	840F	938F	948F
PERM CROPS	10F	10F	12F	13F
PERM PASTURE	850	800*	800*	800*
FOREST&WOODL	15000	15000	15000	15000
OTHER LAND	6330	6430	6330	6319
LEBANON				
TOTAL AREA	1040	1040	1040	1040
LAND AREA	1023	1023	1023	1023
ARAB&PERM CR	276	296	330*	348F
ARABLE LAND	218	230	240*	240F
PERM CROPS	58	66	90F	108F
PERM PASTURE	8	10	10	10*
FOREST&WOODL	94	95	95	78F
OTHER LAND	645	622	588	587
MACAU				
TOTAL AREA	2	2	2	2
LAND AREA	2	2	2	2
OTHER LAND	2	2	2	2
MAL PENINSUL				
TOTAL AREA	13159	13159	13159	13159
LAND AREA	13159	13159	13159	13159
ARAB&PERM CR	2455	2549	2853F	2935F
ARABLE LAND	459	506	593	625F
PERM CROPS	1996	2043	2260F	2310F
PERM PASTURE	30F	30F	30F	30F
FOREST&WOODL	8734	8102	7874	6850F
OTHER LAND	1940	2478	2402	3344
MAL SABAH				
TOTAL AREA	7371	7371	7371	7371
LAND AREA	7371	7371	7371	7371
ARAB&PERM CR	185	208	242	329
ARABLE LAND	48	54	62	79
PERM CROPS	137	154	180	250
PERM PASTURE	5	6	7	8
FOREST&WOODL	6050*	6050*	6050*	6050*
OTHER LAND	1131	1107	1072	984
MAL SARAWAK				
TOTAL AREA	12445	12445	12445	12445

TABLE
TABLEAU 1
CUADRO

LAND USE   UTILISATION DES TERRES   USO DE TIERRAS

1000 HA

	1961-65	1966	1971	1976
MAL SARAWAK CONT.				
LAND AREA	12325	12325	12325	12325
ARAB&PERM CR	2625F	2571*	2688*	2740F
ARABLE LAND	2370F	2400*	2423*	2435F
PERM CROPS	255	171	265F	305F
PERM PASTURE	15*	15*	15*	15*
FOREST&WOOD	9395	9433	9433	9433
OTHER LAND	290	306	189	137
MALDIVES				
TOTAL AREA	30	30	30	30
LAND AREA	30	30	30	30
ARAB&PERM CR	3*	3*	3*	3*
ARABLE LAND	3*	3*	3*	3*
PERM PASTURE	1F	1F	1F	1F
FOREST&WOODL	1F	1F	1F	1F
OTHER LAND	25	25	25	25
MONGOLIA				
TOTAL AREA	156500	156500	156500	156500
LAND AREA	156500	156500	156500	156500
ARAB&PERM CR	683	885	775	950
ARABLE LAND	683	885	775	950
PERM PASTURE	139483	139798	139900	124602
FOREST&WOODL	15000	15000	15000	15641
OTHER LAND	1334	817	825	15307
NEPAL				
TOTAL AREA	14080	14080	14080	14080
LAND AREA	13800*	13800*	13800*	13800*
ARAB&PERM CR	1833	1831	1980	2024F
ARABLE LAND	1823	1821	1968	2010F
PERM CROPS	10F	10F	12F	14F
PERM PASTURE	2000*	2000*	2000	2000*
FOREST&WOODL	4532	4532	4475	4450F
OTHER LAND	5435	5437	5345	5326
OMAN				
TOTAL AREA	21246	21246	21246	21246
LAND AREA	21246	21246	21246	21246
ARAB&PERM CR	36*	36*	36	36*
ARABLE LAND	16*	16*	16	16*
PERM CROPS	20*	20*	20	20*
PERM PASTURE	1000F	1000F	1000F	1000F
OTHER LAND	20210	20210	20210	20210
PAKISTAN				
TOTAL AREA	80394	80394	80394	80394
LAND AREA	77872	77872	77872	77872
ARAB&PERM CR	17874	19692	19279	19420F
ARABLE LAND	17722*	19537*	19114*	19250F
PERM CROPS	152F	155F	165*	170F
PERM PASTURE	5000F	5000F	5000F	5000F
FOREST&WOODL	1838	2084	2720	2857*
OTHER LAND	53160	51096	50873	50595
PHILIPPINES				
TOTAL AREA	30000	30000	30000	30000
LAND AREA	29826	29826	29826	29817
ARAB&PERM CR	6807	6980F	7272F	8000F
ARABLE LAND	4840	4750F	4780F	5200F
PERM CROPS	1967	2230F	2492	2800F
PERM PASTURE	812	830F	700F	665F
FOREST&WOODL	17050	16900F	15875	12500F
OTHER LAND	5157	5116	5979	8652
QATAR				
TOTAL AREA	1100	1100	1100	1100
LAND AREA	1100	1100	1100	1100
ARAB&PERM CR	2F	2F	2F	2F
ARABLE LAND	2F	2F	2F	2F
PERM PASTURE	50F	50F	50F	50F
OTHER LAND	1048	1048	1048	1048
SAUDI ARABIA				
TOTAL AREA	214969	214969	214969	214969
LAND AREA	214969	214969	214969	214969
ARAB&PERM CR	705F	809	878	1110F
ARABLE LAND	670F	765	811	1040F
PERM CROPS	35	44	67	70F
PERM PASTURE	85000	85000	85000	85000*
FOREST&WOODL	1688	1680	1601	1601*
OTHER LAND	127576	127480	127490	127258
SINGAPORE				
TOTAL AREA	58	58	58	58
LAND AREA	57	57	57	57
ARAB&PERM CR	13	13	10	8
ARABLE LAND	4	4	2	2
PERM CROPS	10	9	8	6
FOREST&WOODL	4	4	3	3
OTHER LAND	40	40	44	46
SRI LANKA				
TOTAL AREA	6561	6561	6561	6561

	1961-65	1966	1971	1976
SRI LANKA CONT.				
LAND AREA	6474	6474	6474	6474
ARAB&PERM CR	1737	1875	1979	1979
ARABLE LAND	709	792	895	895
PERM CROPS	1027	1083	1084	1084
PERM PASTURE	213	330F	439	439
FOREST&WOODL	3458	3325	2899	2899
OTHER LAND	1066	944	1157	1157
SYRIA				
TOTAL AREA	18518	18518	18518	18518
LAND AREA	18457	18457	18457	18418
ARAB&PERM CR	6523	6130	5908	5672
ARABLE LAND	6280	5873	5647	5260
PERM CROPS	243	257	261	412
PERM PASTURE	8298	7500*	7550*	8541
FOREST&WOODL	446	477	477	457
OTHER LAND	3190	4350	4522	3748
THAILAND				
TOTAL AREA	51400	51400	51400	51400
LAND AREA	51177	51177	51177	51177
ARAB&PERM CR	12639	12865*	13939	17650F
ARABLE LAND	11279	11415	12431	15750F
PERM CROPS	1360	1450F	1508	1900F
PERM PASTURE	308*	308*	308*	308*
FOREST&WOODL	24070	23100F	21800F	20950F
OTHER LAND	14160	14904	15130	12269
TURKEY				
TOTAL AREA	78058	78058	78058	78058
LAND AREA	77076	77076	77076	77076
ARAB&PERM CR	25775	26384	27614	27699
ARABLE LAND	23541	24000	24978	24858
PERM CROPS	2234	2384	2636	2841
PERM PASTURE	28451	27995	27750F	27500F
FOREST&WOODL	20170	20170	20170	20170
OTHER LAND	2680	2527	1542	1707
U A EMIRATES				
TOTAL AREA	8360	8360	8360	8360
LAND AREA	8360	8360	8360	8360
ARAB&PERM CR	8F	10F	12F	13F
ARABLE LAND	5F	6F	7*	8F
PERM CROPS	3F	4F	5*	5*
PERM PASTURE	200F	200F	200F	200F
FOREST&WOODL			1	2F
OTHER LAND	8152	8150	8147	8145
VIET NAM				
TOTAL AREA	32956	32956	32956	32956
LAND AREA	32536	32536	32536	32536
ARAB&PERM CR	5047*	4814F	5195F	5600F
ARABLE LAND	4693F	4466F	4905F	5350F
PERM CROPS	355F	348F	290F	250F
PERM PASTURE	4916F	4870F	4870F	4870F
FOREST&WOODL	13520*	13320F	12220F	11300F
OTHER LAND	9053	9532	10251	10766
YEMEN AR				
TOTAL AREA	19500	19500	19500	19500
LAND AREA	19000	19000	19000	19000
ARAB&PERM CR	1410F	1445F	1490F	1570F
ARABLE LAND	1380F	1410F	1450F	1520F
PERM CROPS	30F	35F	40F	50F
PERM PASTURE	7000F	7000F	7000F	7000*
FOREST&WOODL	400*	400*	400*	400*
OTHER LAND	10190	10155	10110	10030
YEMEN DEM				
TOTAL AREA	33297	33297	33297	33297
LAND AREA	33297	33297	33297	33297
ARAB&PERM CR	105F	123F	149F	172F
ARABLE LAND	90F	105F	130F	152F
PERM CROPS	15F	18F	19F	20F
PERM PASTURE	9065*	9065	9065*	9065*
FOREST&WOODL	2610	2590	2540F	2490F
OTHER LAND	21517	21519	21543	21570
EUROPE				
TOTAL AREA	486950	486952	487008	487032
LAND AREA	472880	472855	472862	472809
ARAB&PERM CR	152184	149421	144204	142380
ARABLE LAND	137926	134932	129470	127438
PERM CROPS	14259	14489	14734	14942
PERM PASTURE	89653	90104	89323	87606
FOREST&WOODL	143077	146176	150518	153444
OTHER LAND	87966	87154	88817	89379
ALBANIA				
TOTAL AREA	2875	2875	2875	2875
LAND AREA	2740	2740	2740	2740
ARAB&PERM CR	497	540	610F	660F
ARABLE LAND	442	475F	530F	565F
PERM CROPS	55	65F	80F	95F

TABLE
TABLEAU 1
CUADRO

LAND USE   UTILISATION DES TERRES   USO DE TIERRAS

1000 HA

	1961-65	1966	1971	1976
ALBANIA CONT.				
PERM PASTURE	736	700F	638F	580F
FOREST&WOODL	1266	1247F	1242F	1242F
OTHER LAND	241	253	250	258
ANDORRA				
TOTAL AREA	45	45	45	45
LAND AREA	45	45	45	45
ARAB&PERM CR	1F	1F	1F	1F
ARABLE LAND	1F	1F	1F	1F
PERM PASTURE	25F	25F	25F	25F
FOREST&WOODL	10F	10F	10F	10F
OTHER LAND	9	9	9	9
AUSTRIA				
TOTAL AREA	8385	8385	8385	8385
LAND AREA	8271	8271	8271	8272
ARAB&PERM CR	1735	1686	1681	1618
ARABLE LAND	1669	1616	1587	1520
PERM CROPS	66	70	94	98
PERM PASTURE	2275	2249	2213	2071
FOREST&WOODL	3161	3203	3206	3266
OTHER LAND	1100	1133	1171	1317
BELGIUM-LUX				
TOTAL AREA	3310	3310	3310	3310
LAND AREA	3282	3282	3282	3282
ARAB&PERM CR	1007	981	908	900
ARABLE LAND	957	934	874	871
PERM CROPS	50	47	34	29
PERM PASTURE	812	818	813	785
FOREST&WOODL	687	687	687	702
OTHER LAND	776	796	874	895
BULGARIA				
TOTAL AREA	11091	11091	11091	11091
LAND AREA	11055	11055	11055	11055
ARAB&PERM CR	4565	4564	4516	4327
ARABLE LAND	4178	4166	4132	3940
PERM CROPS	387	398	384	387
PERM PASTURE	1158	1238	1493	1871
FOREST&WOODL	3621	3617	3735	3807
OTHER LAND	1711	1636	1311	1050
CZECHOSLOVAK				
TOTAL AREA	12788	12788	12788	12788
LAND AREA	12564	12564	12564	12556
ARAB&PERM CR	5332	5373	5329	5258
ARABLE LAND	5237	5265	5205	5125
PERM CROPS	95	108	124	133
PERM PASTURE	1809	1771	1748	1732
FOREST&WOODL	4428	4450	4458	4511
OTHER LAND	995	970	1029	1055
DENMARK				
TOTAL AREA	4307	4307	4307	4307
LAND AREA	4237	4237	4237	4237
ARAB&PERM CR	2761	2701	2662	2675
ARABLE LAND	2749	2686	2647	2662
PERM CROPS	12	15	15	13
PERM PASTURE	337	326	289	266
FOREST&WOODL	438	472	483F	499
OTHER LAND	701	738	803	797
FAERCE IS				
TOTAL AREA	140	140	140	140
LAND AREA	140	140	140	140
ARAB&PERM CR	3*	3*	3*	3*
ARABLE LAND	3*	3*	3*	3*
OTHER LAND	137	137	137	137
FINLAND				
TOTAL AREA	33701	33701	33701	33701
LAND AREA	30545	30545	30545	30545
ARAB&PERM CR	2635	2662	2668	2613
ARABLE LAND	2635	2662	2668	2613
PERM PASTURE	111	110	150F	148
FOREST&WOODL	21779	21930F	22350F	22650F
OTHER LAND	6020	5843	5377	5134
FRANCE				
TOTAL AREA	54703	54703	54703	54703
LAND AREA	54592	54592	54592	54592
ARAB&PERM CR	21067	20214	18690	18730
ARABLE LAND	19293	18463	17056	17139
PERM CROPS	1774	1751	1634	1591
PERM PASTURE	13221	13632	13933	13337
FOREST&WOODL	11905	12714	14243	14576
OTHER LAND	8399	8032	7726	7949
GERMAN DR				
TOTAL AREA	10818	10818	10818	10818
LAND AREA	10603	10603	10603	10603
ARAB&PERM CR	5046	4996	4824	4998
ARABLE LAND	4814*	4786*	4622	4752

	1961-65	1966	1971	1976
GERMAN DR CONT.				
PERM CROPS	232*	210*	202	246
PERM PASTURE	1427	1442	1463	1295
FOREST&WOODL	2953	2950	2949	2951
OTHER LAND	1177	1215	1367	1359
GERMANY FED				
TOTAL AREA	24858	24858	24858	24858
LAND AREA	24446	24434	24415	24402
ARAB&PERM CR	8425	8228	8087	8050
ARABLE LAND	7835	7609	7548	7532
PERM CROPS	591	619	539	518
PERM PASTURE	5731	5802	5417	5219
FOREST&WOODL	7144	7184	7183	7165
OTHER LAND	3146	3220	3728	3968
GIBRALTAR				
TOTAL AREA	1	1	1	1
LAND AREA	1	1	1	1
OTHER LAND	1	1	1	1
GREECE				
TOTAL AREA	13194	13194	13194	13194
LAND AREA	13080	13080	13080	13080
ARAB&PERM CR	3800	3851	3910	3885F
ARABLE LAND	2941	2995	2991	2923F
PERM CROPS	859	856	919	962F
PERM PASTURE	5100	5239	5245F	5255F
FOREST&WOODL	2545	2608	2610F	2618F
OTHER LAND	1635	1382	1315	1322
HOLY SEE				
HUNGARY				
TOTAL AREA	9303	9303	9303	9303
LAND AREA	9238	9238	9238	9238
ARAB&PERM CR	5632	5642	5578	5471
ARABLE LAND	5205F	5225	5184	5108
PERM CROPS	427F	417	394	363
PERM PASTURE	1371	1285	1277	1287
FOREST&WOODL	1383	1442	1480	1556
OTHER LAND	852	869	903	924
ICELAND				
TOTAL AREA	10300	10300	10300	10300
LAND AREA	10025	10025	10025	10025
ARAB&PERM CR	1	1	1	1
ARABLE LAND	1	1	1	1
PERM PASTURE	2279	2279	2279	2279
FOREST&WOODL	116	120	120	120
OTHER LAND	7629	7625	7625	7625
IRELAND				
TOTAL AREA	7028	7028	7028	7028
LAND AREA	6889	6889	6889	6889
ARAB&PERM CR	1317	1199	1132*	992F
ARABLE LAND	1314	1196	1130*	990F
PERM CROPS	3F	3	2F	2*
PERM PASTURE	4180*	4420*	4540*	4740F
FOREST&WOODL	188	208	255F	305F
OTHER LAND	1204	1062	962	852
ITALY				
TOTAL AREA	30123	30123	30123	30123
LAND AREA	29402	29401	29404	29405
ARAB&PERM CR	15454	15258	12409	12348
ARABLE LAND	12691	12444	9511	9364F
PERM CROPS	2763	2814	2898	2984
PERM PASTURE	5096	5147	5240	5176
FOREST&WOODL	5984	6099	6170	6313
OTHER LAND	2868	2897	5585	5568
LIECHTENSTEN				
TOTAL AREA	16	16	16	16
LAND AREA	16	16	16	16
ARAB&PERM CR	4	4	4	4
ARABLE LAND	4	4	4	4
PERM PASTURE	5	5	5	5
FOREST&WOODL	3	3	3	3
OTHER LAND	4	4	4	4
MALTA				
TOTAL AREA	32	32	32	32
LAND AREA	32	32	32	32
ARAB&PERM CR	16	14	14	15
ARABLE LAND	15	13	13	14
PERM CROPS	1	1	1	1
OTHER LAND	16	18	18	17
MONACC				
NETHERLANDS				
TOTAL AREA	3613	3615	3671	3695
LAND AREA	3354	3343	3369	3381
ARAB&PERM CR	992	946	848	843

TABLE
TABLEAU **1**
CUADRO

LAND USE   UTILISATION DES TERRES   USO DE TIERRAS

1000 HA

	1961-65	1966	1971	1976
NETHERLANDS CONT.				
ARABLE LAND	954	897	808	807
PERM CROPS	38	49	40	36
PERM PASTURE	1293	1299	1280	1230
FOREST&WOODL	280	292	299	309
OTHER LAND	789	806	942	999
NORWAY				
TOTAL AREA	32422	32422	32422	32422
LAND AREA	30825	30825	30825	30810
ARAB&PERM CR	850	841	806	795
ARABLE LAND	837*	828*	793	782
PERM CROPS	13*	13*	13	13*
PERM PASTURE	172	158	125	101
FOREST&WOODL	6883	7300F	8030F	8330
OTHER LAND	22920	22526	21864	21584
POLAND				
TOTAL AREA	31268	31268	31268	31268
LAND AREA	30468	30468	30468	30460
ARAB&PERM CR	15968	15682	15277	15036
ARABLE LAND	15729	15436	14980	14762
PERM CROPS	239	246	297	274
PERM PASTURE	4201	4265	4231	4114
FOREST&WOODL	7882	8140	8578	8631
OTHER LAND	2417	2381	2382	2679
PORTUGAL				
TOTAL AREA	9208	9208	9208	9208
LAND AREA	9164	9164	9164	9164
ARAB&PERM CR	4332	4070F	3720F	3600F
ARABLE LAND	3732	3470F	3130F	3010F
PERM CROPS	600	600F	590F	590F
PERM PASTURE	530*	530*	530*	530*
FOREST&WOODL	3165	3400F	3641	3641*
OTHER LAND	1137	1164	1273	1393
ROMANIA				
TOTAL AREA	23750	23750	23750	23750
LAND AREA	23034	23034	23034	23034
ARAB&PERM CR	10466	10502	10506	10518
ARABLE LAND	9854	9797	9728	9760
PERM CROPS	612	705	778	758
PERM PASTURE	4243	4333	4429	4437
FOREST&WOODL	6394	6371	6313	6316
OTHER LAND	1931	1828	1786	1763
SAN MARINO				
TOTAL AREA	6	6	6	6
LAND AREA	6	6	6	6
ARAB&PERM CR	1F	1F	1F	1F
ARABLE LAND	1F	1F	1F	1F
OTHER LAND	5	5	5	5
SPAIN				
TOTAL AREA	50478	50478	50478	50478
LAND AREA	49978	49978	49978	49957
ARAB&PERM CR	20709	20156	21184	20659
ARABLE LAND	16126	15509	16339	15657
PERM CROPS	4583	4647	4845	5002
PERM PASTURE	12300F	12000F	11500F	10857
FOREST&WOODL	13160	13600F	14300F	15333
OTHER LAND	3809	4222	2994	3108
SWEDEN				
TOTAL AREA	44996	44996	44996	44996
LAND AREA	41141	41141	41141	41148
ARAB&PERM CR	3397	3158	3051	3003
ARABLE LAND	3352	3113	3006	2958
PERM CROPS	45	45F	45F	45F
PERM PASTURE	684	685F	707	732
FOREST&WOODL	26216*	26505*	26424	26424
OTHER LAND	10844	10793	10959	10989
SWITZERLAND				
TOTAL AREA	4129	4129	4129	4129
LAND AREA	3993	3993	3993	3977
ARAB&PERM CR	415	401F	385	396
ARABLE LAND	402	385F	367	376
PERM CROPS	13	16F	18	20
PERM PASTURE	1755	1778F	1789	1625
FOREST&WOODL	981	981	981	1052
OTHER LAND	842	833	838	904
UK				
TOTAL AREA	24482	24482	24482	24482
LAND AREA	24174	24173	24170	24177
ARAB&PERM CR	7407	7480	7226	6975
ARABLE LAND	7299	7382	7146	6907
PERM CROPS	108*	98	80	68
PERM PASTURE	12330	12118	11617	11593
FOREST&WOODL	1761	1831	1510	2043
OTHER LAND	2676	2744	3417	3566
YUGOSLAVIA				
TOTAL AREA	25580	25580	25580	25580

	1961-65	1966	1971	1976
YUGOSLAVIA CONT.				
LAND AREA	25540	25540	25540	25540
ARAB&PERM CR	8349	8266	8173	8005
ARABLE LAND	7656	7570	7465	7291
PERM CROPS	693	696	708	714
PERM PASTURE	6472	6450	6347	6316
FOREST&WOODL	8744	8812	8858	9071
OTHER LAND	1975	2012	2162	2148
OCEANIA				
TOTAL AREA	850956	850956	850956	850956
LAND AREA	842906	842906	842906	842906
ARAB&PERM CR	35449	41632	46671	47124
ARABLE LAND	34620	40752	45766	46212
PERM CROPS	828	880	905	912
PERM PASTURE	459369	460615	468061	469761
FOREST&WOODL	186622	186478	186211	185950
OTHER LAND	161466	154181	141963	140071
AMER SAMOA				
TOTAL AREA	20	20	20	20
LAND AREA	20	20	20	20
ARAB&PERM CR	8	8	8*	8*
ARABLE LAND	4	4	4*	4*
PERM CROPS	4	4	4*	4*
FOREST&WOODL	8	10	10*	10*
OTHER LAND	4	2	2	2
AUSTRALIA				
TOTAL AREA	768685	768685	768685	768685
LAND AREA	761793	761793	761793	761793
ARAB&PERM CR	33713	39797	44752	45170F
ARABLE LAND	33534	39614	44562	45000F
PERM CROPS	179	183	190	170
PERM PASTURE	445908	447208	454768	455527*
FOREST&WOODL	137700*	137700*	137700*	137700
OTHER LAND	144472	137088	124573	123396
CANTON IS				
TOTAL AREA	7	7	7	7
LAND AREA	7	7	7	7
OTHER LAND	7	7	7	7
CHRISTMAS IS				
TOTAL AREA	13	13	13	13
LAND AREA	13	13	13	13
OTHER LAND	13	13	13	13
COCOS IS				
TOTAL AREA	1	1	1	1
LAND AREA	1	1	1	1
OTHER LAND	1	1	1	1
COOK ISLANDS				
TOTAL AREA	23	23	23	23
LAND AREA	23	23	23	23
ARAB&PERM CR	10	9*	11	11F
ARABLE LAND	4	3*	3	3F
PERM CROPS	6	6*	8	8F
FOREST&WOODL	8	6	6	
OTHER LAND	5	8	6	6
FIJI				
TOTAL AREA	1827	1827	1827	1827
LAND AREA	1827	1827	1827	1827
ARAB&PERM CR	213	204	225	231F
ARABLE LAND	146	109	146	147F
PERM CROPS	67	95	79	84F
PERM PASTURE	73	65	65	65*
FOREST&WOODL	1177*	1185*	1185*	1185*
OTHER LAND	364	373	352	346
FR POLYNESIA				
TOTAL AREA	400	400	400	400
LAND AREA	366	366	366	366
ARAB&PERM CR	62	66F	70F	75F
ARABLE LAND	5F	5F	5F	5F
PERM CROPS	57F	61F	65F	70F
PERM PASTURE	20	20*	20*	20*
FOREST&WOODL	115	115*	115*	115*
OTHER LAND	169	165	161	156
GILBERT IS				
TOTAL AREA	71	71	71	71
LAND AREA	71	71	71	71
ARAB&PERM CR	39	41	36	36*
PERM CROPS	39	41	36	36*
FOREST&WOODL	2	2	2	2*
OTHER LAND	30	28	33	33
GUAM				
TOTAL AREA	55	55	55	55
LAND AREA	55	55	55	55
ARAB&PERM CR	12*	12*	12*	12*
ARABLE LAND	6*	6*	6*	6*

TABLE / TABLEAU / CUADRO 1

LAND USE   UTILISATION DES TERRES   USO DE TIERRAS

1000 HA

	1961-65	1966	1971	1976
GUAM CONT.				
PERM CROPS	6*	6*	6*	6*
PERM PASTURE	8*	8*	8*	8*
FOREST&WOODL	10	10	10	10
OTHER LAND	25	25	25	25
NAURU				
TOTAL AREA	2	2	2	2
LAND AREA	2	2	2	2
OTHER LAND	2	2	2	2
NEWCALEDONIA				
TOTAL AREA	1906	1906	1906	1906
LAND AREA	1876	1876	1876	1876
ARAB&PERM CR	28F	31F	31F	33*
ARABLE LAND	25F	28F	28F	30*
PERM CROPS	3*	3*	3*	3*
PERM PASTURE	355	325*	325*	325*
FOREST&WOODL	960*	960*	960*	960*
OTHER LAND	533	560	560	558
NEW HEBRIDES				
TOTAL AREA	1476	1476	1476	1476
LAND AREA	1476	1476	1476	1476
ARAB&PERM CR	82	88F	95F	95F
ARABLE LAND	11	13F	15F	15F
PERM CROPS	71	75F	80F	80F
PERM PASTURE	25	25	25	25
FOREST&WOODL	16	16	16	16
OTHER LAND	1353	1347	1340	1340
NEW ZEALAND				
TOTAL AREA	26868	26868	26868	26868
LAND AREA	26867	26867	26867	26867
ARAB&PERM CR	737	819	834	849F
ARABLE LAND	726	806	820	835F
PERM CROPS	11	13	14	14*
PERM PASTURE	12847	12827	12680	13600F
FOREST&WOODL	7403	7250F	7000F	6750F
OTHER LAND	5880	5971	6353	5668
NIUE ISLAND				
TOTAL AREA	26	26	26	26
LAND AREA	26	26	26	26
ARAB&PERM CR	20	20	19	19*
ARABLE LAND	18	18	17	17*
PERM CROPS	2	2	2	2
PERM PASTURE			1	1
FOREST&WOODL	5	5	5	5
OTHER LAND	1	1	1	1
NORFOLK IS				
TOTAL AREA	4	4	4	4
LAND AREA	4	4	4	4
PERM PASTURE	1	1	1	1
OTHER LAND	3	3	3	3
PACIFIC IS				
TOTAL AREA	178	178	178	178
LAND AREA	178	178	178	178
ARAB&PERM CR	52	53*	58F	59F
ARABLE LAND	22	23*	25F	25F
PERM CROPS	30	30*	33F	34
PERM PASTURE	18	18	20F	24
FOREST&WOODL	40	40	40	40
OTHER LAND	68	67	60	55
PAPUA N GUIN				
TOTAL AREA	46169	46169	46169	46169
LAND AREA	45171	45171	45171	45171
ARAB&PERM CR	309F	315F	348F	353F
ARABLE LAND	15F	15F	18*	16*
PERM CROPS	294F	300F	330F	337F
PERM PASTURE	68F	71	101	110F
FOREST&WOODL	36424	36424	36424	36424
OTHER LAND	8370	8361	8298	8284
SAMOA				
TOTAL AREA	286	286	286	286
LAND AREA	285	285	285	285
ARAB&PERM CR	59	63	63*	64F
ARABLE LAND	39	42	42*	42*
PERM CROPS	19	21	21*	22F
PERM PASTURE	5	6	6	12
FOREST&WOODL	183	184	170F	159
OTHER LAND	38	32	46	50
SOLOMON IS				
TOTAL AREA	2845	2845	2845	2845
LAND AREA	2754	2754	2754	2754
ARAB&PERM CR	54F	54F	54F	54F
ARABLE LAND	50F	50F	50F	50F
PERM CROPS	4F	4F	4F	4F
PERM PASTURE	39*	39*	39*	39*
FOREST&WOODL	2560	2560	2560	2560

	1961-65	1966	1971	1976
SOLOMON IS CONT.				
OTHER LAND	101	101	101	101
TOKELAU				
TOTAL AREA	1	1	1	1
LAND AREA	1	1	1	1
OTHER LAND	1	1	1	1
TONGA				
TOTAL AREA	70	70	70	70
LAND AREA	67	67	67	67
ARAB&PERM CR	49F	50F	53	53
ARABLE LAND	14	15	24	16
PERM CROPS	35F	35F	29	37
PERM PASTURE	2	2	2	4
FOREST&WOODL	11	11	8	8
OTHER LAND	5	4	4	2
TUVALU				
TOTAL AREA	3	3	3	3
LAND AREA	3	3	3	3
OTHER LAND	3	3	3	3
WALLIS ETC				
TOTAL AREA	20	20	20	20
LAND AREA	20	20	20	20
ARAB&PERM CR	2F	2F	2F	2F
ARABLE LAND	1F	1F	1F	1F
PERM CROPS	1F	1F	1F	1F
OTHER LAND	18	18	18	18
USSR				
TOTAL AREA	2240220	2240220	2240220	2240220
LAND AREA	2227200	2227200	2227200	2227200
ARAB&PERM CR	229496	229300F	232609	232306
ARABLE LAND	225080	224700F	227600	227400
PERM CROPS	4416	4600F	5009	4906
FOREST&WOODL	920000	920000*	920000*	920000*
OTHER LAND	706104	704700	699891	701494
DEV.PED M E				
TOTAL AREA	3275948	3275950	3276006	3276030
LAND AREA	3157874	3157849	3157856	3157819
ARAB&PERM CR	380522	382862	397548	394153
ARABLE LAND	364933	366849	381257	377567
PERM CROPS	15590	16013	16291	16586
PERM PASTURE	903237	899041	893375E	889302
FOREST&WOODL	906576	908681	912517	914592
OTHER LAND	967539	967265	954033	959772
N AMERICA				
TOTAL AREA	1933926	1933926	1933926	1933926
LAND AREA	1834796	1834796	1834796	1834796
ARAB&PERM CR	222490	220954	233978	232039
ARABLE LAND	220532	219029	232087	230129
PERM CROPS	1958	1925	1891	1910
PERM PASTURE	282229	278957	268396	264961
FOREST&WOODL	616271	615671	616300	616129
OTHER LAND	713806	719214	716122	721667
W EUROPE				
TOTAL AREA	385057	385059	385115	385139
LAND AREA	373178	373153	373160	373123
ARAB&PERM CR	104678	102122	97564	96112
ARABLE LAND	92467	89782	85089	83426
PERM CROPS	12212	12340	12475	12686
PERM PASTURE	74708	75070	74044	72290
FOREST&WOODL	115150	117959	121763	124430
OTHER LAND	78642	78002	79789	80291
OCEANIA				
TOTAL AREA	795553	795553	795553	795553
LAND AREA	788660	788660	788660	788660
ARAB&PERM CR	34450	40616	45586	46019
ARABLE LAND	34260	40420	45382	45835
PERM CROPS	190	196	204	184
PERM PASTURE	458755	460035	467448	469127
FOREST&WOODL	145103	144950	144700	144450
OTHER LAND	150352	143059	130926	129064
OTH DEV.PED				
TOTAL AREA	161412	161412	161412	161412
LAND AREA	161240	161240	161240	161240
ARAB&PERM CR	18904	19170	20420	19983
ARABLE LAND	17674	17618	18699	18177
PERM CROPS	1230	1552	1721	1806
PERM PASTURE	87545	84979	83870	82924
FOREST&WOODL	30052	30101	29754	29583
OTHER LAND	24739	26990	27196	28750
DEV.PING M E				
TOTAL AREA	6597788	6597788	6597788	6597788
LAND AREA	6444564	6444564	6444564	6444398
ARAB&PERM CR	606310	624617	647847	674057

TABLE
TABLEAU 1
CUADRO

LAND USE   UTILISATION DES TERRES   USO DE TIERRAS

1000 HA

	1961-65	1966	1971	1976
DEV.PING M E CONT.				
ARABLE LAND	550774	566935	586016	608905
PERM CROPS	55532	57682	61831	65152
PERM PASTURE	1409121	1417392	1427712	1435963
FOREST&WOODL	2144002	2123460	2099492	2076826
OTHER LAND	2285131	2279095	2269513	2257552
AFRICA				
TOTAL AREA	2382384	2382384	2382384	2382384
LAND AREA	2329494	2329494	2329494	2329410
ARAB&PERM CR	165523	171675	177636	181990
ARABLE LAND	154360	159555	165161	168950
PERM CROPS	11160	12120	12475	13040
PERM PASTURE	689299	687913	687781	688057
FOREST&WOODL	569965	558623	548793	542966
OTHER LAND	904707	911283	915284	916397
LAT AMERICA				
TOTAL AREA	2060298	2060298	2060298	2060298
LAND AREA	2025219	2025219	2025219	2025185
ARAB&PERM CR	116852	123558	134137	143568
ARABLE LAND	93518	99652	108654	116734
PERM CROPS	23334	23906	25483	26834
PERM PASTURE	494681	503028	515768	523603
FOREST&WOODL	1062041	1052601	1038580	1026433
OTHER LAND	351645	346032	336734	331581
NEAR EAST				
TOTAL AREA	1208353	1208353	1208353	1208353
LAND AREA	1191860	1191860	1191860	1191821
ARAB&PERM CR	75662	76827	79133	81062
ARABLE LAND	72066	72876	74183	75588
PERM CROPS	3596	3951	4950	5474
PERM PASTURE	188476	189488	189946	190417
FOREST&WOODL	139652	139613	139369	138932
OTHER LAND	788070	785932	783412	781410
FAR EAST				
TOTAL AREA	857151	857151	857151	857151
LAND AREA	809547	809547	809547	809538
ARAB&PERM CR	247271	251538	255853	266329
ARABLE LAND	230467	234517	237631	247253
PERM CROPS	16804	17021	18222	19076
PERM PASTURE	36046	36378	33599	33247
FOREST&WOODL	330813	331083	331227	326983
OTHER LAND	195417	190548	188868	182979
OTH DV.PING				
TOTAL AREA	89602	89602	89602	89602
LAND AREA	88444	88444	88444	88444
ARAB&PERM CR	1002	1019	1088	1108
ARABLE LAND	363	335	387	380
PERM CROPS	638	684	701	728
PERM PASTURE	619	585	618	639
FOREST&WOODL	41531	41540	41523	41512
OTHER LAND	45292	45300	45215	45185
CENTR PLANND				
TOTAL AREA	3521423	3521423	3521423	3521423
LAND AREA	3476127	3476127	3476127	3476111
ARAB&PERM CR	406268	409031	417315	419840
ARABLE LAND	398674	401092	408761	411262
PERM CROPS	7595	7939	8554	8578
PERM PASTURE	735574	739932	745379	732818
FOREST&WOODL	1107969	1117379	1134317	1153797
OTHER LAND	1226316	1209785	1179116	1169656
ASIAN CPE				
TOTAL AREA	1179310	1179310	1179310	1179310
LAND AREA	1149225	1149225	1149225	1149225
ARAB&PERM CR	129266	132432	138066	141266
ARABLE LAND	128135	131242	136780	139850
PERM CROPS	1132	1190	1286	1416
PERM PASTURE	349029	351698	355400	344102
FOREST&WOODL	160042	169162	185562	204783
OTHER LAND	510888	495933	470197	459074
E EUR+USSR				
TOTAL AREA	2342113	2342113	2342113	2342113
LAND AREA	2326902	2326902	2326902	2326886
ARAB&PERM CR	277002	276599	279249	278574
ARABLE LAND	270539	269850	271981	271412
PERM CROPS	6463	6749	7268	7162
PERM PASTURE	386545	388234	389979	388716
FOREST&WOODL	947927	948217	948755	949014
OTHER LAND	715428	713852	708919	710582
DEV.PED ALL				
TOTAL AREA	5618061	5618063	5618119	5618143
LAND AREA	5484776	5484751	5484758	5484705
ARAB&PERM CR	657524	659461	676797	672727
ARABLE LAND	635472	636699	653238	648979
PERM CROPS	22053	22762	23559	23748
PERM PASTURE	1289782	1287275	1283737	1278018
FOREST&WOODL	1854503	1856898	1861272	1863606

	1961-65	1966	1971	1976
DEV.PED ALL CONT.				
OTHER LAND	1682967	1681117	1662952	1670354
DEV.PING ALL				
TOTAL AREA	7777098	7777098	7777098	7777098
LAND AREA	7593789	7593789	7593789	7593623
ARAB&PERM CR	735576	757049	785913	815323
ARABLE LAND	678909	698177	722796	748755
PERM CROPS	56664	58872	63117	66568
PERM PASTURE	1758150	1769090	1783112	1780065
FOREST&WOODL	2304044	2292622	2285054	2281609
OTHER LAND	2796019	2775028	2739710	2716626

TABLE
TABLEAU 2
CUADRO

IRRIGATION IRRIGATION RIEGO

AGRIC AREA 1000HA SUP AGRICOLE 1000HA SUP AGRICOLA 1000HA

	1961-65	1966	1971	1976
WORLD	189085	196455	212622	230556
AFRICA	5873	6439	7148	7697
ALGERIA	259	234	280F	330F
BENIN	1	3F	5F	5F
BOTSWANA	2	2	2	1*
BURUNDI	4F	4F	5	5*
CAMEROON	3F	5F	7F	8F
CAPE VERDE	2	2	2	2*
CHAD		1	1	1
EGYPT	2548	2780	2852	2826F
ETHIOPIA	48	48*	55F	55F
GAMBIA	12F	15F	20F	25F
GHANA	11	12*	15F	20F
GUINEA	4F	4F	5F	6
IVORY COAST	5	9	22	25F
KENYA	14*	17F	32F	42
LIBERIA			2*	2
LIBYA	123	142	125	135F
MADAGASCAR	306F	330F	350*	430F
MALAWI	2	2	5	5*
MALI	43F	50F	66	90F
MAURITANIA	3	3F	3F	3F
MAURITIUS	10	13	15	15*
MOROCCO	199	280F	355F	470*
MOZAMBIQUE	46F	55F	62F	68F
NAMIBIA	4*	5F	6F	7F
NIGER	4	5	5	6*
NIGERIA	11	13*	15	15*
REUNION	4	6	7	7
RHODESIA	27	34*	50F	55F
RWANDA			1	1
SENEGAL	77	90F	119	127F
SIERRA LEONE	1	2F	3F	4F
SOMALIA	165*	165*	162*	165F
SOUTH AFRICA	850	900F	1017	1017*
SUDAN	952F	1050F	1300F	1500F
SWAZILAND	20	27	25	26F
TANZANIA	33F	42*	46F	55F
TOGO	2	3	3*	3*
TUNISIA	74	80*	95F	130F
UGANDA	2	4	4	4
UPPER VOLTA			1	2*
ZAMBIA	2	2*	3F	4F
N C AMERICA	19582	20374	21621	23174
BELIZE			1*	2*
CANADA	365F	390F	430F	480F
COSTA RICA	26	26*	26*	26*
CUBA	456	500F	620F	730F
DOMINICAN RP	113	120F	125F	135F
EL SALVADOR	18	20*	20	33
GUADELOUPE	1	2*	2	2*
GUATEMALA	38	45F	60	62F
HAITI	38F	42	65F	70*
HONDURAS	60	66*	70F	80F
JAMAICA	23	24*	30F	32F
MARTINIQUE	1	1	1	2
MEXICO	3700F	3750F	4000F	4816
NICARAGUA	18	18*	45F	70
PANAMA	15	18F	20F	23F
PUERTO RICO	39	39	39	39
ST LUCIA	1	1	1*	1*
ST VINCENT		1	1	1*
TRINIDAD ETC	11	11*	15F	20F
USA	14659	15300F	16050F	16550F
SOUTH AMERIC	5403	5712	6193	6730
ARGENTINA	1587	1650F	1720F	1820F
BOLIVIA	74F	75F	80	120F
BRAZIL	546F	640F	830F	980F
CHILE	1084	1100F	1200F	1280F
COLOMBIA	231	240F	255F	285F
ECUADOR	446F	463*	470F	510F
GUYANA	100	109	115F	122F
PARAGUAY	30F	40F	50*	55F
PERU	1041	1078	1110	1150F
SURINAM	14	20F	27*	30F
URUGUAY	32	42	52*	58F
VENEZUELA	218	255*	284*	320F
ASIA	138753	143439	153885	163637
AFGHANISTAN	2208F	2260F	2360F	2460F
BAHRAIN	1	1	1	1*
BANGLADESH	501	620	1047	1420F
BURMA	681	773	890	984
CHINA	77200F	79000F	82100F	85200F
CYPRUS	95	102	94*	94*
HONG KONG	9	8*	8	6
INDIA	25523	26660	31100	34400F
INDONESIA	4100F	4175F	4490F	4840

	1961-65	1966	1971	1976
IRAN	4800F	4950F	5251	5840
IRAQ	1150*	1150*	1150*	1150*
ISRAEL	142	156	171	187
JAPAN	3176	3129	2626	2690F
JORDAN	57	60	60*	60*
KAMPUCHEA DM	72	100	89	89F
KOREA DPR	500F	500F	500F	500F
KOREA REP	682	731	868	936
KUWAIT			1	1
LAO	13	15*	19	11*
LEBANON	49	65	75F	85F
MAL PENINSUL	224	229	246	310F
MAL SABAH	8	8*	9	15
MAL SARAWAK	1	1	1	3*
NEPAL	77	105	117	190F
PAKISTAN	11139	12029	12986	13600F
PHILIPPINES	896	960	1200F	1430F
SAUDI ARABIA	270F	300F	350F	390F
SRI LANKA	361	398	439	530F
SYRIA	579	508	476	547
THAILAND	1729	1768	2106	2448
TURKEY	1336	1500F	1850F	2000F
U A EMIRATES	3F	4F	5*	5*
VIET NAM	992F	980F	980F	980F
YEMEN AR	175F	190F	215F	230*
YEMEN DEM	4F	4F	5F	5F
EUROPE	8659	9248	10702	12393
ALBANIA	180	217	230F	250F
AUSTRIA	4*	4	4*	4*
BELGIUM-LUX	1*	1*	1*	1*
BULGARIA	848	946	1021	1147
CZECHOSLOVAK	67F	75F	86F	94F
DENMARK	1*	1*	1*	1*
FINLAND	4	8F	20F	45
FRANCE	510F	520F	540F	565F
GERMAN DR	95	128F	145F	160F
GERMANY FED	260F	270*	288F	315F
GREECE	530	601	793	867
HUNGARY	196	163	205	320
ITALY	2420F	2470F	2600F	2820F
MALTA	1	1	1	1
NETHERLANDS	46	50F	50F	65F
NORWAY	19	20F	24F	26F
POLAND	285F	270F	202	198
PORTUGAL	620F	621F	623F	628F
ROMANIA	207	331	957	1729
SPAIN	2089	2275	2625	2854
SWEDEN	21F	24F	37F	58*
SWITZERLAND	21F	23*	28F	30F
UK	106	104	87	91
YUGOSLAVIA	128	125	134	124
OCEANIA	1197	1443	1590	1625
AUSTRALIA	1115	1359	1470F	1475
NEW ZEALAND	82	84	120	150F
USSR	9618	9800F	11483	15300
DEV.PED M E	27170	28436	29740	31044
N AMERICA	15024	15690	16480	17030
W EUROPE	6781	7118	7856	8495
OCEANIA	1197	1443	1590	1625
OTH DEV.PED	4168	4185	3814	3894
DEV.PING M E	71655	75509	84884	93545
AFRICA	1400	1567	1854	2219
LAT AMERICA	9961	10396	11334	12874
NEAR EAST	14350	15066	16170	17329
FAR EAST	45944	48480	55526	61123
CENTR PLANND	90260	92510	97998	105967
ASIAN CPE	78764	80580	83669	86769
E EUR+USSR	11496	11930	14329	19198
DEV.PED ALL	38666	40366	44069	50242
DEV.PING ALL	150419	156089	168553	180314

POPULATION

POBLACION

POPULATION

POBLACION

TABLE 3 TOTAL POPULATION,AGRICULTURAL POPULAT ION AND ECONOMICALLY ACTIVE POPULATION AS ESTIMATED FOR 1965,1970,1975,1976 AND 1977 — POBLACION TOTAL,POBLACION AGRI COLA·Y POBLACION ECONOMICAMEN TE ACTIVA:ESTIMACIONES PARA 1965,1970,1975,1976 Y 1977 — POPULATION TOTALE,POPULATION AGRICOLE ET POPULATION ECONOMIQUEMENT ACTIVE: ESTIMATIONS POUR 1965,1970,1975,1976 ET 1977

CONTINENT AND COUNTRY	YEAR	POPULATION		ECONOMICALLY ACTIVE POPULATION		
		TOTAL	AGRICULTURAL	TOTAL	IN AGRICULTURE	PERCENT IN AGRICULTURE
		......................THOUSANDS......................				
WORLD	1965	3276470	1768740	1391660	753985	54.2
	1970	3596330	1837070	1504570	766581	51.0
	1975	3951100	1914100	1640650	785481	47.9
	1976	4026400	1930560	1669160	789494	47.3
	1977	4103310	1947070	1697800	793362	46.7
AFRICA	1965	306049	220465	122080	90526	74.2
	1970	348002	241684	135847	97234	71.6
	1975	397106	264299	151179	103765	68.6
	1976	408257	269229	154608	105157	68.0
	1977	419801	274225	158123	106544	67.4
ALGERIA	1965	11060	7003	2742	1749	63.8
	1970	13300	8010	3019	1833	60.7
	1975	15530	8596	3462	1916	55.4
	1976	16100	8736	3588	1945	54.2
	1977	16800	8931	3744	1986	53.0
ANGOLA	1965	5101	3395	1429	951	66.6
	1970	5670	3614	1562	995	63.7
	1975	6394	3886	1726	1049	60.8
	1976	6561	3948	1763	1061	60.2
	1977	6733	4010	1801	1073	59.6
BENIN	1965	2365	1232	1190	620	52.1
	1970	2686	1336	1317	655	49.7
	1975	3074	1470	1457	696	47.8
	1976	3160	1499	1487	706	47.4
	1977	3249	1529	1519	715	47.1
BOTSWANA	1965	554	495	284	254	89.3
	1970	617	535	297	258	86.7
	1975	691	579	330	276	83.7
	1976	709	589	338	281	83.1
	1977	729	601	346	285	82.4
BURUNDI	1965	3210	2843	1667	1477	88.6
	1970	3350	2920	1715	1494	87.2
	1975	3765	3208	1866	1590	85.2
	1976	3863	3275	1900	1611	84.8
	1977	3964	3343	1935	1632	84.3
CAMEROON	1965	5356	4611	2689	2315	86.1
	1970	5836	4937	2883	2439	84.6
	1975	6433	5323	3085	2553	82.7
	1976	6571	5411	3130	2578	82.3
	1977	6711	5499	3176	2602	81.9
CAPE VERDE	1965	240	156	69	45	65.2
	1970	268	167	77	48	62.3
	1975	292	173	87	52	59.3
	1976	296	174	89	52	58.6
	1977	301	175	92	53	58.0
CENT AFR EMP	1965	1452	1346	826	766	92.7
	1970	1612	1471	899	821	91.3
	1975	1790	1601	983	880	89.5
	1976	1829	1629	1001	892	89.0
	1977	1870	1657	1021	904	88.6
CHAD	1965	3368	3108	1325	1223	92.3
	1970	3640	3281	1418	1278	90.1
	1975	3947	3443	1526	1331	87.2
	1976	4016	3476	1550	1342	86.6
	1977	4086	3509	1574	1352	85.9
COMOROS	1965	242	166	96	66	68.7
	1970	270	181	104	70	67.1
	1975	306	200	114	75	65.4
	1976	314	204	116	76	65.1
	1977	322	208	118	77	64.7
CONGO	1965	1069	499	393	183	46.7
	1970	1191	498	428	179	41.8
	1975	1345	510	473	179	37.9
	1976	1380	513	483	180	37.2
	1977	1416	516	493	180	36.5
EGYPT	1965	27755	15643	7846	4422	56.4
	1970	31476	17117	8801	4786	54.4
	1975	35455	18575	9931	5203	52.4
	1976	36292	18869	10176	5291	52.0
	1977	37145	19164	10427	5379	51.6
EQ GUINEA	1965	263	215	81	66	81.8
	1970	285	228	87	70	79.9
	1975	313	242	94	73	77.4
	1976	319	245	96	74	76.9
	1977	326	249	97	74	76.4
ETHIOPIA	1965	22078	19007	9759	8401	86.1
	1970	24855	20913	10822	9105	84.1
	1975	28134	23006	11918	9745	81.8
	1976	28854	23447	12150	9874	81.3
	1977	29593	23894	12389	10003	80.7
GABON	1965	484	402	247	205	83.1
	1970	500	407	250	203	81.4
	1975	521	411	255	201	79.0
	1976	526	413	256	201	78.4
	1977	531	414	257	200	77.9
GAMBIA	1965	422	353	226	189	83.6
	1970	463	379	241	197	81.9
	1975	509	407	256	205	80.0
	1976	520	414	260	207	79.6
	1977	530	420	263	208	79.2

TABLE **3** TOTAL POPULATION,AGRICULTURAL POPULAT ION AND ECONOMICALLY ACTIVE POPULATION AS ESTIMATED FOR 1965,1970,1975,1976 AND 1977

POBLACION TOTAL,POBLACION AGRI COLA Y POBLACION ECONOMICAMEN TE ACTIVA:ESTIMACIONES PARA 1965,1970,1975,1976 Y 1977

POPULATION TOTALE,POPULATION AGRICOLE ET POPULATION ECONOMIQUEMENT ACTIVE: ESTIMATIONS POUR 1965,1970,1975,1976 ET 1977

CONTINENT AND COUNTRY	YEAR	POPULATION		ECONOMICALLY ACTIVE POPULATION		
		TOTAL	AGRICULTURAL	TOTAL	IN AGRICULTURE	PERCENT IN AGRICULTURE
		....................THOUSANDS....................				
GHANA	1965	7740	4773	3121	1910	61.2
	1970	8628	5084	3351	1958	58.4
	1975	9873	5443	3710	2027	54.6
	1976	10161	5520	3800	2046	53.8
	1977	10461	5600	3897	2067	53.0
GUINEA	1965	3510	3034	1713	1481	86.4
	1970	3921	3320	1870	1583	84.7
	1975	4416	3644	2037	1681	82.5
	1976	4527	3715	2074	1702	82.1
	1977	4642	3788	2112	1723	81.6
GUIN BISSAU	1965	492	438	161	143	88.9
	1970	487	424	158	137	87.0
	1975	525	445	166	141	84.8
	1976	534	450	168	141	84.3
	1977	544	456	170	142	83.8
IVORY COAST	1965	5260	4556	2879	2494	86.6
	1970	5912	4996	3156	2667	84.5
	1975	6701	5499	3460	2840	82.1
	1976	6878	5608	3527	2876	81.5
	1977	7062	5721	3597	2914	81.0
KENYA	1965	9527	8001	3901	3276	84.0
	1970	11247	9237	4570	3753	82.1
	1975	13251	10594	5196	4154	79.9
	1976	13701	10890	5329	4235	79.5
	1977	14170	11196	5467	4320	79.0
LESOTHO	1965	954	873	529	485	91.5
	1970	1043	935	581	521	89.7
	1975	1148	999	623	542	87.0
	1976	1173	1013	632	546	86.4
	1977	1199	1028	642	550	85.7
LIBERIA	1965	1253	980	515	403	78.2
	1970	1387	1049	552	417	75.6
	1975	1555	1132	596	434	72.8
	1976	1594	1150	606	437	72.2
	1977	1635	1170	616	441	71.6
LIBYA	1965	1714	732	471	201	42.7
	1970	2054	657	550	176	32.0
	1975	2390	546	619	141	22.8
	1976	2464	523	634	135	21.2
	1977	2542	501	650	128	19.7
MADAGASCAR	1965	6079	5539	3160	2879	91.1
	1970	6932	6194	3622	3237	89.4
	1975	8020	6946	4049	3507	86.6
	1976	8263	7104	4138	3558	86.0
	1977	8515	7265	4230	3609	85.3
MALAWI	1965	3932	3570	1868	1696	90.8
	1970	4360	3884	2046	1823	89.1
	1975	4909	4253	2248	1948	86.6
	1976	5035	4334	2293	1974	86.1
	1977	5164	4415	2339	2000	85.5
MALI	1965	4530	4189	2587	2392	92.5
	1970	5047	4592	2850	2593	91.0
	1975	5697	5078	3140	2799	89.1
	1976	5842	5183	3202	2841	88.7
	1977	5991	5290	3266	2884	88.3
MAURITANIA	1965	1174	1049	372	332	89.3
	1970	1299	1136	406	355	87.5
	1975	1435	1224	444	379	85.3
	1976	1465	1243	452	384	84.8
	1977	1496	1262	461	388	84.3
MAURITIUS	1965	761	280	231	85	36.8
	1970	824	280	256	87	34.0
	1975	899	279	300	93	31.0
	1976	914	278	309	94	30.4
	1977	928	277	319	95	29.9
MOROCCO	1965	13139	7806	3654	2171	59.4
	1970	15126	8601	3987	2267	56.9
	1975	17504	9454	4572	2469	54.0
	1976	18038	9638	4718	2521	53.4
	1977	18592	9826	4872	2575	52.9
MOZAMBIQUE	1965	7449	5757	3069	2372	77.3
	1970	8234	6050	3381	2484	73.5
	1975	9223	6376	3648	2522	69.1
	1976	9461	6454	3707	2528	68.2
	1977	9705	6530	3766	2534	67.3
NAMIBIA	1965	660	387	230	135	58.7
	1970	766	426	263	146	55.6
	1975	883	460	293	153	52.1
	1976	909	467	299	154	51.3
	1977	936	474	306	155	50.6
NIGER	1965	3524	3312	1129	1061	94.0
	1970	4016	3726	1280	1188	92.8
	1975	4600	4169	1449	1313	90.6
	1976	4732	4264	1487	1339	90.1
	1977	4869	4361	1525	1366	89.6
NIGERIA	1965	48779	32404	20273	13467	66.4
	1970	55073	34178	22277	13825	62.1
	1975	63049	36390	24665	14236	57.7
	1976	64887	36881	25201	14324	56.8
	1977	66778	37366	25747	14406	56.0

| TABLE 3 TOTAL POPULATION,AGRICULTURAL POPULAT ION AND ECONOMICALLY ACTIVE POPULATION AS ESTIMATED FOR 1965,1970,1975,1976 AND 1977 | | POBLACION TOTAL,POBLACION AGRI COLA Y POBLACION ECONOMICAMEN TE ACTIVA:ESTIMACIONES PARA 1965,1970,1975,1976 Y 1977 | | POPULATION TOTALE,POPULATION AGRICOLE ET POPULATION ECONOMIQUEMENT ACTIVE: ESTIMATIONS POUR 1965,1970,1975,1976 ET 1977 | |

CONTINENT AND COUNTRY	YEAR	POPULATION		ECONOMICALLY ACTIVE POPULATION		
		TOTAL	AGRICULTURAL	TOTAL	IN AGRICULTURE	PERCENT IN AGRICULTURE
		....................THOUSANDS....................				
REUNION	1965	393	163	104	43	41.5
	1970	447	168	118	44	37.5
	1975	501	164	139	46	32.7
	1976	511	162	144	46	31.7
	1977	521	161	149	46	30.8
RHODESIA	1965	4393	2921	1596	1061	66.5
	1970	5308	3391	1875	1198	63.9
	1975	6272	3845	2139	1311	61.3
	1976	6493	3947	2198	1336	60.8
	1977	6723	4051	2258	1361	60.3
RWANDA	1965	3198	3016	1752	1652	94.3
	1970	3679	3428	1988	1853	93.2
	1975	4233	3875	2253	2062	91.5
	1976	4362	3976	2314	2109	91.2
	1977	4495	4080	2376	2156	90.8
SENEGAL	1965	3941	3216	1797	1466	81.6
	1970	4432	3531	1963	1564	79.7
	1975	4989	3848	2135	1647	77.1
	1976	5111	3915	2172	1664	76.6
	1977	5236	3982	2210	1681	76.0
SIERRA LEONE	1965	2198	1641	902	673	74.6
	1970	2455	1756	980	701	71.5
	1975	2770	1895	1071	733	68.4
	1976	2840	1924	1091	739	67.8
	1977	2914	1955	1112	746	67.1
SOMALIA	1965	2500	2156	1014	874	86.3
	1970	2789	2361	1084	918	84.7
	1975	3170	2614	1240	1023	82.5
	1976	3258	2671	1275	1046	82.0
	1977	3350	2730	1310	1068	81.5
SOUTH AFRICA	1965	18337	5692	6678	2073	31.0
	1970	21500	6637	8032	2480	30.9
	1975	24663	7347	9145	2724	29.8
	1976	25375	7491	9379	2769	29.5
	1977	26124	7641	9624	2815	29.2
SUDAN	1965	11259	9442	3667	3075	83.9
	1970	13050	10701	4183	3430	82.0
	1975	15190	12084	4788	3809	79.5
	1976	15674	12386	4924	3891	79.0
	1977	16178	12697	5065	3975	78.5
SWAZILAND	1965	361	307	178	151	85.1
	1970	409	331	198	161	81.1
	1975	469	362	220	170	77.1
	1976	483	368	225	172	76.2
	1977	497	375	230	174	75.4
TANZANIA	1965	11616	10181	5121	4489	87.7
	1970	13273	11411	5768	4959	86.0
	1975	15388	12872	6488	5427	83.7
	1976	15872	13196	6646	5525	83.1
	1977	16371	13526	6808	5625	82.6
TOGO	1965	1697	1297	752	575	76.4
	1970	1960	1437	859	630	73.3
	1975	2248	1589	953	673	70.7
	1976	2312	1622	973	682	70.1
	1977	2379	1656	994	692	69.6
TUNISIA	1965	4620	2464	1180	630	53.3
	1970	5137	2558	1217	606	49.8
	1975	5747	2594	1364	616	45.1
	1976	5893	2604	1404	620	44.2
	1977	6049	2615	1446	625	43.2
UGANDA	1965	8578	7520	3776	3310	87.7
	1970	9806	8427	4262	3662	85.9
	1975	11353	9487	4799	4010	83.6
	1976	11701	9717	4915	4082	83.0
	1977	12062	9953	5035	4155	82.5
UPPER VOLTA	1965	4498	4010	2558	2280	89.2
	1970	4985	4326	2777	2410	86.8
	1975	5585	4707	3033	2556	84.3
	1976	5716	4786	3088	2586	83.7
	1977	5849	4865	3145	2616	83.2
ZAIRE	1965	18835	15324	8719	7094	81.4
	1970	21638	17159	9720	7708	79.3
	1975	24450	18796	10663	8197	76.9
	1976	25098	19165	10877	8305	76.4
	1977	25673	19470	11054	8383	75.8
ZAMBIA	1965	3723	2819	1476	1118	75.7
	1970	4295	3128	1669	1215	72.8
	1975	5004	3494	1882	1314	69.8
	1976	5167	3575	1929	1335	69.2
	1977	5335	3658	1978	1356	68.6
N C AMERICA	1965	293033	51427	111535	17367	15.6
	1970	317109	50680	123406	16728	13.6
	1975	341294	51021	135959	16754	12.3
	1976	346298	51197	138478	16802	12.1
	1977	351401	51394	140991	16851	12.0
BARBADOS	1965	235	54	89	21	23.2
	1970	239	48	90	18	20.0
	1975	245	44	98	18	18.1
	1976	247	44	100	18	17.8
	1977	248	43	102	18	17.5

TABLE **3** TOTAL POPULATION,AGRICULTURAL POPULAT ION AND ECONOMICALLY ACTIVE POPULATION AS ESTIMATED FOR 1965,1970,1975,1976 AND 1977 | POBLACION TOTAL,POBLACION AGRI COLA Y POBLACION ECONOMICAMEN TE ACTIVA:ESTIMACIONES PARA 1965,1970,1975,1976 Y 1977 | POPULATION TOTALE,POPULATION AGRICOLE ET POPULATION ECONOMIQUEMENT ACTIVE: ESTIMATIONS POUR 1965,1970,1975,1976 ET 1977

CONTINENT AND COUNTRY	YEAR	POPULATION		ECONOMICALLY ACTIVE POPULATION		
		TOTAL	AGRICULTURAL	TOTAL	IN AGRICULTURE	PERCENT IN AGRICULTURE
		................THOUSANDS................				
CANADA	1965	19644	2092	7384	786	10.7
	1970	21320	1755	8562	705	8.2
	1975	22831	1474	9552	617	6.5
	1976	23143	1424	9741	599	6.2
	1977	23320	1367	9869	579	5.9
COSTA RICA	1965	1495	697	447	208	46.7
	1970	1727	728	530	223	42.2
	1975	1968	759	634	244	38.5
	1976	2018	764	656	248	37.9
	1977	2065	768	677	252	37.2
CUBA	1965	7802	2707	2510	871	34.7
	1970	8565	2620	2609	798	30.6
	1975	9481	2538	2849	763	26.8
	1976	9682	2522	2913	759	26.0
	1977	9889	2505	2982	756	25.3
DOMINICAN RP	1965	3703	2364	1021	651	63.8
	1970	4343	2660	1164	713	61.2
	1975	5118	3004	1355	795	58.7
	1976	5291	3078	1398	814	58.2
	1977	5471	3155	1444	833	57.7
EL SALVADOR	1965	2954	1766	925	544	58.9
	1970	3516	2007	1079	606	56.1
	1975	4108	2228	1270	677	53.3
	1976	4239	2275	1314	692	52.7
	1977	4375	2323	1359	708	52.1
GUADELOUPE	1965	279	104	93	35	37.4
	1970	304	78	99	26	25.7
	1975	328	68	112	23	20.7
	1976	333	67	115	23	20.0
	1977	338	65	118	23	19.3
GUATEMALA	1965	4122	2630	1273	812	63.8
	1970	4765	2905	1458	889	61.0
	1975	5513	3195	1685	976	58.0
	1976	5678	3256	1735	995	57.3
	1977	5848	3317	1785	1013	56.7
HAITI	1965	3950	3048	2146	1656	77.2
	1970	4235	3142	2145	1591	74.2
	1975	4552	3211	2299	1622	70.5
	1976	4626	3228	2338	1632	69.8
	1977	4703	3245	2377	1640	69.0
HONDURAS	1965	2021	1382	626	428	68.4
	1970	2336	1553	714	475	66.5
	1975	2779	1794	828	535	64.6
	1976	2876	1845	854	548	64.2
	1977	2976	1898	880	561	63.8
JAMAICA	1965	1760	603	630	216	34.2
	1970	1882	554	636	187	29.5
	1975	2029	504	672	167	24.8
	1976	2058	493	683	164	24.0
	1977	2087	482	696	161	23.1
MARTINIQUE	1965	278	89	89	29	32.1
	1970	306	72	100	23	23.4
	1975	329	62	109	21	18.9
	1976	334	61	112	20	18.3
	1977	338	59	114	20	17.6
MEXICO	1965	42859	21541	12519	6292	50.3
	1970	50313	22752	14490	6552	45.2
	1975	59204	23960	17069	6908	40.5
	1976	61196	24206	17653	6983	39.6
	1977	63266	24454	18257	7057	38.7
NICARAGUA	1965	1701	950	516	294	56.9
	1970	1970	990	591	303	51.3
	1975	2318	1066	691	325	47.0
	1976	2396	1082	714	330	46.2
	1977	2476	1098	737	334	45.4
PANAMA	1965	1261	581	420	194	46.1
	1970	1458	606	495	206	41.6
	1975	1678	638	567	216	38.0
	1976	1725	644	582	217	37.4
	1977	1774	651	598	219	36.7
PUERTO RICO	1965	2626	420	767	123	16.0
	1970	2743	225	797	65	8.2
	1975	2902	150	885	46	5.2
	1976	2936	142	906	44	4.8
	1977	2970	133	927	42	4.5
TRINIDAD ETC	1965	908	182	285	57	20.1
	1970	955	178	314	58	18.6
	1975	1009	174	354	61	17.2
	1976	1019	173	363	62	17.0
	1977	1030	172	372	62	16.7
USA	1965	194303	9909	79412	4050	5.1
	1970	204879	7519	87115	3197	3.7
	1975	213540	5872	94470	2598	2.8
	1976	215118	5616	95832	2502	2.6
	1977	216820	5382	97213	2413	2.5
SOUTH AMERIC	1965	165074	69507	53454	21990	41.1
	1970	188454	73390	60304	22995	38.1
	1975	214935	76638	68773	24022	34.9
	1976	220666	77229	70635	24217	34.3
	1977	226555	77806	72547	24407	33.6

TABLE 3 TOTAL POPULATION,AGRICULTURAL POPULAT ION AND ECONOMICALLY ACTIVE POPULATION AS ESTIMATED FOR 1965,1970,1975,1976 AND 1977

POBLACION TOTAL,POBLACION AGRI COLA Y POBLACION ECONOMICAMEN TE ACTIVA:ESTIMACIONES PARA 1965,1970,1975,1976 Y 1977

POPULATION TOTALE,POPULATION AGRICOLE ET POPULATION ECONOMIWUEMENT ACTIVE: ESTIMATIONS POUR 1965,1970,1975,1976 ET 1977

CONTINENT AND COUNTRY	YEAR	POPULATION		ECONOMICALLY ACTIVE POPULATION		
		TOTAL	AGRICULTURAL	TOTAL	IN AGRICULTURE	PERCENT IN AGRICULTURE
		....................THOUSANDS....................				
ARGENTINA	1965	22179	4025	8659	1572	18.2
	1970	23748	3883	9217	1507	16.4
	1975	25384	3701	9791	1427	14.6
	1976	25719	3662	9905	1410	14.2
	1977	26056	3624	10018	1393	13.9
BOLIVIA	1965	3548	2065	1190	693	58.2
	1970	3995	2216	1326	735	55.5
	1975	4521	2384	1493	787	52.7
	1976	4639	2421	1530	798	52.2
	1977	4761	2458	1567	809	51.6
BRAZIL	1965	82541	40255	26116	12737	48.8
	1970	95204	43432	30046	13707	45.6
	1975	109730	46054	34609	14525	42.0
	1976	112890	46525	35597	14670	41.2
	1977	116139	46981	36609	14809	40.5
CHILE	1965	8510	2336	2687	722	26.9
	1970	9369	2277	2885	685	23.8
	1975	10253	2201	3259	684	21.0
	1976	10441	2186	3349	685	20.5
	1977	10633	2171	3440	686	19.9
COLOMBIA	1965	17110	7614	5114	2276	44.5
	1970	20207	7648	6006	2273	37.9
	1975	23700	7629	7027	2262	32.2
	1976	24453	7621	7253	2260	31.2
	1977	25226	7611	7487	2259	30.2
ECUADOR	1965	5095	2758	1630	882	54.1
	1970	6031	3072	1912	974	50.9
	1975	7090	3381	2243	1069	47.7
	1976	7319	3442	2316	1089	47.0
	1977	7555	3504	2393	1110	46.4
GUYANA	1965	633	206	187	61	32.5
	1970	709	199	204	57	28.1
	1975	791	196	242	60	24.8
	1976	808	195	251	61	24.2
	1977	827	195	261	61	23.6
PARAGUAY	1965	2016	1099	644	351	54.5
	1970	2301	1211	728	383	52.6
	1975	2647	1344	844	429	50.8
	1976	2724	1373	871	439	50.4
	1977	2805	1403	899	449	50.0
PERU	1965	11440	5839	3453	1680	48.7
	1970	13248	6253	3851	1726	44.8
	1975	15326	6645	4472	1835	41.0
	1976	15777	6721	4615	1860	40.3
	1977	16242	6795	4763	1884	39.5
SURINAM	1965	332	87	90	23	26.2
	1970	371	83	92	21	22.4
	1975	422	84	104	21	19.9
	1976	434	84	107	21	19.5
	1977	447	85	111	21	19.0
URUGUAY	1965	2524	450	990	177	17.8
	1970	2662	405	1028	156	15.2
	1975	2800	377	1077	145	13.4
	1976	2827	371	1088	143	13.1
	1977	2855	366	1099	141	12.8
VENEZUELA	1965	9105	2761	2681	813	30.3
	1970	10559	2701	2995	766	25.6
	1975	12213	2630	3591	773	21.5
	1976	12575	2615	3733	776	20.8
	1977	12947	2598	3878	778	20.1
ASIA	1965	1819100	1237690	785388	534975	68.1
	1970	2021790	1312400	857783	555714	64.8
	1975	2249280	1384370	938925	575514	61.3
	1976	2298350	1398870	956211	579408	60.6
	1977	2348620	1413370	973826	583244	59.9
AFGHANISTAN	1965	15097	12604	5402	4510	83.5
	1970	16978	13866	5951	4860	81.7
	1975	19280	15387	6586	5256	79.8
	1976	19796	15723	6729	5345	79.4
	1977	20330	16067	6879	5437	79.0
BANGLADESH	1965	58795	50775	20619	17807	86.4
	1970	67692	58141	23401	20099	85.9
	1975	73746	62743	25339	21559	85.1
	1976	75529	64098	25920	21997	84.9
	1977	77601	65675	26594	22507	84.6
BHUTAN	1965	939	890	475	450	94.8
	1970	1045	986	521	491	94.3
	1975	1173	1101	576	540	93.9
	1976	1202	1127	588	551	93.8
	1977	1232	1154	601	563	93.7
BURMA	1965	24754	15815	11181	7144	63.9
	1970	27748	16529	11898	7088	59.6
	1975	31240	17351	12880	7154	55.5
	1976	31992	17521	13001	7175	54.8
	1977	32762	17694	13327	7198	54.0
CHINA	1965	710324	506247	337759	240720	71.3
	1970	771840	522998	364617	247064	67.8
	1975	838803	535659	393398	251223	63.9
	1976	852565	537549	399211	251705	63.1
	1977	866376	539194	405001	252054	62.2

TABLE **3** TOTAL POPULATION,AGRICULTURAL POPULAT
ION AND ECONOMICALLY ACTIVE POPULATION
AS ESTIMATED FOR 1965,1970,1975,1976
AND 1977

POBLACION TOTAL,POBLACION AGRI
COLA Y POBLACION ECONOMICAMEN
TE ACTIVA:ESTIMACIONES PARA
1965,1970,1975,1976 Y 1977

POPULATION TOTALE,POPULATION AGRICOLE
ET POPULATION ECONOMIQUEMENT ACTIVE:
ESTIMATIONS POUR 1965,1970,1975,1976
ET 1977

CONTINENT AND COUNTRY	YEAR	POPULATION		ECONOMICALLY ACTIVE POPULATION		
		TOTAL	AGRICULTURAL	TOTAL	IN AGRICULTURE	PERCENT IN AGRICULTURE
		.....................THOUSANDS.....................				
CYPRUS	1965	594	239	247	100	40.3
	1970	633	244	265	102	38.6
	1975	673	245	289	105	36.4
	1976	681	244	294	106	35.9
	1977	690	245	299	106	35.4
EAST TIMOR	1965	549	373	174	118	67.9
	1970	604	393	189	123	65.1
	1975	672	417	208	129	62.1
	1976	688	423	212	130	61.4
	1977	704	428	217	132	60.8
HONG KONG	1965	3692	225	1407	86	6.1
	1970	3942	171	1621	70	4.3
	1975	4225	140	1872	62	3.3
	1976	4283	135	1922	61	3.2
	1977	4342	130	1972	59	3.0
INDIA	1965	482541	345885	199482	142988	71.7
	1970	543333	376747	218149	151263	69.3
	1975	613439	408366	240345	159997	66.6
	1976	628834	414734	245267	161760	66.0
	1977	644695	421065	250342	163504	65.3
INDONESIA	1965	105070	74074	37405	26370	70.5
	1970	119467	79159	42208	27967	66.3
	1975	136044	85207	47030	29456	62.6
	1976	139635	86451	48081	29768	61.9
	1977	143316	87683	49182	30090	61.2
IRAN	1965	23154	11327	6770	3381	49.9
	1970	26624	11980	7721	3552	46.0
	1975	30909	12743	8778	3706	42.2
	1976	31880	12907	9011	3737	41.5
	1977	32891	13074	9253	3769	40.7
IRAQ	1965	7976	3976	2086	1040	49.9
	1970	9356	4362	2410	1124	46.6
	1975	11067	4803	2787	1209	43.4
	1976	11453	4897	2870	1227	42.8
	1977	11853	4993	2955	1245	42.1
ISRAEL	1965	2567	308	913	110	12.0
	1970	2978	288	1077	104	9.7
	1975	3458	282	1252	102	8.2
	1976	3551	280	1284	101	7.9
	1977	3640	278	1313	100	7.6
JAPAN	1965	98881	25669	49035	12921	26.4
	1970	104345	20173	53393	10492	19.7
	1975	111570	16251	57492	8520	14.8
	1976	112770	15518	58100	8135	14.0
	1977	113860	14791	58636	7752	13.2
JORDAN	1965	1955	759	492	191	38.8
	1970	2280	770	565	191	33.8
	1975	2688	796	652	193	29.6
	1976	2779	802	671	193	28.8
	1977	2874	807	691	194	28.1
KAMPUCHEA DM	1965	6142	4915	2543	2035	80.0
	1970	7060	5518	2849	2227	78.2
	1975	8110	6174	3201	2437	76.1
	1976	8349	6321	3280	2483	75.7
	1977	8599	6473	3361	2530	75.3
KOREA DPR	1965	12100	7056	5308	3095	58.3
	1970	13892	7598	5993	3278	54.7
	1975	15852	7970	6945	3492	50.3
	1976	16256	8025	7154	3532	49.4
	1977	16665	8077	7366	3570	48.5
KOREA REP	1965	28089	16440	9579	5606	58.5
	1970	31365	15996	10978	5599	51.0
	1975	34663	15497	12687	5672	44.7
	1976	35340	15374	13056	5680	43.5
	1977	36031	15240	13434	5682	42.3
KUWAIT	1965	441	7	171	3	1.6
	1970	705	12	229	4	1.8
	1975	1007	17	287	5	1.7
	1976	1071	18	300	5	1.7
	1977	1136	19	312	5	1.7
LAO	1965	2652	2146	1399	1133	80.9
	1970	2962	2334	1495	1178	78.8
	1975	3303	2523	1616	1235	76.4
	1976	3381	2565	1645	1248	75.9
	1977	3462	2608	1674	1261	75.3
LEBANON	1965	2151	626	572	166	29.1
	1970	2469	487	644	127	19.7
	1975	2869	404	745	105	14.1
	1976	2959	391	768	102	13.2
	1977	3053	378	793	98	12.4
MAL:SABAH	1965	544	319	181	107	59.4
	1970	650	356	220	122	55.5
	1975	770	391	261	135	51.7
	1976	812	405	275	140	50.9
	1977	843	414	286	143	50.1
MAL:SARAWAK	1965	838	490	278	165	59.4
	1970	1001	546	338	188	55.5
	1975	1200	609	406	210	51.7
	1976	1249	624	423	215	50.9
	1977	1295	637	439	220	50.1

TABLE **3** TOTAL POPULATION,AGRICULTURAL POPULAT POBLACION TOTAL,POBLACION AGRI POPULATION TOTALE,POPULATION AGRICOLE
ION AND ECONOMICALLY ACTIVE POPULATION COLA Y POBLACION ECONOMICAMEN ET POPULATION ECONOMIQUEMENT ACTIVE:
AS ESTIMATED FOR 1965,1970,1975,1976 TE ACTIVA:ESTIMACIONES PARA ESTIMATIONS POUR 1965,1970,1975,1976
AND 1977 1965,1970,1975,1976 Y 1977 ET 1977

| CONTINENT AND COUNTRY | YEAR | POPULATION | | ECONOMICALLY ACTIVE POPULATION | | |
		TOTAL	AGRICULTURAL	TOTAL	IN AGRICULTURE	PERCENT IN AGRICULTURE
		................................THOUSANDS......................				
MAL:PENINSUL	1965	7698	4497	2557	1519	59.4
	1970	8815	4804	2977	1652	55.5
	1975	10123	5134	3428	1771	51.7
	1976	10393	5192	3520	1791	50.9
	1977	10688	5258	3621	1815	50.1
MONGOLIA	1965	1070	706	438	289	66.0
	1970	1248	772	487	301	61.8
	1975	1446	802	549	304	55.5
	1976	1489	806	563	305	54.1
	1977	1532	808	577	304	52.8
NEPAL	1965	10100	9516	4948	4662	94.2
	1970	11232	10545	5471	5136	93.9
	1975	12572	11724	6070	5660	93.3
	1976	12877	11991	6201	5774	93.1
	1977	13196	12270	6336	5891	93.0
OMAN	1965	565	394	155	108	69.8
	1970	657	442	178	120	67.3
	1975	766	494	203	131	64.5
	1976	791	506	208	133	64.0
	1977	817	518	214	135	63.4
PAKISTAN	1965	52415	31360	15646	9361	59.8
	1970	60449	35574	17367	10220	58.9
	1975	70560	39653	19651	11043	56.2
	1976	72859	40517	20174	11219	55.6
	1977	75250	41413	20718	11402	55.0
PHILIPPINES	1965	30320	17247	11594	6624	57.1
	1970	35596	18848	13017	6925	53.2
	1975	42066	20763	14862	7373	49.6
	1976	43468	21148	15271	7468	48.9
	1977	44904	21529	15695	7564	48.2
SAUDI ARABIA	1965	5406	3716	1503	1033	68.7
	1970	6199	4094	1699	1122	66.0
	1975	7180	4534	1909	1205	63.1
	1976	7398	4627	1954	1222	62.5
	1977	7624	4723	2001	1240	62.0
SINGAPORE	1965	1880	105	611	34	5.6
	1970	2075	71	727	25	3.4
	1975	2248	62	853	23	2.7
	1976	2284	61	878	23	2.7
	1977	2321	60	903	23	2.6
SRI LANKA	1965	11164	6224	3780	2107	55.8
	1970	12514	6898	4187	2308	55.1
	1975	13986	7593	4741	2574	54.3
	1976	14282	7726	4864	2631	54.1
	1977	14576	7855	4989	2689	53.9
SYRIA	1965	5320	2799	1441	758	52.6
	1970	6247	3195	1659	848	51.1
	1975	7259	3580	1890	932	49.3
	1976	7490	3666	1942	951	48.9
	1977	7734	3757	1998	971	48.6
THAILAND	1965	30641	25064	14830	12131	81.8
	1970	35745	28550	16661	13307	79.9
	1975	42093	32706	19194	14914	77.7
	1976	43490	33593	19762	15265	77.2
	1977	44927	34495	20344	15620	76.8
TURKEY	1965	31151	22740	14647	10692	73.0
	1970	35232	23866	15590	10561	67.7
	1975	39882	24450	17054	10455	61.3
	1976	40908	24522	17392	10426	59.9
	1977	41968	24581	17740	10390	58.6
VIET NAM	1965	34835	27492	17762	14018	78.9
	1970	39106	29889	18771	14347	76.4
	1975	43451	31989	20344	14977	73.6
	1976	44412	32434	20714	15127	73.0
	1977	45415	32896	21093	15278	72.4
YEMEN AR	1965	3993	3242	1176	955	81.2
	1970	4591	3638	1334	1057	79.3
	1975	5308	4096	1503	1160	77.2
	1976	5467	4195	1540	1182	76.7
	1977	5632	4297	1578	1204	76.3
YEMEN DEM	1965	1252	846	347	234	67.6
	1970	1436	929	393	254	64.7
	1975	1660	1024	441	272	61.7
	1976	1710	1045	451	276	61.1
	1977	1762	1066	462	280	60.5
EUROPE	1965	444901	107571	197412	48606	24.6
	1970	458981	92338	201392	41783	20.7
	1975	473098	81081	210258	37427	17.8
	1976	474893	79006	211689	36596	17.3
	1977	476689	76922	213047	35730	16.8
ALBANIA	1965	1903	1309	821	565	68.8
	1970	2169	1437	923	611	66.3
	1975	2482	1572	1057	669	63.3
	1976	2549	1599	1087	682	62.7
	1977	2618	1626	1117	694	62.1
AUSTRIA	1965	7255	1399	3249	626	19.3
	1970	7426	1101	3133	464	14.8
	1975	7520	878	3249	380	11.7
	1976	7514	838	3272	365	11.1
	1977	7520	799	3299	351	10.6

TABLE 3 TOTAL POPULATION,AGRICULTURAL POPULAT
ION AND ECONOMICALLY ACTIVE POPULATION
AS ESTIMATED FOR 1965,1970,1975,1976
AND 1977

POBLACION TOTAL,POBLACION AGRI
COLA Y POBLACION ECONOMICAMEN
TE ACTIVA:ESTIMACIONES PARA
1965,1970,1975,1976 Y 1977

POPULATION TOTALE,POPULATION AGRICOLE
ET POPULATION ECONOMIQUEMENT ACTIVE:
ESTIMATIONS POUR 1965,1970,1975,1976
ET 1977

CONTINENT AND COUNTRY	YEAR	POPULATION		ECONOMICALLY ACTIVE POPULATION		
		TOTAL	AGRICULTURAL	TOTAL	IN AGRICULTURE	PERCENT IN AGRICULTURE
		...................THOUSANDS....................				
BELGIUM-LUX	1965	9796	645	3713	245	6.6
	1970	9995	495	3773	186	4.9
	1975	10153	398	3893	152	3.9
	1976	10178	381	3915	147	3.7
	1977	10200	365	3935	141	3.6
BULGARIA	1965	8201	4227	4408	2272	51.5
	1970	8490	3957	4534	2113	46.6
	1975	8722	3466	4658	1851	39.7
	1976	8760	3358	4673	1791	38.3
	1977	8795	3252	4681	1731	37.0
CZECHOSLOVAK	1965	14159	3026	6791	1451	21.4
	1970	14339	2425	7032	1189	16.9
	1975	14757	1955	7399	980	13.2
	1976	14846	1871	7458	940	12.6
	1977	14934	1790	7506	899	12.0
DENMARK	1965	4758	686	2211	319	14.4
	1970	4929	549	2312	258	11.1
	1975	5060	449	2395	212	8.9
	1976	5073	430	2406	204	8.5
	1977	5090	413	2420	196	8.1
FINLAND	1965	4564	1375	2098	603	28.7
	1970	4606	1035	2120	452	21.3
	1975	4711	834	2248	376	16.7
	1976	4726	799	2271	363	16.0
	1977	4737	764	2292	349	15.2
FRANCE	1965	48758	8733	20459	3664	17.9
	1970	50770	6961	20998	2876	13.7
	1975	52786	5764	22281	2433	10.9
	1976	52915	5530	22438	2345	10.5
	1977	53050	5303	22583	2258	10.0
GERMAN DR	1965	17019	2601	8380	1280	15.3
	1970	17058	2214	8538	1108	13.0
	1975	16850	1880	8577	957	11.2
	1976	16786	1817	8582	929	10.8
	1977	16730	1757	8595	903	10.5
GERMANY FED	1965	59012	6356	26945	2902	10.8
	1970	60710	4529	26822	2001	7.5
	1975	61829	3395	28039	1540	5.5
	1976	61531	3189	28129	1458	5.2
	1977	61400	3002	28296	1383	4.9
GREECE	1965	8664	4411	3823	1946	50.9
	1970	8793	4041	3750	1724	46.0
	1975	9047	3742	3825	1582	41.4
	1976	9167	3712	3872	1568	40.5
	1977	9280	3679	3915	1552	39.6
HUNGARY	1965	10153	3511	4916	1557	31.7
	1970	10338	2848	5030	1258	25.0
	1975	10540	2330	5229	1042	19.9
	1976	10600	2241	5265	1002	19.0
	1977	10650	2151	5286	960	18.2
ICELAND	1965	192	41	74	16	21.3
	1970	204	37	80	14	18.0
	1975	218	32	89	13	14.6
	1976	220	31	91	13	13.9
	1977	223	30	93	12	13.3
IRELAND	1965	2876	904	1110	349	31.4
	1970	2940	778	1112	294	26.5
	1975	3131	737	1189	280	23.5
	1976	3166	729	1204	277	23.0
	1977	3199	720	1218	274	22.5
ITALY	1965	51944	12872	20269	5023	24.8
	1970	53660	10077	19994	3755	18.8
	1975	55810	8125	20823	3031	14.6
	1976	56156	7774	20988	2905	13.8
	1977	56446	7424	21125	2778	13.2
MALTA	1965	320	27	100	8	8.4
	1970	326	22	110	8	6.9
	1975	329	19	115	7	5.9
	1976	330	19	115	7	5.7
	1977	332	18	116	6	5.5
NETHERLANDS	1965	12292	1158	4523	426	9.4
	1970	13032	1058	4858	394	8.1
	1975	13653	903	5170	342	6.6
	1976	13770	871	5233	331	6.3
	1977	13853	839	5285	320	6.1
NORWAY	1965	3723	583	1435	225	15.7
	1970	3877	462	1470	175	11.9
	1975	4007	384	1518	146	9.6
	1976	4027	370	1526	140	9.2
	1977	4044	357	1533	135	8.8
POLAND	1965	31496	13742	15442	6738	43.6
	1970	32473	12668	16756	6537	39.0
	1975	34022	11757	18406	6360	34.6
	1976	34362	11578	18726	6309	33.7
	1977	34698	11395	19026	6249	32.8
PORTUGAL	1965	9235	3554	3608	1388	38.5
	1970	8628	2875	3387	1129	33.3
	1975	8762	2597	3414	1012	29.6
	1976	8804	2550	3424	992	29.0
	1977	8842	2502	3434	972	28.3

TABLE **3** TOTAL POPULATION,AGRICULTURAL POPULAT
ION AND ECONOMICALLY ACTIVE POPULATION
AS ESTIMATED FOR 1965,1970,1975,1976
AND 1977

POBLACION TOTAL,POBLACION AGRI
COLA Y POBLACION ECONOMICAMEN
TE ACTIVA:ESTIMACIONES PARA
1965,1970,1975,1976 Y 1977

POPULATION TOTALE,POPULATION AGRICOLE
ET POPULATION ECONOMIQUEMENT ACTIVE:
ESTIMATIONS POUR 1965,1970,1975,1976
ET 1977

CONTINENT AND COUNTRY	YEAR	POPULATION		ECONOMICALLY ACTIVE POPULATION		
		TOTAL	AGRICULTURAL	TOTAL	IN AGRICULTURE	PERCENT IN AGRICULTURE
		....................THOUSANDS....................				
ROMANIA	1965	19027	11454	10885	6553	60.2
	1970	20252	11341	11347	6354	56.0
	1975	21245	10967	11867	6126	51.6
	1976	21446	10882	11959	6068	50.7
	1977	21630	10784	12027	5996	49.9
SPAIN	1965	31913	10547	11718	3980	34.0
	1970	33615	8475	11742	3052	26.0
	1975	35263	7261	12321	2620	21.3
	1976	35610	7060	12468	2554	20.5
	1977	35959	6857	12613	2485	19.7
SWEDEN	1965	7734	951	3431	379	11.0
	1970	8043	746	3586	298	8.3
	1975	8196	595	3628	235	6.5
	1976	8219	569	3631	224	6.2
	1977	8255	545	3640	214	5.9
SWITZERLAND	1965	5857	552	2723	256	9.4
	1970	6190	482	2985	233	7.8
	1975	6410	408	3119	198	6.4
	1976	6346	387	3094	189	6.1
	1977	6280	367	3070	180	5.9
UK	1965	54380	1843	25493	864	3.4
	1970	55599	1557	25787	722	2.8
	1975	56086	1330	25895	614	2.4
	1976	56074	1288	25888	595	2.3
	1977	56041	1246	25880	575	2.2
YUGOSLAVIA	1965	19434	11040	8734	4962	56.8
	1970	20371	10145	9175	4569	49.8
	1975	21352	9284	9792	4258	43.5
	1976	21560	9111	9911	4188	42.3
	1977	21720	8914	9999	4103	41.0
OCEANIA	1965	17478	4118	7254	1852	25.5
	1970	19236	4281	8151	1928	23.7
	1975	20999	4468	8947	2007	22.4
	1976	21278	4504	9067	2021	22.3
	1977	21540	4540	9178	2034	22.2
AUSTRALIA	1965	11387	1097	4669	450	9.6
	1970	12510	1011	5300	428	8.1
	1975	13502	924	5763	394	6.8
	1976	13643	903	5822	385	6.6
	1977	13780	882	5879	376	6.4
FIJI	1965	464	238	128	66	51.4
	1970	520	247	151	72	47.5
	1975	577	253	180	79	43.8
	1976	589	254	187	80	43.0
	1977	600	254	193	82	42.3
NEW ZEALAND	1965	2628	347	1004	132	13.2
	1970	2811	333	1096	130	11.9
	1975	3070	322	1221	128	10.5
	1976	3095	317	1235	126	10.2
	1977	3105	310	1242	124	10.0
PAPUA N GUIN	1965	2148	1878	1143	1000	87.5
	1970	2413	2074	1255	1079	86.0
	1975	2716	2287	1382	1164	84.2
	1976	2783	2333	1410	1182	83.8
	1977	2854	2382	1439	1201	83.5
USSR	1965	230936	77964	114544	38670	33.8
	1970	242768	62294	117693	30200	25.7
	1975	254390	52226	126609	25993	20.5
	1976	256674	50536	128471	25294	19.7
	1977	258700	48824	130091	24552	18.9
DEV.PED M E	1965	690590	112815	294864	48712	16.5
	1970	724205	93164	311808	40147	12.9
	1975	757114	79626	331959	34523	10.4
	1976	762239	77209	335333	33493	10.0
	1977	767283	74819	338584	32457	9.6
N AMERICA	1965	213947	12002	86796	4836	5.6
	1970	226199	9274	95677	3902	4.1
	1975	236371	7346	104022	3215	3.1
	1976	238261	7040	105572	3101	2.9
	1977	240140	6749	107081	2992	2.8
W EUROPE	1965	342843	67702	145769	28190	19.3
	1970	353862	55448	147233	22612	15.4
	1975	364480	47154	153065	19441	12.7
	1976	365544	45660	153942	18875	12.3
	1977	366634	44167	154810	18298	11.8
OCEANIA	1965	14015	1443	5673	582	10.3
	1970	15321	1344	6396	558	8.7
	1975	16572	1246	6983	522	7.5
	1976	16738	1220	7057	512	7.3
	1977	16885	1192	7121	500	7.0
OTH DEV.PED	1965	119785	31669	56627	15103	26.7
	1970	128823	27098	62502	13075	20.9
	1975	139691	23880	67889	11346	16.7
	1976	141696	23290	68762	11006	16.0
	1977	143624	22710	69573	10667	15.3
DEV.PING M E	1965	1488520	991680	566806	386031	68.1
	1970	1691100	1077950	628199	409848	65.2
	1975	1923320	1165730	700453	434547	62.0
	1976	1975080	1184340	716689	439835	61.4
	1977	2028690	1203230	733494	445186	60.7

TABLE **3**   TOTAL POPULATION,AGRICULTURAL POPULAT   POBLACION TOTAL,POBLACION AGRI   POPULATION TOTALE,POPULATION AGRICOLE
ION AND ECONOMICALLY ACTIVE POPULATION   COLA Y POBLACION ECONOMICAMEN   ET POPULATION ECONOMIQUEMENT ACTIVE:
AS ESTIMATED FOR 1965,1970,1975,1976   TE ACTIVA:ESTIMACIONES PARA   ESTIMATIONS POUR 1965,1970,1975,1976
AND 1977   1965,1970,1975,1976 Y 1977   ET 1977

| CONTINENT AND COUNTRY | YEAR | POPULATION | | ECONOMICALLY ACTIVE POPULATION | | |
		TOTAL	AGRICULTURAL	TOTAL	IN AGRICULTURE	PERCENT IN AGRICULTURE
		................................THOUSANDS.....................				
AFRICA	1965	246984	188957	103418	80755	78.1
	1970	279922	206573	114282	86363	75.6
	1975	319408	225749	126696	91888	72.5
	1976	328452	229962	129495	93072	71.9
	1977	337812	234224	132358	94247	71.2
LAT AMERICA	1965	244067	108920	78155	34515	44.2
	1970	279260	114784	87990	35817	40.7
	1975	319743	120302	100660	37550	37.3
	1976	328586	121376	103490	37905	36.6
	1977	337697	122440	106404	38252	35.9
NEAR EAST	1965	140614	89394	47227	30953	65.5
	1970	160972	96710	52442	32409	61.8
	1975	184742	104166	58771	33974	57.8
	1976	190011	105716	60185	34305	57.0
	1977	195467	107290	61646	34641	56.2
FAR EAST	1965	853302	601722	336387	238534	70.9
	1970	966936	656937	371687	253885	68.3
	1975	1094890	712288	412312	269639	65.4
	1976	1123380	724000	421458	273031	64.8
	1977	1152940	735927	430977	276498	64.2
OTH DV.PING	1965	3556	2687	1620	1275	78.7
	1970	4019	2949	1799	1375	76.4
	1975	4542	3233	2014	1496	74.3
	1976	4657	3294	2062	1522	73.8
	1977	4774	3357	2110	1548	73.4
CENTR PLANND	1965	1097360	664248	529997	319243	60.2
	1970	1181030	665956	564569	316586	56.1
	1975	1270670	668745	608238	316411	52.0
	1976	1289060	669016	617138	316166	51.2
	1977	1307340	669026	625725	315719	50.5
ASIAN CPE	1965	764471	546415	363810	260157	71.5
	1970	833146	566773	392717	267216	68.0
	1975	907662	582593	424436	272432	64.2
	1976	923071	585135	430920	273151	63.4
	1977	938587	587448	437397	273736	62.6
E EUR+USSR	1965	332894	117833	166187	59086	35.6
	1970	347887	99184	171852	49371	28.7
	1975	363008	86153	183802	43979	23.9
	1976	366023	83881	186218	43015	23.1
	1977	368755	81579	188328	41984	22.3
DEV.PED ALL	1965	1023480	230648	461051	107798	23.4
	1970	1072090	192347	483660	89518	18.5
	1975	1120120	165778	515761	78503	15.2
	1976	1128260	161090	521551	76509	14.7
	1977	1136030	156397	526912	74441	14.1
DEV.PING ALL	1965	2252990	1538090	930616	646188	69.4
	1970	2524240	1644720	1020910	677064	66.3
	1975	2830980	1748320	1124880	706979	62.8
	1976	2898150	1769470	1147600	712986	62.1
	1977	2967270	1790670	1170890	718922	61.4

FAO INDEX NUMBERS OF AGRICULTURAL PRODUCTION

NOMBRES-INDICES FAO DE LA PRODUCTION AGRICOLE

NUMEROS INDICES DE LA FAO DE LA PRODUCCION AGRICOLA

TABLE
TABLEAU 4
CUADRO

FOOD   ALIMENTAIRES   ALIMENTICIOS

PROD INDICES   1969-71=100   TOTAL

	1966	1967	1968	1969	1970	1971	1972	1973	1974	1975	1976	1977
WORLD	90	94	97	97	100	103	103	108	110	113	115	117
AFRICA	86	92	93	98	99	103	104	100	108	108	111	111
ALGERIA	69	86	100	96	104	101	106	95	100	103	112	105
ANGOLA	90	93	94	99	101	100	97	102	102	98	100	104
BENIN	81	90	92	99	103	98	98	98	99	88	102	113
BOTSWANA	86	97	91	97	92	111	102	112	121	118	133	130
BURUNDI	97	98	99	97	101	102	105	113	113	111	113	116
CAMEROON	82	90	93	97	100	102	104	104	113	112	113	116
CENT AFR EMP	84	91	92	96	100	104	108	113	115	114	116	120
CHAD	100	98	101	101	98	101	83	79	89	90	91	93
CONGO	92	100	99	100	98	101	97	98	98	104	113	114
EGYPT	88	86	96	98	99	103	105	106	107	110	112	113
ETHIOPIA	92	96	95	99	101	99	98	100	101	103	105	103
GABON	94	94	96	97	100	102	104	105	106	106	108	109
GAMBIA	109	96	98	100	98	102	108	97	112	114	114	100
GHANA	85	92	91	95	99	106	106	108	119	99	91	91
GUINEA	88	94	98	98	100	102	94	89	92	95	102	102
IVORY COAST	80	85	87	94	97	109	107	111	127	134	136	146
KENYA	88	91	95	98	102	100	100	99	98	102	108	112
LESOTHO	92	99	97	100	97	103	87	120	105	93	124	111
LIBERIA	86	88	93	94	101	105	108	112	120	119	125	124
LIBYA	115	112	99	123	83	95	140	142	146	173	192	185
MADAGASCAR	86	94	93	98	100	102	99	99	110	109	114	115
MALAWI	82	100	93	98	90	112	118	116	117	110	120	121
MALI	91	97	94	101	101	97	82	70	76	91	98	99
MAURITANIA	95	97	98	102	101	97	92	76	74	73	82	83
MAURITIUS	91	108	96	107	95	99	114	113	111	85	122	129
MOROCCO	80	85	114	93	98	109	107	97	109	92	105	81
MOZAMBIQUE	85	88	94	98	100	101	105	110	106	93	100	100
NAMIBIA	79	90	94	92	100	108	119	123	127	128	130	132
NIGER	98	107	95	108	96	96	91	69	83	78	107	99
NIGERIA	86	86	89	100	102	98	100	93	102	106	108	108
REUNION	98	102	106	114	94	92	111	110	103	107	111	116
RHODESIA	88	86	85	97	93	110	121	103	129	125	124	125
RWANDA	75	89	85	94	102	104	101	105	103	117	124	126
SENEGAL	98	119	94	107	82	111	73	91	120	139	126	96
SIERRA LEONE	91	93	93	98	99	104	102	102	102	108	113	114
SOMALIA	90	95	97	99	100	101	111	104	94	99	108	110
SOUTH AFRICA	82	107	89	94	94	112	120	98	121	113	111	118
SUDAN	78	95	85	95	100	104	107	110	124	129	124	128
SWAZILAND	77	89	93	95	102	103	117	112	121	110	125	129
TANZANIA	92	90	91	96	104	100	103	104	102	107	111	116
TOGO	97	90	94	88	101	111	82	85	67	71	73	75
TUNISIA	90	89	88	82	97	121	122	128	136	153	146	144
UGANDA	88	94	97	100	100	100	105	106	107	107	112	116
UPPER VOLTA	96	99	101	100	102	97	94	90	101	111	107	99
ZAIRE	89	92	95	98	101	102	103	106	108	111	113	114
ZAMBIA	91	94	95	99	95	106	108	109	119	125	132	132
N C AMERICA	93	97	98	98	97	105	104	105	107	114	117	121
BARBADOS	105	122	99	95	107	98	87	96	88	84	90	106
CANADA	106	95	102	99	93	108	103	104	97	108	121	120
COSTA RICA	80	80	92	93	99	107	109	113	114	130	130	131
CUBA	74	89	87	85	123	93	85	88	91	97	96	102
DOMINICAN RP	81	81	83	93	100	107	109	110	112	105	115	111
EL SALVADOR	87	86	92	91	101	108	102	118	116	131	128	144
GUATEMALA	83	87	92	96	100	104	109	112	113	116	130	132
HAITI	94	96	94	98	100	102	105	105	108	107	110	111
HONDURAS	78	91	97	101	96	103	112	108	99	88	99	109
JAMAICA	103	106	99	94	99	108	107	103	108	108	111	108
MEXICO	92	93	95	95	100	105	107	108	116	118	114	120
NICARAGUA	81	86	92	96	100	103	102	104	108	119	126	131
PANAMA	81	86	97	103	94	103	103	107	111	117	115	121
PUERTO RICO	115	109	104	101	101	98	99	95	97	97	103	107
TRINIDAD ETC	90	90	100	94	104	103	109	104	102	101	104	103
USA	92	98	98	98	97	105	104	105	107	114	118	121
SOUTH AMERIC	87	92	92	97	101	102	102	106	114	119	127	130
ARGENTINA	92	101	94	102	101	96	94	104	107	110	120	118
BOLIVIA	87	89	96	95	101	104	112	117	121	130	134	129
BRAZIL	82	88	91	94	102	104	109	112	123	129	143	147
CHILE	97	97	101	96	103	101	95	89	103	109	106	116
COLOMBIA	86	88	92	94	99	107	110	110	116	126	129	135
ECUADOR	91	93	101	96	101	103	100	102	114	120	115	119
GUYANA	88	90	87	99	96	104	98	97	108	107	110	108
PARAGUAY	90	93	92	94	103	103	102	101	113	115	122	135
PERU	87	92	83	94	102	104	102	103	108	109	111	110
URUGUAY	91	80	91	100	107	93	91	97	105	106	117	96
VENEZUELA	81	85	89	97	101	102	103	105	111	118	114	127
ASIA	88	92	95	97	101	102	102	107	109	115	116	120
AFGHANISTAN	99	106	108	111	98	91	103	111	114	116	122	119
BANGLADESH	86	99	101	107	101	92	91	102	97	109	102	110
BHUTAN	91	93	96	98	100	102	105	107	110	112	114	115
BURMA	83	93	94	96	102	103	98	104	107	107	112	115
CHINA	89	92	92	95	100	104	103	108	112	115	118	119
CYPRUS	73	90	90	97	90	113	110	82	97	101	99	109
HONG KONG	91	68	104	91	104	104	107	79	114	96	88	87
INDIA	80	87	91	95	102	102	97	107	100	114	111	117
INDONESIA	89	85	96	95	100	105	107	117	124	126	128	131

TABLE
TABLEAU **4**
CUADRO

FOOD   ALIMENTAIRES   ALIMENTICIOS

PROD INDICES   1969-71=100   TOTAL

	1966	1967	1968	1969	1970	1971	1972	1973	1974	1975	1976	1977
IRAN	87	86	95	99	101	100	111	115	119	123	132	137
IRAQ	83	92	103	101	99	100	130	95	93	83	102	92
ISRAEL	77	89	93	93	98	109	121	120	128	130	135	142
JAPAN	97	106	109	106	101	93	101	101	104	110	100	109
JORDAN	156	148	100	108	80	112	123	81	142	86	86	78
KAMPUCHEA DM	84	87	103	90	115	94	78	61	52	67	71	71
KOREA DPR	87	89	91	93	100	107	114	117	128	135	145	155
KOREA REP	92	86	87	100	99	101	102	104	111	121	127	135
LAO	85	93	92	100	102	98	100	105	109	111	111	105
LEBANON	91	110	104	90	99	111	119	110	123	114	126	129
MAL PENINSUL	80	81	88	94	100	106	111	118	126	127	133	138
MAL SABAH	73	78	90	94	97	110	125	121	128	149	145	159
MAL SARAWAK	80	83	97	105	97	99	104	105	105	104	112	115
MONGOLIA	110	114	96	97	101	102	105	112	117	122	118	120
NEPAL	91	94	95	98	102	100	97	107	108	111	110	104
PAKISTAN	78	84	93	97	102	101	104	110	112	114	120	127
PHILIPPINES	87	93	92	97	101	102	101	114	119	128	143	141
SAUDI ARABIA	92	96	91	99	102	98	77	84	109	117	104	112
SINGAPORE	20	63	53	83	96	121	101	125	136	125	118	175
SRI LANKA	81	93	96	95	103	102	105	101	115	120	125	138
SYRIA	88	113	104	116	88	96	144	87	153	163	194	180
THAILAND	91	84	90	96	99	104	105	123	120	126	132	133
TURKEY	89	90	95	95	100	105	108	100	110	121	127	125
VIET NAM	92	94	92	96	102	102	105	107	108	112	112	116
YEMEN AR	106	111	106	101	84	114	117	116	105	125	116	115
YEMEN DEM	93	98	98	99	92	109	105	116	121	127	128	127
EUROPE	92	96	99	98	99	103	103	108	112	112	111	113
ALBANIA	86	96	95	97	100	103	100	108	110	112	124	128
AUSTRIA	89	97	98	102	99	99	96	102	106	109	108	107
BELGIUM-LUX	77	90	93	94	100	105	101	106	112	109	101	104
BULGARIA	102	102	99	100	98	101	107	106	98	104	115	115
CZECHOSLOVAK	89	93	99	98	98	104	105	114	118	116	115	123
DENMARK	101	102	107	102	96	102	98	98	111	105	100	104
FINLAND	87	90	90	97	98	105	106	99	102	106	119	106
FRANCE	91	98	102	96	100	104	104	110	112	108	109	109
GERMAN DR	98	105	105	100	100	100	107	111	121	120	115	117
GERMANY FED	88	95	100	98	100	102	98	100	104	101	100	102
GREECE	91	94	89	93	103	103	108	109	120	125	129	127
HUNGARY	91	88	95	101	91	108	116	117	121	126	116	131
ICELAND	123	114	97	101	97	102	115	124	117	125	121	121
IRELAND	88	99	98	96	96	108	103	101	117	139	125	136
ITALY	93	96	96	99	100	100	95	101	106	107	105	104
MALTA	73	79	86	90	108	102	108	104	106	94	116	115
NETHERLANDS	78	89	92	93	103	104	101	107	115	119	117	121
NORWAY	93	100	105	98	100	103	104	103	122	108	106	119
POLAND	98	100	102	97	102	101	106	112	114	116	111	107
PORTUGAL	87	98	97	97	106	97	95	100	102	102	98	91
ROMANIA	103	102	102	100	89	110	125	117	121	125	154	149
SPAIN	89	90	94	95	101	105	111	117	120	126	129	127
SWEDEN	94	102	105	94	102	104	105	101	122	108	116	116
SWITZERLAND	89	96	101	98	99	102	101	104	105	108	114	112
UK	95	96	95	96	101	104	106	108	114	108	104	112
YUGOSLAVIA	95	95	96	105	93	103	99	106	118	117	121	123
OCEANIA	94	87	101	98	99	103	104	115	110	119	126	123
AUSTRALIA	98	86	104	97	99	104	103	119	114	123	129	125
FIJI	86	85	101	96	105	99	93	99	95	94	97	99
FR POLYNESIA	102	92	89	106	98	96	90	80	75	99	100	100
NEW HEBRIDES	97	114	99	105	95	100	72	85	115	110	112	114
NEW ZEALAND	84	89	93	100	99	101	106	107	100	110	124	117
PAPUA N GUIN	91	93	95	98	100	102	107	111	115	119	117	118
SAMOA	87	95	95	100	97	102	93	94	94	96	103	109
USSR	93	95	99	94	102	104	99	117	112	109	114	115
DEV.PED M E	92	97	99	98	99	103	102	105	109	112	112	115
N AMERICA	93	97	98	98	97	105	104	105	106	114	118	121
W EUROPE	90	96	98	97	100	103	101	106	111	110	109	110
OCEANIA	94	87	101	98	99	103	104	116	110	119	128	123
OTH DEV.PED	94	106	106	104	100	96	104	101	107	111	103	112
DEV.PING M E	85	90	93	97	101	102	102	106	109	115	118	121
AFRICA	87	90	94	98	100	102	102	100	106	106	110	108
LAT AMERICA	88	92	93	96	102	102	103	106	113	117	122	126
NEAR EAST	89	93	97	99	99	103	110	104	113	119	126	124
FAR EAST	83	87	92	97	102	102	99	108	106	116	116	122
OTH DV.PING	91	93	96	98	100	101	102	106	109	113	112	114
CENTR PLANND	92	95	97	95	101	104	103	112	113	113	116	118
ASIAN CPE	89	92	92	95	101	104	103	108	111	115	117	119
E EUR+USSR	94	96	100	96	101	104	103	115	113	112	115	117
DEV.PED ALL	93	97	99	97	99	103	103	109	110	112	113	116
DEV.PING ALL	87	91	93	96	101	103	103	107	109	115	118	120

TABLE
TABLEAU **5**
CUADRO

AGRICULTURE   AGRICULTURE   AGRICULTURA

PROD INDICES   1969-71=100   TOTAL

	1966	1967	1968	1969	1970	1971	1972	1973	1974	1975	1976	1977
WORLD	90	94	97	97	100	103	103	108	110	113	115	117
AFRICA	86	91	93	98	100	103	104	100	107	107	109	110
ALGERIA	69	86	100	96	103	101	106	95	100	103	112	105
ANGOLA	92	95	92	99	99	102	99	100	104	76	77	79
BENIN	79	88	91	98	103	100	100	99	98	86	103	113
BOTSWANA	87	97	91	97	92	111	102	112	120	118	133	130
BURUNDI	96	97	99	96	101	103	104	113	114	110	113	116
CAMEROON	81	90	93	97	100	103	104	104	113	109	110	113
CENT AFR EMP	84	90	93	100	100	100	106	110	114	110	113	116
CHAD	101	98	105	102	97	101	85	82	94	98	96	99
CONGO	91	100	99	100	98	102	97	98	98	105	113	115
EGYPT	88	86	94	99	99	102	104	104	103	103	105	108
ETHIOPIA	92	96	95	99	102	99	99	99	101	103	106	103
GABON	94	94	96	98	100	102	104	105	106	106	107	109
GAMBIA	109	96	98	100	98	102	108	97	112	114	114	100
GHANA	85	92	91	95	99	106	106	108	119	99	91	92
GUINEA	88	94	99	99	100	101	94	89	91	94	101	101
IVORY COAST	87	77	94	92	101	106	108	115	117	129	135	141
KENYA	89	89	92	97	102	100	105	106	106	108	116	127
LESOTHO	93	98	97	101	98	102	85	115	101	93	120	110
LIBERIA	85	86	91	91	102	106	108	112	117	115	117	118
LIBYA	114	111	99	122	84	95	138	140	144	170	188	181
MADAGASCAR	88	95	94	98	101	101	100	102	111	113	117	119
MALAWI	82	97	91	96	92	112	119	118	118	116	125	132
MALI	89	95	93	101	101	98	83	71	77	94	102	103
MAURITANIA	95	97	98	102	101	97	92	76	74	73	82	83
MAURITIUS	91	107	96	106	95	99	114	114	112	86	123	129
MOROCCO	80	85	114	93	98	109	108	98	109	92	105	82
MOZAMBIQUE	86	89	95	99	101	100	105	108	105	91	97	97
NAMIBIA	81	90	94	93	100	108	118	122	127	127	129	132
NIGER	98	107	95	108	96	96	91	68	83	78	106	99
NIGERIA	86	86	88	100	102	98	100	94	102	106	109	109
REUNION	98	102	106	114	94	92	111	110	103	107	111	116
RHODESIA	97	92	83	97	94	108	119	99	124	128	128	123
RWANDA	74	88	85	94	102	105	101	105	105	119	126	128
SENEGAL	97	118	94	107	82	111	73	92	122	139	127	98
SIERRA LEONE	92	92	93	98	99	103	104	104	101	108	112	115
SOMALIA	90	95	97	99	100	101	110	104	94	99	108	110
SOUTH AFRICA	83	106	90	95	95	110	117	97	118	111	108	117
SUDAN	76	91	83	96	100	105	105	103	122	121	107	119
SWAZILAND	78	91	93	94	102	104	118	114	125	115	129	133
TANZANIA	96	92	91	96	105	98	102	103	100	104	109	113
TOGO	96	91	95	89	102	109	82	85	68	72	74	76
TUNISIA	91	90	89	82	97	121	122	128	136	152	145	143
UGANDA	86	91	89	103	101	96	101	106	100	101	103	106
UPPER VOLTA	95	98	100	100	103	97	94	90	100	110	109	101
ZAIRE	89	92	95	97	101	102	103	106	108	110	112	114
ZAMBIA	92	93	96	99	94	107	108	109	118	124	131	131
N C AMERICA	93	96	98	98	98	105	104	106	107	113	116	121
BARBADOS	105	123	99	95	107	98	87	96	88	85	90	106
CANADA	105	93	101	99	94	107	101	103	96	107	117	118
COSTA RICA	82	84	91	96	97	108	106	114	112	123	125	124
CUBA	77	92	89	86	122	93	86	90	93	98	97	103
DOMINICAN RP	83	83	84	94	100	106	109	115	115	107	118	116
EL SALVADOR	89	92	89	96	97	107	107	110	118	128	122	138
GUATEMALA	85	86	91	95	102	103	109	113	118	116	127	132
HAITI	95	97	94	97	100	102	105	105	108	108	108	110
HONDURAS	80	95	97	102	95	103	112	108	102	95	103	116
JAMAICA	104	106	99	94	99	107	106	103	108	108	110	108
MEXICO	95	95	98	95	100	105	107	108	116	115	111	118
NICARAGUA	90	97	97	99	98	103	106	109	121	127	129	142
PANAMA	81	87	97	103	94	103	103	107	110	117	115	120
PUERTO RICO	116	111	104	100	103	97	99	97	96	96	102	106
TRINIDAD ETC	90	90	102	94	102	104	109	104	101	102	103	103
USA	93	97	98	98	97	105	104	106	107	113	117	121
SOUTH AMERIC	89	93	93	98	100	102	103	105	114	117	121	126
ARGENTINA	93	100	94	102	102	96	94	103	107	110	120	118
BOLIVIA	86	88	95	95	101	105	114	125	126	134	135	131
BRAZIL	85	91	91	97	98	105	111	108	122	124	126	136
CHILE	97	97	101	96	103	101	95	89	102	109	105	115
COLOMBIA	88	89	94	96	100	104	106	109	112	123	124	132
ECUADOR	94	95	101	96	101	103	100	102	116	120	117	120
GUYANA	89	90	87	99	96	104	98	97	108	107	110	108
PARAGUAY	88	93	93	95	102	103	104	103	114	118	126	141
PERU	91	92	86	95	102	104	101	103	106	105	107	106
URUGUAY	92	82	92	100	106	94	88	93	99	100	112	93
VENEZUELA	83	86	89	97	101	102	101	106	110	119	113	125
ASIA	88	93	95	97	101	102	103	108	109	115	116	119
AFGHANISTAN	98	106	108	111	99	90	102	111	114	118	122	119
BANGLADESH	88	100	101	108	102	90	92	101	95	105	101	109
BHUTAN	91	93	96	98	100	102	105	107	110	113	114	115
BURMA	82	92	94	96	102	103	99	106	107	107	113	115
CHINA	89	93	92	95	100	104	104	110	113	116	118	119
CYPRUS	73	91	90	97	90	113	110	82	96	100	99	108
HONG KONG	91	68	104	91	104	104	107	79	114	96	88	87
INDIA	81	88	91	95	102	103	98	107	101	113	111	117
INDONESIA	89	85	95	96	100	104	106	115	121	123	125	128

TABLE
TABLEAU 5
CUADRO

AGRICULTURE   AGRICULTURE   AGRICULTURA

PROD INDICES   1969-71=100   TOTAL

	1966	1967	1968	1969	1970	1971	1972	1973	1974	1975	1976	1977
IRAN	86	86	95	99	101	100	112	115	120	121	130	135
IRAQ	82	92	103	100	99	100	128	94	93	82	101	91
ISRAEL	76	88	93	94	98	108	120	119	128	130	135	143
JAPAN	98	107	109	106	101	93	100	101	103	109	100	108
JORDAN	154	147	100	108	81	111	123	81	141	87	86	79
KAMPUCHEA DM	87	91	107	95	114	92	78	61	53	67	71	71
KOREA DPR	87	90	91	93	100	107	114	117	127	134	144	153
KOREA REP	92	86	87	100	99	101	104	107	113	123	129	138
LAO	85	93	93	100	101	98	100	104	108	110	110	104
LEBANON	91	108	103	90	99	111	119	111	124	116	125	123
MAL PENINSUL	78	80	88	96	99	105	108	118	124	122	131	132
MAL SABAH	77	80	90	96	98	106	116	120	122	139	138	150
MAL SARAWAK	86	86	96	110	95	95	101	111	107	104	116	119
MONGOLIA	108	112	96	96	101	103	105	111	117	121	116	120
NEPAL	91	94	95	98	102	100	97	107	107	110	109	104
PAKISTAN	79	85	93	97	101	103	105	109	111	111	114	123
PHILIPPINES	87	93	92	97	101	102	101	114	119	128	143	141
SAUDI ARABIA	92	96	91	99	102	98	77	84	109	117	104	112
SINGAPORE	22	64	55	83	96	120	100	124	135	124	117	172
SRI LANKA	88	96	99	98	101	101	102	96	105	109	110	120
SYRIA	90	108	103	113	90	97	136	90	144	150	177	166
THAILAND	92	85	91	97	100	103	103	120	117	122	128	129
TURKEY	89	91	95	95	99	106	109	101	112	120	128	125
VIET NAM	92	94	92	96	102	103	105	107	108	112	112	116
YEMEN AR	105	111	106	101	84	115	118	117	107	126	118	116
YEMEN DEM	92	97	94	99	92	109	103	115	120	125	123	123
EUROPE	92	97	99	98	99	103	104	108	112	112	112	113
ALBANIA	91	99	94	97	100	103	102	110	111	112	122	126
AUSTRIA	89	97	99	102	99	99	96	102	106	109	108	107
BELGIUM-LUX	78	91	93	94	100	105	101	106	112	109	101	103
BULGARIA	104	103	99	99	99	102	111	108	102	109	117	119
CZECHOSLOVAK	89	93	99	98	98	104	105	114	117	115	115	122
DENMARK	101	102	107	102	96	102	98	98	111	105	100	104
FINLAND	87	90	90	97	98	105	106	99	102	106	119	107
FRANCE	91	98	103	96	100	104	104	110	112	108	109	109
GERMAN DR	98	105	105	100	99	100	107	110	121	120	115	117
GERMANY FED	88	95	100	98	100	102	98	100	104	101	100	102
GREECE	92	96	89	94	103	103	108	108	118	125	129	127
HUNGARY	91	89	95	101	91	108	115	116	120	125	116	130
ICELAND	125	116	100	104	96	100	112	121	115	123	119	120
IRELAND	88	99	98	96	96	108	103	101	117	138	125	135
ITALY	93	97	96	99	100	100	95	101	106	107	106	104
MALTA	73	79	86	90	108	102	108	104	105	94	116	115
NETHERLANDS	79	90	93	93	103	104	101	108	116	120	118	122
NORWAY	93	100	105	98	100	102	103	103	121	107	106	119
POLAND	98	100	102	97	102	101	106	111	113	116	111	106
PORTUGAL	88	99	98	97	106	97	95	99	102	102	98	91
ROMANIA	103	102	102	100	89	110	125	116	121	125	154	148
SPAIN	90	90	95	95	101	104	111	117	120	125	128	127
SWEDEN	94	102	105	94	102	104	105	101	122	108	116	116
SWITZERLAND	89	96	101	99	99	102	101	104	105	108	114	112
UK	95	97	95	96	101	104	106	108	114	108	104	112
YUGOSLAVIA	96	96	96	105	93	102	99	107	117	118	121	123
OCEANIA	92	88	99	98	100	102	103	109	104	113	118	114
AUSTRALIA	94	86	101	97	100	103	102	111	106	115	118	115
FIJI	86	85	101	96	105	99	93	99	95	95	97	99
FR POLYNESIA	102	92	90	106	98	96	90	80	75	99	100	101
NEW HEBRIDES	97	114	100	105	95	100	72	85	115	109	112	114
NEW ZEALAND	87	91	94	100	99	101	104	104	97	106	119	112
PAPUA N GUIN	89	92	94	98	100	102	108	112	118	121	120	122
SAMOA	88	95	95	100	97	102	94	94	96	97	104	110
USSR	93	95	99	94	102	104	99	116	112	110	114	116
DEV.PED M E	92	97	99	98	99	103	103	105	108	111	112	115
N AMERICA	94	97	98	98	97	105	104	106	106	113	117	121
W EUROPE	90	96	98	97	100	103	101	106	111	110	109	110
OCEANIA	92	87	99	98	100	102	103	109	104	113	118	114
OTH DEV.PED	95	106	106	104	100	96	103	101	106	110	102	111
DEV.PING M E	86	90	93	97	101	102	103	106	109	114	116	119
AFRICA	87	90	94	98	100	102	103	100	106	106	110	108
LAT AMERICA	89	93	94	97	101	102	103	105	113	116	118	123
NEAR EAST	89	93	96	99	99	103	109	104	113	117	123	123
FAR EAST	83	88	92	97	101	102	100	108	106	115	115	121
OTH DV.PING	90	92	95	98	100	102	103	107	112	114	115	117
CENTR PLANND	92	95	97	95	101	104	103	113	113	114	116	118
ASIAN CPE	90	93	92	95	101	104	104	109	112	116	118	120
E EUR+USSR	94	96	100	96	101	104	103	115	113	113	115	117
DEV.PED ALL	93	97	99	97	100	103	103	109	110	112	113	115
DEV.PING ALL	87	91	93	97	101	103	103	107	110	114	117	119

TABLE
TABLEAU 6
CUADRO

FOOD   ALIMENTAIRES   ALIMENTICIOS

PROD INDICES   1969-71=100   PER CAPUT

	1966	1967	1968	1969	1970	1971	1972	1973	1974	1975	1976	1977
WORLD	97	100	100	99	100	101	99	102	102	103	103	103
AFRICA	95	99	98	100	99	100	99	93	97	95	95	92
ALGERIA	78	94	107	99	104	97	99	87	88	88	92	83
ANGOLA	98	99	98	101	101	98	93	95	93	87	87	88
BENIN	90	97	97	101	103	96	93	91	89	76	87	93
BOTSWANA	94	104	95	99	93	109	97	105	110	105	116	110
BURUNDI	100	100	102	99	101	100	101	107	103	99	98	98
CAMEROON	88	95	96	99	100	101	100	98	104	102	100	101
CENT AFR EMP	91	97	96	99	100	102	103	106	106	103	103	103
CHAD	107	103	104	103	98	99	81	76	83	83	83	83
CONGO	100	107	104	102	98	99	92	92	89	92	98	96
EGYPT	97	93	101	100	99	101	100	98	97	98	97	96
ETHIOPIA	101	103	100	102	101	97	93	93	91	91	90	86
GABON	96	96	97	98	100	102	103	102	103	102	102	103
GAMBIA	118	102	101	102	98	100	104	91	103	104	102	87
GHANA	92	98	95	98	99	104	101	100	107	86	77	76
GUINEA	96	100	103	100	100	99	90	83	84	85	89	87
IVORY COAST	88	91	91	96	97	106	102	104	115	118	117	122
KENYA	101	100	102	101	102	96	94	89	86	87	88	89
LESOTHO	99	104	101	102	97	101	84	114	98	85	110	97
LIBERIA	94	93	97	96	101	103	104	105	110	106	109	105
LIBYA	133	124	106	127	82	91	131	129	129	148	159	148
MADAGASCAR	95	102	98	101	100	99	94	91	98	95	96	94
MALAWI	89	106	97	100	90	110	113	108	106	98	104	103
MALI	100	103	98	104	101	95	78	65	69	80	85	83
MAURITANIA	103	103	102	104	101	96	88	72	69	66	72	72
MAURITIUS	97	113	98	108	95	97	110	108	104	78	111	115
MOROCCO	89	93	121	96	98	106	101	89	97	79	88	66
MOZAMBIQUE	92	93	98	100	100	99	100	103	97	83	87	85
NAMIBIA	91	98	100	95	100	105	113	113	113	111	109	108
NIGER	109	116	100	111	96	93	86	63	74	68	90	82
NIGERIA	94	93	93	103	102	96	95	86	92	92	92	89
REUNION	109	110	111	117	94	90	106	102	94	95	97	99
RHODESIA	103	97	92	101	93	107	114	93	113	106	102	99
RWANDA	84	97	90	97	102	102	96	97	92	102	105	104
SENEGAL	108	128	99	110	82	108	69	85	109	123	109	81
SIERRA LEONE	99	99	97	100	99	101	97	95	92	95	97	96
SOMALIA	99	101	102	101	100	99	105	97	85	87	92	92
SOUTH AFRICA	93	118	95	97	94	109	113	90	109	98	94	97
SUDAN	88	104	90	98	100	101	101	101	110	111	103	104
SWAZILAND	85	96	97	97	102	100	111	103	109	96	106	106
TANZANIA	103	98	96	98	105	97	97	96	91	93	93	94
TOGO	108	99	100	91	102	108	78	78	60	62	62	62
TUNISIA	99	95	93	84	97	119	117	121	125	137	127	122
UGANDA	98	102	102	103	100	97	99	97	95	92	94	94
UPPER VOLTA	105	106	106	103	102	95	90	84	92	99	94	85
ZAIRE	99	100	100	100	101	99	98	98	98	98	97	96
ZAMBIA	102	102	101	102	95	103	102	100	105	107	110	107
N C AMERICA	99	101	101	99	97	103	101	101	101	106	107	109
BARBADOS	107	124	100	95	107	98	87	95	86	82	87	102
CANADA	113	99	105	101	93	107	100	100	92	101	111	109
COSTA RICA	90	87	98	96	100	104	103	105	103	114	112	110
CUBA	80	95	90	86	123	91	81	83	84	88	85	89
DOMINICAN RP	92	90	89	96	101	103	102	100	98	89	94	88
EL SALVADOR	100	95	98	94	101	104	96	107	102	112	106	116
GUATEMALA	93	95	98	99	100	101	103	103	101	101	109	108
HAITI	99	100	96	100	100	101	102	100	102	100	101	100
HONDURAS	88	99	103	104	96	100	105	98	86	74	81	86
JAMAICA	109	110	102	95	99	106	104	99	102	101	101	97
MEXICO	104	103	101	98	100	102	101	98	102	101	93	96
NICARAGUA	91	95	98	100	100	100	95	95	94	101	103	104
PANAMA	91	94	102	106	94	100	97	99	99	102	98	99
PUERTO RICO	119	112	105	102	101	97	97	92	93	91	96	99
TRINIDAD ETC	94	93	102	95	104	102	107	101	98	96	97	96
USA	96	101	100	99	97	104	102	103	103	110	112	115
SOUTH AMERIC	97	100	97	100	101	99	97	98	102	104	109	108
ARGENTINA	98	105	97	104	101	95	92	100	102	103	111	107
BOLIVIA	96	96	101	98	101	101	107	108	109	115	115	108
BRAZIL	92	96	96	97	102	101	103	103	109	112	121	121
CHILE	104	103	105	97	103	99	92	84	96	100	95	102
COLOMBIA	99	97	98	98	99	103	103	100	102	107	107	108
ECUADOR	105	103	108	99	101	99	94	93	100	102	95	95
GUYANA	97	96	91	101	96	102	94	91	99	96	96	93
PARAGUAY	100	101	97	97	103	100	97	93	101	100	103	111
PERU	98	100	88	97	102	101	97	95	96	94	93	90
URUGUAY	94	83	93	101	107	92	89	94	101	100	110	89
VENEZUELA	91	92	95	100	101	99	97	96	98	102	96	104
ASIA	95	98	99	99	101	100	98	100	100	104	102	103
AFGHANISTAN	109	114	114	114	98	88	98	103	102	102	104	99
BANGLADESH	97	107	106	109	101	90	87	97	92	100	91	96
BHUTAN	100	100	100	100	100	100	100	100	100	100	99	97
BURMA	91	100	98	98	102	100	93	97	97	95	98	97
CHINA	95	97	95	97	101	103	100	103	104	106	107	106
CYPRUS	77	94	92	98	90	112	107	79	92	95	92	100
HONG KONG	96	70	106	92	105	103	104	76	108	89	81	79
INDIA	88	93	95	98	102	100	93	100	91	101	96	99
INDONESIA	99	92	101	98	100	102	102	108	112	110	110	109

TABLE
TABLEAU 6
CUADRO

FOOD   ALIMENTAIRES   ALIMENTICIOS

PROD INDICES   1969-71=100   PER CAPUT

	1966	1967	1968	1969	1970	1971	1972	1973	1974	1975	1976	1977
IRAN	97	94	100	101	101	97	105	105	106	106	110	111
IRAQ	95	101	110	105	99	97	121	85	81	70	83	73
ISRAEL	87	98	99	96	99	106	113	109	113	112	113	116
JAPAN	102	110	112	107	101	91	98	97	99	103	93	100
JORDAN	177	163	106	111	80	109	116	73	125	73	70	62
KAMPUCHEA DM	93	95	109	93	115	92	74	57	47	59	60	59
KOREA DPR	97	97	96	96	100	104	108	108	115	118	124	129
KOREA REP	100	92	90	102	99	99	98	98	102	109	112	118
LAO	93	99	96	102	102	96	96	98	100	100	97	90
LEBANON	102	120	110	93	99	108	112	101	110	99	106	104
MAL PENINSUL	89	88	93	97	100	103	105	109	113	110	113	114
MAL SABAH	85	87	97	97	97	106	116	109	110	126	116	123
MAL SARAWAK	92	93	104	109	97	95	97	94	90	86	89	89
MONGOLIA	124	126	103	100	101	99	99	102	104	105	99	98
NEPAL	99	100	99	101	102	97	92	100	99	99	96	89
PAKISTAN	88	92	99	100	102	98	98	100	100	98	100	102
PHILIPPINES	99	102	98	100	101	99	95	103	104	108	117	112
SAUDI ARABIA	102	104	97	102	102	95	73	77	97	101	87	91
SINGAPORE	21	66	55	85	96	119	98	119	127	116	108	157
SRI LANKA	89	100	100	97	103	100	100	94	105	108	109	119
SYRIA	100	125	111	120	88	92	135	79	136	140	161	145
THAILAND	103	92	95	99	100	101	98	112	105	107	109	106
TURKEY	98	97	99	97	100	103	102	93	100	107	110	105
VIET NAM	101	100	97	98	102	100	101	101	99	101	99	100
YEMEN AR	118	121	112	104	85	111	110	106	94	108	98	94
YEMEN DEM	104	106	104	102	92	106	99	106	108	110	108	104
EUROPE	94	98	100	98	100	102	102	105	110	109	108	109
ALBANIA	96	104	100	100	100	101	95	100	99	98	106	106
AUSTRIA	90	98	99	103	99	98	95	100	104	107	107	106
BELGIUM-LUX	79	91	93	94	100	105	100	105	111	108	99	102
BULGARIA	105	105	100	101	98	101	106	104	96	102	111	111
CZECHOSLOVAK	90	94	99	98	98	104	104	112	115	113	111	118
DENMARK	103	103	109	103	96	101	97	96	108	102	97	101
FINLAND	88	90	90	97	98	105	105	98	100	104	116	104
FRANCE	94	101	104	97	100	103	102	108	109	104	104	104
GERMAN DR	98	105	105	100	100	100	108	111	122	121	117	120
GERMANY FED	90	96	101	98	100	101	97	98	102	99	98	101
GREECE	92	95	89	94	104	103	107	108	118	122	124	120
HUNGARY	92	89	95	102	91	108	115	116	119	124	113	127
ICELAND	128	117	99	102	97	101	112	119	111	117	112	111
IRELAND	90	101	99	97	96	107	101	97	111	130	116	125
ITALY	95	98	97	100	100	100	93	99	103	103	101	99
MALTA	75	80	87	91	108	102	110	105	105	93	114	113
NETHERLANDS	82	92	95	94	103	103	99	104	111	113	111	114
NORWAY	96	103	107	98	100	102	102	101	118	104	102	114
POLAND	101	102	103	97	102	101	104	110	111	111	105	100
PORTUGAL	82	94	94	96	106	98	95	100	101	101	97	89
ROMANIA	109	108	105	101	89	109	123	113	117	119	146	139
SPAIN	93	93	96	96	101	104	109	114	116	120	122	119
SWEDEN	96	104	107	95	102	103	104	100	120	106	113	113
SWITZERLAND	93	100	103	99	99	102	97	100	101	104	111	110
UK	96	97	95	96	101	103	105	107	113	107	103	111
YUGOSLAVIA	99	98	98	106	93	102	97	103	113	112	115	115
OCEANIA	101	92	105	100	99	101	100	109	102	109	114	109
AUSTRALIA	105	91	109	99	99	102	100	114	107	114	118	114
FIJI	94	91	106	98	105	97	89	93	87	85	85	8/
FR POLYNESIA	117	103	95	109	98	93	85	73	66	84	83	8:
NEW HEBRIDES	108	123	105	108	94	98	67	78	102	96	95	93
NEW ZEALAND	89	92	95	102	99	99	103	102	93	101	113	106
PAPUA N GUIN	100	100	100	100	100	100	103	103	105	105	102	100
SAMOA	98	102	100	103	98	99	88	85	84	82	85	88
USSR	96	97	101	95	102	103	97	113	108	104	108	108
DEV.PED M E	95	100	101	99	99	102	101	102	105	107	107	109
N AMERICA	97	100	101	99	97	104	102	102	102	109	112	114
W EUROPE	92	97	99	98	100	102	100	103	108	106	106	106
OCEANIA	101	92	105	100	99	101	100	111	103	110	117	112
OTH DEV.PED	100	111	109	106	100	94	100	97	100	102	93	100
DEV.PING M E	95	97	98	100	101	99	97	98	98	101	101	10I
AFRICA	96	98	99	100	100	99	97	92	95	93	94	90
LAT AMERICA	98	100	98	99	102	99	98	98	101	102	104	104
NEAR EAST	100	101	102	101	99	100	104	96	101	104	106	102
FAR EAST	91	94	97	99	102	99	95	101	96	103	100	102
OTH DV.PING	100	100	101	101	100	99	97	99	99	100	97	96
CENTR PLANND	98	99	100	97	101	102	100	108	106	105	106	106
ASIAN CPE	96	97	96	97	101	102	100	103	104	106	106	106
E EUR+USSR	97	99	101	96	101	103	101	112	110	108	109	110
DEV.PED ALL	96	99	101	98	99	102	101	106	106	107	108	109
DEV.PING ALL	95	97	97	99	101	100	98	100	100	103	103	102

TABLE
TABLEAU **7**
CUADRO

AGRICULTURE   AGRICULTURE   AGRICULTURA

PROD INDICES   1969-71=100   PER CAPUT

	1966	1967	1968	1969	1970	1971	1972	1973	1974	1975	1976	1977
WORLD	97	100	100	99	100	101	99	102	102	103	102	103
AFRICA	96	99	98	100	100	100	99	93	97	94	93	91
ALGERIA	79	94	107	99	104	97	99	87	88	88	92	83
ANGOLA	101	101	96	101	99	100	95	93	94	68	67	67
BENIN	88	95	95	100	103	97	95	91	88	75	88	94
BOTSWANA	95	104	95	99	92	109	97	105	110	105	116	110
BURUNDI	99	100	101	98	101	101	101	107	105	98	98	98
CAMEROON	87	95	97	99	100	101	100	98	105	99	97	99
CENT AFR EMP	91	96	97	102	100	97	101	103	105	99	99	100
CHAD	108	102	108	104	97	99	82	78	88	90	87	88
CONGO	100	107	104	102	98	100	93	92	89	93	98	96
EGYPT	97	92	99	101	99	100	99	96	93	92	91	91
ETHIOPIA	101	103	100	102	102	96	94	92	91	91	91	87
GABON	97	96	97	98	100	101	102	102	102	102	102	103
GAMBIA	118	102	101	102	98	100	104	91	103	104	102	87
GHANA	92	98	95	98	99	104	101	100	107	87	78	76
GUINEA	96	101	104	101	100	99	90	83	83	84	88	86
IVORY COAST	95	83	99	95	102	104	103	107	106	114	116	118
KENYA	101	98	98	101	102	97	98	96	93	92	95	101
LESOTHO	100	104	101	103	98	100	82	109	93	84	107	95
LIBERIA	92	92	95	93	103	104	104	105	107	102	102	100
LIBYA	131	123	106	125	83	91	129	127	127	145	156	146
MADAGASCAR	97	103	100	101	101	98	95	93	99	97	99	97
MALAWI	90	104	95	98	92	109	114	110	108	103	109	112
MALI	98	102	97	103	101	96	79	66	70	83	88	86
MAURITANIA	103	103	102	104	101	96	88	72	69	66	72	72
MAURITIUS	97	112	98	107	95	98	111	108	105	79	111	115
MOROCCO	90	93	120	96	98	106	102	90	97	80	88	67
MOZAMBIQUE	93	94	98	101	101	98	100	101	96	81	84	82
NAMIBIA	92	99	99	95	100	105	111	112	113	111	109	108
NIGER	109	116	100	110	96	93	86	63	74	68	90	82
NIGERIA	95	93	93	102	102	96	95	87	92	93	92	90
REUNION	109	110	111	116	94	90	106	102	94	95	97	100
RHODESIA	112	102	89	101	94	105	111	89	109	108	104	97
RWANDA	83	96	90	96	102	102	95	97	94	103	107	105
SENEGAL	107	127	99	109	82	108	70	86	111	124	110	83
SIERRA LEONE	100	98	98	100	99	101	99	97	92	96	97	97
SOMALIA	99	101	102	101	100	99	105	96	85	87	92	92
SOUTH AFRICA	94	116	96	98	95	107	111	89	106	97	92	96
SUDAN	85	100	88	99	100	102	99	94	108	104	89	96
SWAZILAND	86	99	98	97	102	101	112	105	112	100	109	110
TANZANIA	107	100	96	99	105	95	97	94	89	90	91	91
TOGO	107	99	101	92	102	107	77	78	61	63	63	63
TUNISIA	99	96	93	84	97	119	117	120	125	137	127	122
UGANDA	95	98	94	105	101	94	95	97	89	87	86	86
UPPER VOLTA	103	105	105	103	103	95	90	84	92	98	95	86
ZAIRE	99	100	100	100	101	99	98	98	98	98	97	96
ZAMBIA	103	102	101	102	95	103	102	100	105	107	109	106
N C AMERICA	99	101	101	99	98	103	101	101	101	105	107	109
BARBADOS	107	124	100	95	107	98	87	95	86	82	87	102
CANADA	112	97	104	101	94	105	99	100	91	100	108	108
COSTA RICA	92	92	97	99	97	104	100	106	101	108	108	104
CUBA	83	97	92	88	122	91	83	85	86	89	86	90
DOMINICAN RP	94	91	89	97	100	103	102	105	101	91	97	92
EL SALVADOR	103	103	95	99	97	103	100	100	105	110	102	111
GUATEMALA	95	94	97	98	102	100	103	104	105	101	107	107
HAITI	101	101	97	99	100	101	103	101	102	101	99	99
HONDURAS	90	103	103	106	95	100	105	97	88	80	84	91
JAMAICA	109	111	102	96	99	106	103	99	102	100	101	97
MEXICO	108	104	105	99	100	102	101	98	102	98	91	94
NICARAGUA	101	106	104	102	98	100	99	99	107	108	106	113
PANAMA	91	94	102	106	94	100	97	98	98	102	97	99
PUERTO RICO	120	114	106	101	103	96	97	94	92	91	96	98
TRINIDAD ETC	94	93	104	95	102	103	107	101	97	97	96	96
USA	97	100	100	99	97	103	102	103	104	109	112	115
SOUTH AMERIC	99	101	98	101	100	99	98	97	102	103	103	105
ARGENTINA	98	104	97	104	102	94	92	99	101	103	110	107
BOLIVIA	95	94	100	97	101	102	109	116	114	119	116	110
BRAZIL	95	99	96	100	98	103	105	99	109	108	107	112
CHILE	105	103	105	98	103	99	91	84	95	99	95	102
COLOMBIA	101	98	100	99	100	101	99	99	99	105	103	105
ECUADOR	107	105	108	100	101	99	93	93	102	102	97	96
GUYANA	97	97	91	101	97	102	94	91	99	96	96	93
PARAGUAY	98	101	98	97	102	101	98	94	102	103	106	116
PERU	102	101	92	97	102	101	95	94	95	91	90	86
URUGUAY	96	85	94	101	106	93	87	90	95	95	105	87
VENEZUELA	93	94	94	100	101	99	96	97	98	103	95	102
ASIA	96	99	99	99	101	100	98	101	100	103	102	103
AFGHANISTAN	108	113	113	114	98	88	97	102	103	103	104	99
BANGLADESH	99	109	106	110	102	88	88	96	89	96	90	95
BHUTAN	99	100	100	100	100	100	100	100	100	100	99	98
BURMA	90	99	99	98	102	100	94	98	98	95	98	98
CHINA	96	97	95	97	101	103	100	104	106	107	107	106
CYPRUS	77	94	92	99	90	112	107	79	91	95	92	79
HONG KONG	96	70	106	92	105	103	104	76	108	89	81	79
INDIA	89	94	95	98	102	101	94	100	92	101	96	98
INDONESIA	99	92	100	99	100	101	100	106	109	108	107	107

TABLE
TABLEAU 7
CUADRO

AGRICULTURE   AGRICULTURE   AGRICULTURA

PROD INDICES   1969-71=100   PER CAPUT

	1966	1967	1968	1969	1970	1971	1972	1973	1974	1975	1976	1977
IRAN	97	94	101	102	101	97	105	105	106	104	109	109
IRAQ	93	101	110	104	99	97	119	85	81	70	82	72
ISRAEL	87	97	99	97	99	105	112	108	113	112	113	117
JAPAN	102	110	112	107	101	92	98	97	98	102	92	99
JORDAN	174	161	106	111	81	108	115	74	124	74	71	63
KAMPUCHEA DM	98	99	113	98	113	89	74	56	48	58	60	59
KOREA DPR	97	97	96	96	100	104	108	108	115	117	123	128
KOREA REP	100	92	91	102	99	99	100	100	104	111	114	120
LAO	93	99	97	102	101	96	96	97	99	99	96	89
LEBANON	101	118	109	93	99	108	113	102	110	100	104	100
MAL PENINSUL	87	87	92	98	100	102	102	109	111	106	111	109
MAL SABAH	88	89	97	99	98	102	108	107	105	118	111	115
MAL SARAWAK	99	96	103	114	95	92	93	99	92	87	93	92
MONGOLIA	123	123	102	99	101	100	99	102	104	104	98	97
NEPAL	99	100	99	101	102	97	92	100	98	98	95	89
PAKISTAN	88	93	99	99	101	100	99	99	98	95	95	99
PHILIPPINES	99	102	99	100	101	99	95	103	104	108	117	112
SAUDI ARABIA	102	104	97	102	102	96	73	77	97	101	87	91
SINGAPORE	24	67	56	85	96	119	97	118	126	115	106	154
SRI LANKA	96	102	103	100	101	99	97	90	96	97	97	103
SYRIA	102	118	110	117	90	94	128	82	127	129	147	134
THAILAND	105	93	96	100	100	100	97	109	103	104	105	103
TURKEY	99	98	100	97	100	104	103	94	101	106	110	105
VIET NAM	101	101	97	98	102	100	100	101	99	101	99	100
YEMEN AR	118	121	112	104	84	111	111	107	95	109	99	95
YEMEN DEM	103	106	99	102	92	106	97	105	107	108	103	101
EUROPE	94	99	100	98	100	102	102	105	110	109	108	109
ALBANIA	101	108	99	99	100	101	97	102	100	98	104	104
AUSTRIA	90	98	99	103	99	98	95	100	105	107	107	106
BELGIUM-LUX	79	91	94	95	100	105	100	105	111	108	99	101
BULGARIA	107	105	100	99	99	101	110	106	100	106	114	115
CZECHOSLOVAK	90	94	100	99	98	104	104	112	115	112	111	117
DENMARK	103	104	109	103	96	101	97	96	108	102	97	101
FINLAND	88	90	90	97	98	105	105	98	100	104	116	114
FRANCE	94	101	104	97	100	103	102	108	109	104	104	104
GERMAN DR	98	105	105	100	99	100	107	111	122	121	117	120
GERMANY FED	90	96	101	98	100	101	97	99	102	99	99	101
GREECE	93	97	89	94	103	103	107	106	116	122	124	121
HUNGARY	92	90	96	102	91	108	115	115	118	123	113	126
ICELAND	130	119	101	104	96	99	110	117	109	115	110	110
IRELAND	90	101	99	97	96	107	101	97	111	130	116	125
ITALY	95	98	97	100	100	100	94	99	103	103	101	99
MALTA	75	80	87	91	108	102	110	104	105	93	114	113
NETHERLANDS	82	93	95	94	103	103	99	105	112	115	112	115
NORWAY	96	102	107	98	100	102	102	101	118	104	102	114
POLAND	100	102	103	98	102	100	104	109	109	110	105	100
PORTUGAL	83	95	95	97	106	98	95	99	101	101	97	89
ROMANIA	109	108	105	101	89	109	123	113	116	119	146	139
SPAIN	94	93	97	96	101	103	109	113	116	120	121	119
SWEDEN	96	104	107	95	102	103	104	100	120	106	113	113
SWITZERLAND	93	100	103	99	99	102	97	100	101	104	111	110
UK	97	98	95	96	101	103	105	107	113	107	104	111
YUGOSLAVIA	100	99	98	106	93	101	98	104	113	112	115	115
OCEANIA	99	93	103	100	100	100	99	103	97	103	107	102
AUSTRALIA	102	91	105	99	100	101	99	106	99	106	108	104
FIJI	94	91	106	98	105	97	89	93	87	85	86	86
FR POLYNESIA	118	103	96	109	98	93	85	73	66	84	83	81
NEW HEBRIDES	108	124	105	108	94	98	67	78	102	95	95	93
NEW ZEALAND	91	94	96	102	99	99	101	99	91	97	108	102
PAPUA N GUIN	98	98	98	100	100	100	103	105	107	107	104	103
SAMOA	98	102	101	103	98	99	89	86	85	83	86	89
USSR	96	98	101	95	102	103	98	113	108	105	108	109
DEV.PED M E	96	100	101	99	99	102	101	102	104	106	106	108
N AMERICA	98	100	100	99	97	104	102	103	103	108	111	114
W EUROPE	93	98	99	98	100	102	100	104	108	107	106	107
OCEANIA	99	92	103	100	100	100	99	104	97	104	108	104
OTH DEV.PED	101	111	110	106	100	94	100	96	99	101	93	99
DEV.PING M E	95	98	98	100	101	100	97	98	98	100	99	100
AFRICA	96	97	98	100	100	99	97	93	95	93	93	90
LAT AMERICA	100	101	99	100	101	100	98	97	101	101	100	102
NEAR EAST	99	101	102	101	99	100	104	96	101	102	104	101
FAR EAST	92	95	97	99	101	100	95	101	96	101	99	101
OTH DV.PING	99	99	100	101	100	99	98	100	101	101	99	98
CENTR PLANND	98	99	100	97	101	102	100	108	106	106	107	107
ASIAN CPE	96	98	96	97	101	102	100	104	105	106	107	106
E EUR+USSR	97	99	101	96	101	103	101	112	109	108	109	110
DEV.PED ALL	96	99	101	98	100	102	101	106	106	107	107	109
DEV.PING ALL	96	98	97	99	101	100	98	100	100	102	102	102

STATISTICAL SUMMARY

SOMMAIRE STATISTIQUE

RESUMEN ESTADISTICO

TABLE / TABLEAU / CUADRO **8** ST.SUMMARY

WORLD — PRODUCTION / PRODUCTION / PRODUCCION (1000 MT)
DEVELOPED MARKET EC — PRODUCTION / PRODUCTION / PRODUCCION (1000 MT)

	WORLD 1972	1973	1974	1975	1976	1977	DEVELOPED MARKET EC 1972	1973	1974	1975	1976	1977
TOTAL CEREALS	1279834	1376104	1335485	1362153	1469877	1459012	451914	466058	442101	480811	489188	502283
WHEAT	347342	376690	360344	354748	418001	386596	121979	132852	138366	142264	153928	140193
RICE PADDY	305820	332145	332792	359653	350171	366505	20988	22082	23206	25021	22505	23386
MAIZE	305402	311377	294158	324257	333079	349676	179545	180409	159805	188623	194716	205567
BARLEY	148131	164699	165400	150003	184434	173094	67148	67312	65918	67146	64694	75002
ROOT CROPS	524054	574432	555545	553065	563261	570211	80524	80198	84129	71856	71680	80645
POTATOES	281225	316972	295384	286401	291422	292938	76969	77218	81071	69062	68993	77997
TOTAL PULSES	45438	46249	46898	44451	49726	47959	3676	3427	3840	3479	3202	3105
VEGETABLES AND MELONS	281608	293038	299277	307021	312293	318906	85563	88215	89637	92096	87472	91412
FRUITS	228894	251315	251834	256670	259695	257068	89470	105106	100396	101853	99532	92710
GRAPES	50570	63347	61236	60440	60161	57005	33102	43291	40929	39638	38365	34663
CITRUS FRUIT	42159	45325	47397	49546	48693	50328	24308	26072	25703	27068	26554	27368
BANANAS	33011	33558	34925	34209	35772	36868	651	747	685	637	626	649
APPLES	19576	22611	21481	24314	23018	21348	13848	16653	14970	17178	15169	13449
TOTAL NUTS	3296	3520	3516	3614	3606	3595	1341	1446	1458	1450	1520	1429
OIL CROPS (OIL EQUIV.)	36550	40135	39462	42695	40979	45220	11243	12697	10851	12851	10769	14356
SUGAR (CENTRIFUGAL,RAW)	73289	77887	77372	80870	85676	92109	22925	22516	21439	24437	25850	26826
COCOA BEANS	1427	1333	1525	1535	1362	1408						
COFFEE GREEN	4570	4138	4736	4462	3659	4340	1	1	1	1	1	1
TEA	1484	1538	1557	1605	1633	1758	97	103	97	107	102	107
VEGETABLE FIBRES	20618	21173	20508	19100	19164	21384	3524	3281	3050	2317	2763	3700
COTTON LINT	13610	13794	13894	12362	12264	14290	3283	3066	2824	2100	2550	3483
JUTE AND SUBSTITUTES	3983	4552	3756	3815	4115	4312	1	1	1	1	1	1
TOBACCO	4865	4935	5298	5447	5680	5647	1412	1471	1551	1713	1706	1597
NATURAL RUBBER	3024	3445	3426	3310	3596	3613						
TOTAL MEAT	111407	112333	118156	119853	122804	126086	54303	54170	57216	56975	60048	60991
TOTAL MILK	412351	418692	426117	430108	437451	450713	206893	205806	206131	207524	212944	216667
HEN EGGS	22255	22415	23073	23596	23922	24700	11565	11288	11372	11406	11549	11600
WOOL GREASY	2729	2569	2531	2648	2607	2588	1563	1390	1323	1418	1391	1327

N AMERICA DEVELOPED — PRODUCTION / PRODUCTION / PRODUCCION (1000 MT)
WESTERN EUROPE — PRODUCTION / PRODUCTION / PRODUCCION (1000 MT)

	N AMERICA DEVELOPED 1972	1973	1974	1975	1976	1977	WESTERN EUROPE 1972	1973	1974	1975	1976	1977
TOTAL CEREALS	263650	274338	235458	285954	302395	303525	148129	150723	159109	146808	142363	153331
WHEAT	56596	62720	61792	74843	81894	74785	56073	55432	62866	52961	57227	53407
RICE PADDY	3875	4208	5098	5826	5246	4501	1411	1784	1729	1703	1547	1352
MAIZE	144262	146845	121997	151706	162943	165788	25436	28934	26439	27411	24073	29643
BARLEY	20478	19314	15306	17672	18624	20571	44217	45077	47495	45666	42588	51255
ROOT CROPS	15873	16225	18656	17431	19191	19042	56419	56535	58534	47297	45041	53991
POTATOES	15316	15669	18046	16827	18578	18470	56272	56395	58390	47158	44893	53848
TOTAL PULSES	1135	1035	1323	1166	1140	1046	2039	1966	2058	1890	1628	1698
VEGETABLES AND MELONS	23509	24543	25172	26849	24949	26491	43655	45757	46754	47262	44446	46184
FRUITS	20292	23985	23726	25556	25345	25817	55392	67284	63293	62341	61046	53142
GRAPES	2396	3869	3881	4038	4070	3920	28418	37388	35101	33357	32054	28468
CITRUS FRUIT	11031	12604	12167	13237	13442	13859	6479	6531	6666	6697	6721	6615
BANANAS	3	3	3	3	2	3	406	480	426	385	362	399
APPLES	3059	3216	3351	3876	3321	3455	8964	11569	9908	11492	10110	8218
TOTAL NUTS	318	422	389	458	440	541	961	956	1005	926	1010	813
OIL CROPS (OIL EQUIV.)	8612	9941	8114	9857	8259	11498	2227	2418	2260	2628	2196	2405
SUGAR (CENTRIFUGAL,RAW)	5898	5329	5048	6441	6168	5568	11596	12248	11167	12844	13745	15124
COFFEE GREEN	1	1	1	1	1	1						
TEA												
VEGETABLE FIBRES	2993	2842	2533	1828	2327	3184	277	225	269	242	208	256
COTTON LINT	2984	2825	2513	1807	2304	3156	199	155	191	176	149	197
JUTE AND SUBSTITUTES												
TOBACCO	878	909	1021	1096	1043	981	334	350	329	398	435	383
TOTAL MEAT	25632	24622	26118	25413	27613	27819	22205	22783	24721	24864	25174	25644
TOTAL MILK	62468	60052	60062	60067	62246	63523	122551	124315	125485	126675	129210	131643
HEN EGGS	4422	4241	4202	4101	4108	4132	4904	4801	4870	5019	5090	5089
WOOL GREASY	81	73	65	60	54	51	160	163	166	162	163	160

TABLE
TABLEAU **8**
CUADRO

ST.SUMMARY

OCEANIA DEVELOPED / OTHER DEVELOPED MARKET EC

PRODUCTION
PRODUCTION — 1000 MT
PRODUCCION

	OCEANIA DEVELOPED (1000 MT)						OTHER DEVELOPED MARKET EC (1000 MT)					
	1972	1973	1974	1975	1976	1977	1972	1973	1974	1975	1976	1977
TOTAL CEREALS	11673	17804	17094	18575	18390	15563	28462	23193	30439	29474	26039	29864
WHEAT	6979	12385	11605	12185	12140	9720	2331	2315	2102	2276	2667	2281
RICE PADDY	248	309	409	388	417	530	15454	15781	15970	17104	15295	17003
MAIZE	330	241	236	338	363	400	9516	4389	11133	9168	7337	9736
BARLEY	2063	2655	2804	3513	3191	2876	391	266	313	295	292	300
ROOT CROPS	1074	1003	868	977	958	998	7158	6435	6072	6151	6489	6614
POTATOES	1064	991	855	967	947	986	4317	4162	3779	4110	4576	4694
TOTAL PULSES	129	92	125	163	188	122	374	334	333	259	246	238
VEGETABLES AND MELONS	1220	1179	1300	1230	1272	1310	17179	16737	16411	16755	16805	17427
FRUITS	2519	2318	2119	2324	2129	2172	11268	11519	11258	11632	11011	11579
GRAPES	841	623	576	752	737	772	1447	1411	1370	1491	1504	1501
CITRUS FRUIT	435	401	433	459	426	456	6362	6536	6437	6676	5965	6439
BANANAS	124	125	118	57	115	92	118	139	137	152	147	156
APPLES	511	574	487	527	430	439	1314	1294	1185	1282	1308	1337
TOTAL NUTS	2	2	2	1	2	2	60	67	62	65	69	73
OIL CROPS (OIL EQUIV.)	111	85	93	98	73	88	293	253	384	268	240	365
SUGAR (CENTRIFUGAL,RAW)	2835	2526	2848	2854	3296	3342	2596	2412	2376	2299	2641	2793
TEA							96	103	97	107	101	106
VEGETABLE FIBRES	49	37	36	38	30	33	206	178	213	209	198	227
COTTON LINT	44	31	31	33	25	28	57	55	90	84	72	102
JUTE AND SUBSTITUTES							1	1	1	1	1	1
TOBACCO	19	20	20	18	18	19	182	192	181	200	210	214
TOTAL MEAT	3544	3628	3180	3519	4018	3986	2922	3137	3197	3178	3242	3542
TOTAL MILK	13514	13155	12645	12712	12980	12532	8359	8283	7938	8070	8509	8969
HEN EGGS	256	252	277	255	236	232	1985	1994	2023	2032	2115	2146
WOOL GREASY	1202	1044	986	1088	1066	1005	119	110	105	109	108	111

DEVELOPING MARKET EC / AFRICA DEVELOPING

PRODUCTION
PRODUCTION — 1000 MT
PRODUCCION

	DEVELOPING MARKET EC (1000 MT)						AFRICA DEVELOPING (1000 MT)					
	1972	1973	1974	1975	1976	1977	1972	1973	1974	1975	1976	1977
TOTAL CEREALS	359261	378801	379634	413407	423587	426914	43922	38153	44212	43450	47429	42976
WHEAT	78054	70523	72670	80485	94871	84830	5786	4443	4871	4650	5862	4124
RICE PADDY	153134	171874	165120	186574	179732	191604	4580	4682	4535	5295	5563	5567
MAIZE	65826	68962	73054	73706	72155	77412	12924	11071	13054	13114	13318	13511
BARLEY	17520	13475	15324	17657	20650	15603	4133	2634	3886	3226	4689	2734
ROOT CROPS	162789	163895	168632	171333	178077	181637	69648	71307	74269	73542	76026	76528
POTATOES	21265	21605	23399	24910	27095	27758	2077	2217	2269	2502	2523	2565
TOTAL PULSES	23574	22738	22295	23336	25403	24434	4158	3923	4402	4555	4755	4391
VEGETABLES AND MELONS	94634	95112	99663	105160	111453	113470	9240	9137	9982	10619	10668	11061
FRUITS	113765	115286	122654	123651	126495	130130	24069	24849	24792	24255	24688	25608
GRAPES	11475	11155	12464	12078	12429	12639	1189	1236	1268	995	699	889
CITRUS FRUIT	16591	17912	20242	21007	20679	21403	2310	2413	2437	2190	2350	2441
BANANAS	31249	31638	33136	32600	34135	35188	3455	3729	3931	3819	4006	4135
APPLES	2963	2776	3637	3576	3699	3857	43	49	49	57	53	58
TOTAL NUTS	1415	1490	1496	1569	1488	1554	485	547	550	498	484	516
OIL CROPS (OIL EQUIV.)	16675	17390	18781	20418	20592	21073	3650	3537	3767	3993	3929	3769
SUGAR (CENTRIFUGAL,RAW)	33606	37396	39656	39995	43562	46297	2875	2941	2949	2738	3086	3168
COCOA BEANS	1427	1333	1525	1535	1362	1408	1014	942	1016	997	850	906
COFFEE GREEN	4561	4127	4725	4452	3648	4329	1311	1357	1273	1189	1260	1274
TEA	1016	1043	1061	1086	1106	1206	148	153	150	152	155	189
VEGETABLE FIBRES	10839	10862	10313	9470	9128	10289	920	904	900	811	828	825
COTTON LINT	5815	5689	6074	5203	4745	5721	528	513	488	498	548	560
JUTE AND SUBSTITUTES	2995	3275	2286	2336	2592	2776	19	19	19	19	19	19
TOBACCO	1921	1824	2079	2048	2211	2307	192	174	199	232	243	253
NATURAL RUBBER	2976	3393	3367	3255	3518	3530	237	244	244	239	223	237
TOTAL MEAT	20338	20700	21276	22140	23176	23813	3415	3389	3416	3503	3641	3704
TOTAL MILK	79965	80780	83313	87165	89936	91551	6401	6183	6127	6352	6685	6842
HEN EGGS	3160	3328	3483	3665	3795	4000	413	422	439	461	483	501
WOOL GREASY	574	570	569	582	601	615	60	66	62	65	67	68

TABLE
TABLEAU **8**
CUADRO

ST.SUMMARY

LATIN AMERICA / NEAR EAST DEVELOPING

	PRODUCTION PRODUCTION PRODUCCION (1000 MT)						PRODUCTION PRODUCTION PRODUCCION (1000 MT)					
	1972	1973	1974	1975	1976	1977	1972	1973	1974	1975	1976	1977
TOTAL CEREALS	67835	74472	78311	79039	85443	83668	47320	40657	44908	51883	56001	53556
WHEAT	12432	12084	13467	14970	19321	11557	25957	21221	24349	28418	31345	30080
RICE PADDY	10925	11795	11910	13765	15332	14771	4583	4446	4304	4602	4742	4848
MAIZE	35082	37380	39375	37997	37194	42247	4265	4536	4842	5028	5486	5260
BARLEY	1778	1665	1253	1560	1881	1532	7275	5197	6238	7850	8949	8056
ROOT CROPS	48340	44706	44590	45513	46225	47251	4311	4581	4606	4953	5612	5819
POTATOES	8383	8585	9947	9196	9668	10245	3946	4245	4229	4524	5173	5381
TOTAL PULSES	4871	4527	4646	4706	3980	4703	1827	1533	1747	1638	1894	1875
VEGETABLES AND MELONS	13214	13632	13971	14015	14249	14406	23970	23125	25363	26903	28685	28866
FRUITS	45372	45105	50092	49780	51457	52722	14411	14425	15216	15554	15615	16035
GRAPES	4628	4415	5389	5177	5904	5530	5364	5196	5496	5586	5498	5882
CITRUS FRUIT	9834	10852	12973	13829	13271	13649	2762	2896	3054	3108	3070	3322
BANANAS	18213	17818	17907	16753	17988	18564	277	283	289	294	296	306
APPLES	912	632	1284	1152	1141	1416	1279	1237	1359	1313	1436	1276
TOTAL NUTS	150	145	123	150	137	138	582	600	627	714	650	678
OIL CROPS (OIL EQUIV.)	3238	3581	4229	4436	4683	5236	1556	1265	1543	1417	1545	1455
SUGAR (CENTRIFUGAL,RAW)	21054	23339	24508	24062	25965	27437	2190	2221	2322	2455	2857	2794
COCOA BEANS	368	350	456	478	454	447						
COFFEE GREEN	2894	2415	3081	2880	1979	2623	6	6	6	6	6	6
TEA	41	40	44	49	44	45	69	66	67	77	82	82
VEGETABLE FIBRES	2392	2391	2574	2287	1975	2441	1723	1635	1794	1487	1409	1657
COTTON LINT	1677	1672	1857	1528	1318	1772	1699	1608	1763	1455	1375	1624
JUTE AND SUBSTITUTES	81	115	78	110	113	110	2	3	2	2	2	2
TOBACCO	565	563	679	671	711	774	241	214	238	252	375	279
NATURAL RUBBER	32	28	24	25	26	31						
TOTAL MEAT	10606	10871	11194	11795	12504	12843	2442	2518	2635	2731	2824	2937
TOTAL MILK	26839	26247	27883	30575	32082	32999	11524	11854	12279	12704	13144	13270
HEN EGGS	1539	1635	1706	1776	1780	1912	376	394	411	460	522	515
WOOL GREASY	309	299	291	299	307	315	146	147	155	155	159	162

ASIA AND FAR EAST DEV / OTHER DEVELOPING MARKET EC

	PRODUCTION PRODUCTION PRODUCCION (1000 MT)						PRODUCTION PRODUCTION PRODUCCION (1000 MT)					
	1972	1973	1974	1975	1976	1977	1972	1973	1974	1975	1976	1977
TOTAL CEREALS	200155	225491	212166	238995	234675	246679	29	27	37	39	38	35
WHEAT	33880	32774	29984	32447	38343	39070						
RICE PADDY	133025	150933	143943	162878	154068	166394	20	18	28	30	27	25
MAIZE	13549	15969	15776	17561	16152	16388	6	6	6	6	6	6
BARLEY	4334	3979	3947	5021	5131	3281						
ROOT CROPS	39121	41902	43742	45884	48755	50557	1369	1399	1425	1441	1460	1482
POTATOES	6854	6552	6948	8684	9727	9563	5	6	5	5	5	5
TOTAL PULSES	12697	12734	11478	12413	14750	13441	21	22	22	23	24	24
VEGETABLES AND MELONS	47960	48961	50083	53355	57576	58858	250	256	263	268	274	280
FRUITS	28884	29855	31484	32977	33614	34626	1029	1052	1070	1085	1120	1135
GRAPES	294	308	311	320	329	339						
CITRUS FRUIT	1670	1730	1755	1864	1929	1972	15	21	19	16	19	20
BANANAS	8462	8952	10139	10849	10931	11250	842	856	871	885	913	928
APPLES	729	857	946	1054	1069	1108						
TOTAL NUTS	195	195	192	205	214	218	3	3	3	3	4	4
OIL CROPS (OIL EQUIV.)	7954	8716	8911	10230	10114	10287	277	292	331	342	321	326
SUGAR (CENTRIFUGAL,RAW)	7184	8594	9604	10467	11347	12522	303	301	273	272	307	376
COCOA BEANS	14	17	21	25	24	25	30	25	32	36	34	31
COFFEE GREEN	320	312	321	336	360	380	30	37	45	41	44	47
TEA	757	781	796	803	819	884	2	3	4	5	6	6
VEGETABLE FIBRES	5804	5931	5045	4884	4916	5365	1	1	1	1	1	1
COTTON LINT	1911	1896	1966	1721	1504	1765						
JUTE AND SUBSTITUTES	2894	3138	2187	2205	2458	2645						
TOBACCO	922	872	962	893	881	1001						
NATURAL RUBBER	2701	3114	3092	2986	3263	3256	6	6	6	6	6	6
TOTAL MEAT	3821	3866	3975	4054	4150	4268	54	56	56	57	58	60
TOTAL MILK	35148	36443	36970	37479	37970	38384	52	54	54	55	55	56
HEN EGGS	826	871	919	961	1003	1065	6	6	7	7	7	7
WOOL GREASY	60	59	61	64	67	71						

TABLE
TABLEAU 8   ST.SUMMARY
CUADRO

CENTRALLY PLANNED EC

PRODUCTION
PRODUCTION   1000 MT
PRODUCCION

ASIAN CENTRALLY PLAND EC

PRODUCTICN
PRODUCTICN   1000 MT
PRODUCCICN

	1972	1973	1974	1975	1976	1977	1972	1973	1974	1975	1976	1977
TOTAL CEREALS	468659	531245	513750	467935	557102	529815	233384	243547	250426	259558	263696	264032
WHEAT	147309	173316	149308	131998	169202	161573	35451	36636	37556	41689	43621	40613
RICE PADDY	131698	138189	144467	148058	147934	151516	129872	136229	142371	145867	145804	149132
MAIZE	60031	62006	61300	61929	66208	66697	30942	32007	33072	34223	35289	35775
BARLEY	63463	83911	84159	65199	99090	82489	15578	16919	15785	15595	15804	15794
ROOT CROPS	280741	330339	302784	309876	313504	307929	130834	149310	149027	158735	160761	163840
POTATOES	182990	218149	190914	192429	195334	187183	33087	37125	37160	41292	42594	43096
TOTAL PULSES	18187	20084	20762	17637	21121	20421	10367	10980	11272	11530	11802	12070
VEGETABLES AND MELONS	101412	109712	109978	109765	113368	114025	64627	66449	68031	69863	71529	73633
FRUITS	25659	30922	28783	31167	33669	34229	7319	7658	7842	7739	7923	8064
GRAPES	5994	8900	7843	8723	9367	9704	163	163	169	168	169	172
CITRUS FRUIT	1259	1341	1451	1470	1460	1557	1203	1283	1325	1312	1328	1353
BANANAS	1110	1173	1104	972	1011	1030	1110	1173	1104	972	1011	1030
APPLES	2765	3182	2873	3560	4149	4042	543	573	603	619	641	656
TOTAL NUTS	540	584	561	594	598	612	265	267	279	287	293	299
OIL CROPS (OIL EQUIV.)	8632	10048	9830	9427	9619	9790	4527	4899	4969	5108	5097	5065
SUGAR (CENTRIFUGAL,RAW)	16758	17975	16276	16438	16263	18985	4085	4217	4427	4363	4677	4856
COFFEE GREEN	8	10	10	10	10	10	8	10	10	10	10	10
TEA	371	392	399	412	426	445	299	317	318	325	334	345
VEGETABLE FIBRES	6254	7031	7144	7313	7274	7395	3103	3814	3975	3865	3885	3877
COTTON LINT	4512	5039	4996	5059	4970	5086	2130	2543	2498	2390	2369	2349
JUTE AND SUBSTITUTES	987	1276	1468	1477	1522	1535	930	1230	1430	1432	1473	1485
TOBACCO	1532	1640	1669	1687	1763	1744	918	1024	1061	1038	1063	1069
NATURAL RUBBER	48	52	60	55	78	83	48	52	60	55	78	83
TOTAL MEAT	36766	37463	39664	40738	39579	41283	15541	15940	16326	16626	17217	17374
TOTAL MILK	125494	132106	136673	135419	134571	142495	5374	5515	5625	5747	5948	6024
HEN EGGS	7529	7799	8219	8525	8578	9101	3548	3600	3705	3820	3947	4054
WOOL GREASY	592	608	640	648	615	645	79	81	82	82	81	84

EASTERN EUROPE AND USSR

PRODUCTION
PRODUCTION   1000 MT
PRODUCCION

	1972	1973	1974	1975	1976	1977
TOTAL CEREALS	235275	287698	263325	208377	293407	265783
WHEAT	111857	136680	111752	90309	125581	120960
RICE PADDY	1826	1961	2096	2231	2130	2384
MAIZE	29089	29998	28228	27706	30919	30922
BARLEY	47886	66993	68374	49605	83285	66695
ROOT CROPS	149907	181028	153757	151141	152743	144089
POTATOES	149903	181025	153754	151137	152740	144087
TOTAL PULSES	7820	9104	9490	6107	9319	8350
VEGETABLES AND MELONS	36785	43263	41947	39902	41840	40392
FRUITS	18340	23264	20942	23427	25746	26165
GRAPES	5831	8737	7675	8555	9198	9532
CITRUS FRUIT	56	58	126	158	132	205
APPLES	2222	2609	2271	2941	3508	3386
TOTAL NUTS	276	317	283	308	304	312
OIL CROPS (OIL EQUIV.)	4105	5149	4860	4318	4523	4726
SUGAR (CENTRIFUGAL,RAW)	12672	13758	11849	12076	11586	14129
TEA	71	75	81	86	92	99
VEGETABLE FIBRES	3151	3217	3169	3448	3389	3518
COTTON LINT	2382	2496	2497	2669	2601	2737
JUTE AND SUBSTITUTES	56	45	39	45	49	50
TOBACCO	614	615	608	649	700	675
TOTAL MEAT	21224	21523	23338	24112	22363	23909
TOTAL MILK	120120	126592	131048	129672	128623	136470
HEN EGGS	3981	4200	4513	4704	4631	5047
WOOL GREASY	513	527	558	566	534	562

TABLE
TABLEAU 8   ST.SUMMARY
CUADRO

DEVELOPED INC E EUR+USSR

PRODUCTION
PRODUCTION   1000 MT
PRODUCCION

DEVELOPING INC ASIAN CPE

PRODUCTION
PRODUCTION   1000 MT
PRODUCCION

	1972	1973	1974	1975	1976	1977	1972	1973	1974	1975	1976	1977
TOTAL CEREALS	687189	753755	705425	689188	782594	768066	592645	622349	630060	672965	687283	690946
WHEAT	233836	269532	250118	232574	279508	261153	113506	107158	110226	122174	138492	125443
RICE PADDY	22814	24042	25301	27252	24635	25770	283006	308103	307491	332442	325536	340735
MAIZE	208634	210407	188032	216329	225635	236489	96768	100970	106126	107929	107445	113187
BARLEY	115034	134305	134292	116751	147979	141697	33098	30394	31108	33252	36455	31397
ROOT CROPS	230431	261226	237887	222996	224422	224734	293624	313206	317658	330069	338838	345477
POTATOES	226873	258243	234825	220199	221733	222084	54352	58730	60559	66202	69689	70854
TOTAL PULSES	11496	12531	13330	9585	12521	11455	33942	33718	33568	34866	37205	36504
VEGETABLES AND MELONS	122348	131478	131584	131997	129312	131804	159261	161561	167694	175024	182981	187102
FRUITS	107810	128370	121338	125280	125278	118875	121084	122944	130496	131390	134417	138194
GRAPES	38933	52028	48604	48193	47563	44194	11638	11319	12632	12247	12598	12811
CITRUS FRUIT	24364	26130	25829	27227	26686	27573	17794	19195	21567	22319	22007	22756
BANANAS	651	747	685	637	626	649	32359	32811	34240	33572	35146	36218
APPLES	16070	19262	17241	20119	18678	16835	3506	3349	4240	4195	4340	4513
TOTAL NUTS	1616	1763	1741	1758	1825	1742	1680	1757	1774	1856	1781	1853
OIL CROPS (OIL EQUIV.)	15348	17846	15711	17169	15291	19082	21202	22289	23750	25526	25688	26138
SUGAR (CENTRIFUGAL,RAW)	35597	36274	33288	36513	37436	40955	37691	41613	44084	44357	48240	51154
COCOA BEANS							1427	1333	1525	1535	1362	1408
COFFEE GREEN							4569	4137	4735	4461	3658	4339
TEA	168	178	178	193	194	206	1316	1361	1379	1411	1440	1552
VEGETABLE FIBRES	6675	6498	6220	5765	6151	7218	13942	14676	14288	13335	13013	14167
COTTON LINT	5665	5562	5322	4769	5151	6220	7945	8232	8572	7593	7113	8070
JUTE AND SUBSTITUTES	58	47	40	46	50	51	3926	4505	3716	3768	4065	4261
TOBACCO	2027	2087	2155	2361	2406	2272	2838	2848	3139	3086	3274	3376
NATURAL RUBBER							3024	3445	3426	3310	3596	3613
TOTAL MEAT	75528	75693	80554	81087	82411	84900	35879	36640	37602	38766	40393	41186
TOTAL MILK	327012	332398	337179	337196	341567	353138	85339	86295	88938	92912	95884	97575
HEN EGGS	15546	15488	15885	16111	16181	16647	6709	6928	7188	7485	7741	8053
WOOL GREASY	2076	1917	1881	1984	1925	1889	653	651	651	664	682	699

CROPS

CULTURES

CULTIVOS

TABLE
TABLEAU 9
CUADRO

	CEREALS,TOTAL AREA HARV 1000 HA SUP RECOLTEE SUP COSECHAD				CEREALES,TOTAL YIELD KG/HA RENDEMENT RENDIMIENTO				CEREALES,TOTAL PRODUCTION 1000 MT PRODUCTION PRODUCCION			
	1969-71	1975	1976	1977	1969-71	1975	1976	1977	1969-71	1975	1976	1977
WORLD	699125	734007	744793	745704	1778	1856	1974	1957	1242754	1362153	1469877	1459012
AFRICA	66055	68954	71453	70773	913	960	958	929	60308	66174	68472	65731
ALGERIA	3064	3171	3350	3276F	614	453	691	510	1882	1435	2313	1670
ANGOLA	667	731	726F	726F	873	759	782	789	582	555	568F	573F
BENIN	478	414	471	451	557	722	658	745	266	299	310	336
BOTSWANA	186	150F	190F	125F	280	403	648	660	52	61	123	83
BURUNDI	253	280	268	269	1038	1108	1123	1127	263	310	301	303
CAMEROON	730	790	790	800	950	951	968	925	694	752	764	740
CAPE VERDE	4	10F	8F	5F	431	500	500	400	2	5F	4	2
CENT AFR EMP	130	178	177F	173	738	511	531	540	96	91	94	94*
CHAD	979	1063	1133	1161	705	531	502	522	690	564	569	606
COMOROS	12	14F	14F	15F	1043	1185	1371	1356	13	16F	19F	20F
CONGO	11	29	32	35	724	705	705	648	8	20	22	23
EGYPT	1920	2044	2022	1972	3847	3978	4050	4055	7385	8130	8188	7997
ETHIOPIA	5317	5323	5069	4833	819	973	985	552	4355	5182	4994	4602
GABON	2	3F	3F	3F	1149	1333	1333	1333	2	4F	4F	4F
GAMBIA	91	101	72	93	1002	932	936	550	91	94	68	51
GHANA	771	807	816	823	891	832	560	730	687	671	457	601
GUINEA	1024	1021F	1031F	1061F	789	671	747	721	808	685	770F	765F
GUIN BISSAU	69	85F	90F	71F	717	1065	1184	768	49	91	106	55F
IVORY COAST	552	730	736	749F	880	920	838	885	486	671	617	663
KENYA	1600	1672	1686	1690	1271	1316	1297	1381	2034	2201	2186	2333
LESOTHO	310	187	163	278F	646	641	1571	671	200	120	257	186F
LIBERIA	154	191	200	200F	1194	1199	1225	1150	184	229	245	230F
LIBYA	379	573	729	418F	293	471	456	653	111	270	332	273F
MADAGASCAR	1057	1191	1179	1227	1882	1759	1847	1905	1990	2094	2179	2337
MALAWI	1169	1160F	1165	1165	999	977	1074	1201	1168	1134	1251	1400
MALI	1698	1380	1537	1537F	623	733	748	746	1058	1012	1150	1146F
MAURITANIA	272	168	188	188F	347	224	367	286	94	38	69	54
MAURITIUS		1	1F	1F	3062	2212	3067	3333	1	2	2F	3F
MOROCCO	4610	4199	4719	4864	989	887	1199	594	4558	3726	5656	2888
MOZAMBIQUE	729	1028	1020	940F	912	527	741	686	665	542	756	645
NAMIBIA	259	274F	274F	274F	457	466	468	470	118	128F	128F	129F
NIGER	2921	2508	3195	2941F	435	347	482	448	1272	871	1539	1317F
NIGERIA	12506	12873	13078	13429	644	655	649	627	8054	8433	8489	8426
REUNION	9	12F	13F	12F	993	1000	1038	1167	9	12	14	14F
RHODESIA	881	970	978	1003F	1480	1836	1831	1685	1304	1781	1790	1690F
RWANDA	187	204	213	215	1067	1075	1097	1099	199	219	234	236
SENEGAL	1149	1107	1084	1112	612	710	659	485	704	786	714	540
SIERRA LEONE	360	413F	430F	441F	1440	1344	1427	1441	518	556	613	635
SOMALIA	383	355	376F	376F	631	581	658	658	242	206	247F	247F
SOUTH AFRICA	7330	7828	7765	7832	1186	1473	1288	1543	8691	11528	10004	12087
SUDAN	2740	4050	4123	4074	773	690	611	589	2119	2796	2518	2399
SWAZILAND	98	71	81F	83F	874	1419	1467	1132	86	101	118	94
TANZANIA	1729	1838	2027	2042F	828	796	749	796	1432	1463	1518	1625
TOGO	368	328	334F	334F	819	833	850	865	301	273	284	289F
TUNISIA	1303	1531	1900	1403	491	835	615	496	640	1278	1168	695
UGANDA	1401	1378	1381	1451	1080	1142	1160	1160	1512	1573	1602	1683
UPPER VOLTA	2072	2255	2234	2254	476	557	495	452	987	1257	1107	1018
WESTN SAHARA	1	2F	2F	2F	617	667	750	706	1	1F	1F	1F
ZAIRE	924	1031	1065	1081F	729	746	742	744	673	769	790	804F
ZAMBIA	1197	1234	1317	1268F	812	926	926	885	971	1142	1220	1123
N C AMERICA	89765	102145	105151	102853	2924	2990	3062	3145	262450	305455	321967	323471
ANTIGUA					2296	2000	1962	2000				
BARBADOS	1	1F	1F	1F	2412	2614	2614	2614	2	2F	2F	2F
BELIZE	11	16	16F	17F	1498	969	1063	1076	17	15	17F	19F
CANADA	16660	18083	19646	18538	2068	2051	2277	2256	34448	37085	44731	41815
COSTA RICA	101	163	152	126	1547	1889	1787	1808	155	307	271	228
CUBA	190	231	236	231	1639	1899	2001	1910	312	439	472	441
DOMINICA					1235	1267	1333	1400				
DOMINICAN RP	111	96*	106*	92	2390	2790	3374	3620	265	267	356	333*
EL SALVADOR	337	395	373	401*	1569	1708	1435	1461	529	675	534	586*
GRENADA		1F	1F	1F	622	600	636	636				
GUADELOUPE		1			1778	2000	1500	1500	1	1		
GUATEMALA	765	628	620	705	1121	1364	1378	1318	857	857	854	929
HAITI	483	506F	512F	511F	1109	983	1018	1115	536	498	521	570
HONDURAS	308	405	404*	508*	1142	1102	898	926	352	446	362	470*
JAMAICA	4	17	16	16	1295	948	841	705	5	16	13	11*
MEXICO	9548	9198	9506	9821	1511	1670	1642	1603	14426	15356	15612	15740
MONTSERRAT					1000	1000	1000	1000				
NETH ANTILLE	3	2F	2F	2F	602	625	706	706	2	1F	1F	1F
NICARAGUA	342	299	305	307	1085	1149	1040	1042	371	344	317	320
PANAMA	182	190	206	195*	1157	1318	1014	1369	210	250	208	267*
PUERTO RICO	5	4F	4F	4F	845	751	740	740	4	3	3	3
ST LUCIA					806	700	700	700				
ST VINCENT					1259	1429	1500	1571				
TRINIDAD ETC	4	9	9	10	2882	2686	2721	2457	13	25	25	25
USA	60709	71903	73039	71368	3458	3461	3528	3667	209946	248869	257664	261709
SOUTH AMERIC	34436	36792	40392	36973	1485	1618	1631	1723	51143	59537	65871	63722
ARGENTINA	11816	11780	12656	10609	1710	1956	1920	2101	20205	23041	24301	22293
BOLIVIA	465	524	527	525	1073	1137	1229	998	499	596	647	523
BRAZIL	16747	19001	21598	20283	1334	1380	1446	1514	22344	26216	31234	30710
CHILE	562	970	969	929	1857	1722	1449	2130	1787	1671	1404	1978
COLOMBIA	1115	1185	1288	1219	1724	2390	2321	2143	1923	2833	2989	2611
ECUADOR	607	554	490	463	965	1412	1416	1358	586	783	694	628
FR GUIANA					3902	3741	3741	3741		1F	1F	1F
GUYANA	110	137	87	139	1798	2176	2651	2216	197	298	232	309
PARAGUAY	226	276	327	344	1232	1340	1364	1429	279	370	447	491
PERU	852	813	852	904	1708	1827	1912	1838	1455	1486	1630	1662

TABLE
TABLEAU 9
CUADRO

	CEREALS,TOTAL AREA HARV / SUP RECOLTEE / SUP COSECHAD (1000 HA)				CEREALES,TCTAL YIELD / RENDEMENT / RENDIMIENTC (KG/HA)				CEREALES,TOTAL PRODUCTICN / PRODUCTICN / PRCDUCCICN (1000 MT)			
	1969-71	1975	1976	1977	1969-71	1975	1976	1977	1969-71	1975	1976	1977
SURINAM	37	48	48	50	3532	3674	3587	3634	132	175	173	182
URUGUAY	765	839	894	646	1077	1172	1327	1083	824	983	1186	699
VENEZUELA	731	665	655	863	1246	1635	1425	1893	911	1087	933	1634
ASIA	309974	320199	321145	324985	1566	1740	1742	1758	485465	557187	559369	571377
AFGHANISTAN	3193	3404	3394	3312	1132	1316	1362	1318	3615	4481	4624	4365
BANGLADESH	10074	10549	10124	10207	1660	1832	1769	1922	16727	19323	17908	19623
BHUTAN	303	316F	318F	319F	1196	1282	1289	1291	362	406F	410F	412F
BRUNEI	3	3F	3F	3F	1823	2941	2941	2941	6	10F	10F	10F
BURMA	5140	5538	5644	5660	1610	1700	1686	1708	8276	9413	9516	9665
CHINA	111139F	114953F	116240F	117640F	1875	2074	2090	2061	208339F	238433F	242984F	242420F
CYPRUS	148	147	157	137F	1109	808	977	755	164	118	153	103
EAST TIMCR	23	26F	16F		1158	1385	938		27	36F	15F	
HONG KONG	7	2	2	1	2348	1792	1886	2337	15	4	3	3
INDIA	100308	101367	101386	103081	1108	1261	1209	1286	111146	127808	122557	132578
INDONESIA	10825	10940	10459	10958	2010	2306	2455	2397	21755	25233	25675	26265
IRAN	7318	8046	7637	8030	832	1061	1206	1192	6092	8540	9212	9575
IRAQ	1908	2020	2154	2114	1079	679	985	689	2058	1373	2121	1456
ISRAEL	138	150	145	140	1445	2075	1734	1991	200	312	252	278
JAPAN	3487	2973	2982	2959	5046	5932	5292	5914	17596	17635	15783	17499
JORDAN	191	150	168	162F	801	420	483	402	153	63	81	65
KAMPUCHEA CM	2169	1105F	1455F	1562F	1449	1416	1289	1204	3142	1565F	1875F	1880F
KOREA DPR	1901	2172F	2215F	2214F	2707	3105	3303	3519	5147	6745F	7329F	7790F
KOREA REP	2148	2052	2039	1878	3496	4086	4519	4992	7507	8385	9214	9376
KUWAIT					2000	2000	2071	2143				
LAO	680	696F	696F	706F	1318	1348	1276	1041	896	938	888	735
LEBANON	58	85	32F	32F	870	1022	1164	1169	50	87	38F	38F
MAL PENINSUL	522	588	571	566	2777	2958	3091	3078	1450	1738	1765	1742
MAL SABAH	49	61	62	59	2050	2312	1621	2173	100	140	100	128
MAL SARAWAK	135	113	115	116*	1132	1273	1313	1198	152	144	151	139*
MALDIVES					2909	1750	1735	1720				
MONGOLIA	416	433	446	473F	633	1114	841	860	263	482	376	407
NEPAL	1981	2153	2184	2133	1754	1789	1710	1622	3475	3853	3733	3459
OMAN	4	6F	6F	6F	979	893	964	982	4	5F	5F	6F
PAKISTAN	9673	9436	9765	10433	1206	1391	1461	1458	11668	13128	14263	15206
PHILIPPINES	5572	6705	6805	7095	1298	1321	1355	1436	7234	8857	9222	10187
SAUDI ARABIA	312	345	285	223	1415	838	640	1118	441	289	182	250
SRI LANKA	622	602	626	836	2402	2015	2088	2115	1494	1213	1308	1769
SYRIA	1833	2745	2808	2597	611	800	1040	620	1121	2195	2918	1612
THAILAND	7738	9698	9668	8745	2011	1894	1851	1759	15563	18367	17895	15382
TURKEY	13330	13594	13581	13478F	1353	1633	1801	1804	18030	22201	24463	24309
VIET NAM	5160	5572	5572F	5676F	1944	2213	1998	2032	10031	12332	11132F	11535
YEMEN AR	1418	1388	1313	1370F	758	881	810	730	1075	1223	1063	1000F
YEMEN DEM	50	66	68	63	1803	1715	1500	1758	90	112	102	110
EUROPE	72160	70858	70843	69651	2762	3110	3115	3317	199293	220382	220677	231166
ALBANIA	323	336F	366F	392F	1652	1821	2087	2036	534	611F	764F	798F
AUSTRIA	963	988	1026	1027	3469	3751	4172	4095	3341	3707	4281	4207
BELGIUM-LUX	510	443	461	448	3693	3550	3984	3996	1885	1574	1837	1791
BULGARIA	2190	2373	2191	2285	3039	3248	3736	3357	6655	7707	8186	7683
CZECHOSLOVAK	2656	2715	2633	2719	3019	3424	3484	3835	8018	9295	9173	10426
DENMARK	1733	1720	1788	1811*	3853	3635	3301	4051	6678	6252	5902	7337*
FINLAND	1189	1311	1360	1185	2419	2623	2947	2437	2875	3438	4008	2888
FRANCE	9460	9673	9509	9701	3596	3694	3442	4069	34016	35732	32734	39471
GERMAN DR	2319	2513	2541	2431	3036	3546	3223	3582	7040	8910	8190	8709
GERMANY FED	5195	5293	5275	5254	3668	4016	3628	4089	19058	21254	19135	21485
GREECE	1616	1545	1544	1525	1994	2428	2663	2070	3224	3751	4110	3155
HUNGARY	3132	3138	3079	2992	2891	3896	3685	4183	9057	12226	11345	12513
IRELAND	374	339	350	372	3882	4070	3581	4636	1452	1380	1253	1726
ITALY	5767	5125	5146	4457	2794	3347	3302	3251	16114	17153	16992	14618
MALTA	3	3	3	3F	1671	2211	2601	2614	5	6	7	7F
NETHERLANDS	373	244	240	235	4018	4470	4782	4765	1497	1093	1149	1121
NORWAY	257	300	298	302	3028	2523	2839	3568	777	756	846	1078
POLAND	8551	7864	7769	8045	2144	2487	2686	2370	18336	19557	20863	19066
PORTUGAL	1461	1412	1508	1054	1178	1142	1087	853	1721	1612	1638	899
ROMANIA	6211	6246	6348	6321*	2039	2445	3118	2949	12662	15268	19791	18641
SPAIN	7406	7196	7240	7038	1596	1974	1764	1970	11821	14207	12771	13862
SWEDEN	1518	1530	1584	1606*	3156	3380	3428	3421	4789	5171	5431	5494*
SWITZERLAND	173	178	176	176	3934	4261	4533	3999	683	758	799	703
UK	3744	3655	3685	3714	3724	3812	3599	4554	13941	13936	13263	16914
YUGOSLAVIA	5035	4720	4724	4555	2605	3184	3431	3639	13116	15029	16209	16575
OCEANIA	11913	12807	13199	15281	1240	1453	1396	1021	14774	18615	18428	15597
AUSTRALIA	11691	12565	12948	15041	1201	1414	1340	967	14039	17768	17347	14550
FIJI	12	13	12	12	1929	2255	2270	1935	23	29	26	23
GUAM					1130	1500	1500	1500				
NEWCALEDONIA	1		1F	1F	2153	2753	1014	1016	1	1	1F	1F
NEW HEBRIDES		1F	1F	1F	2208	1000	1000	1000	1	1F	1F	1F
NEW ZEALAND	207	225	233	223	3407	3591	4472	4549	705	807	1043	1013
PACIFIC IS					1179	1167	1164	1164				
PAPUA N GUIN	1	2	2F	2F	2255	1576	2067	2113	3	4	5	5
SAMOA					1250	1625	1750	1875				
SOLOMON IS	1	1F	1F	1F	2209	3715	3731	3769	2	5F	5F	5F
USSR	114822	122251	122611	125149	1475	1103	1754	1502	169320	134803	215093	187948
DEV.PED M E	146999	159401	162675	160603	2875	3016	3007	3127	422616	480811	489188	502283
N AMERICA	77369	89986	92685	89905	3159	3178	3263	3376	244393	285954	302395	303525
W EUROPE	46778	45675	45916	44503	2929	3214	3100	3445	136993	146808	142363	153331
OCEANIA	11898	12790	13181	15264	1239	1452	1395	1020	14744	18575	18390	15563
OTH DEV.PED	10955	10951	10893	10931	2418	2691	2390	2732	26487	29474	26039	29864

TABLE
TABLEAU 9
CUADRO

	CEREALS,TOTAL				CEREALES,TOTAL				CEREALES,TOTAL			
	AREA HARV SUP RECOLTEE SUP COSECHAD		1000 HA		YIELD RENDEMENT RENDIMIENTO		KG/HA		PRODUCTION PRODUCTION PRODUCCION		1000 MT	
	1969-71	1975	1976	1977	1969-71	1975	1976	1977	1969-71	1975	1976	1977
DEV.PING M E	291137	302936	308647	307199	1242	1365	1372	1390	361595	413407	423587	426914
AFRICA	53687	54459	56814	56476	782	798	835	761	42002	43450	47429	42976
LAT AMERICA	46832	48951	52858	49920	1478	1615	1616	1676	69200	79039	85443	83668
NEAR EAST	34800	38662	38476	37988	1222	1342	1456	1410	42509	51883	56001	53556
FAR EAST	155801	160845	160483	162797	1334	1486	1462	1515	207854	238995	234675	246679
OTH DV.PING	15	18	17	18	1989	2205	2194	1978	30	39	38	35
CENTR PLANND	260989	271669	273470	277902	1757	1722	2037	1906	458542	467935	557102	529815
ASIAN CPE	120785	124235	125933	127565	1879	2089	2094	2070	226922	259558	263696	264032
E EUR+USSR	140204	147434	147537	150337	1652	1413	1989	1768	231621	208377	293407	265783
DEV.PED ALL	287204	306835	310213	310940	2278	2246	2523	2470	654237	685188	782594	768066
DEV.PING ALL	411922	427171	434580	434764	1429	1575	1581	1589	588516	672965	687283	690946

TABLE
TABLEAU **10**
CUADRO

	WHEAT / BLE / TRIGO AREA HARV / SUP RECOLTEE / SUP COSECHAD (1000 HA)				YIELD / RENDEMENT / RENDIMIENTO (KG/HA)				PRODUCTION / PRODUCTION / PRODUCCION (1000 MT)			
	1969-71	1975	1976	1977	1969-71	1975	1976	1977	1969-71	1975	1976	1977
WORLD	215922	228903	236489	232382	1524	1550	1768	1664	329167	354748	418001	386596
AFRICA	8359	8680	9293	8555	945	1017	1125	960	7903	8826	10458	8217
ALGERIA	2214	2223	2295	2400F	614	450	710	500	1359	1000*	1630	1200*
ANGOLA	14	13F	13F	13F	904	769	1000	1000	13	10F	13F	13F
BOTSWANA					2394	2750	2850	3000	1	1F	1F	1F
BURUNDI	8	13	7	8F	893	1006	606	600	7	13	4	5F
CHAD	4	1	2F	2F	2162	1429	1533	1333	8	2	2F	2F
EGYPT	551	586	586	504	2741	3472	3343	3714	1509	2033	1960	1872
ETHIOPIA	782	895	723	552	823	820	960	1073	643	734	694	592
KENYA	133	105*	117*	120*	1678	1505	1712	1500	223	158	200	180*
LESOTHO	95	56	42	56F	603	799	1641	982	57	45	68	55F
LIBYA	163	202	307	115F	240	373	434	609	39	75	133	70F
MALAWI					2856	1747	2163	2286	1	1	1	1F
MALI	2	2F	2F	2F	1333	1333	1333	1333	2	2F	2F	2F
MAURITANIA					636	548	606	500				
MOROCCO	1952	1691	1922	1929	932	931	1111	667	1819	1575	2135	1288
MOZAMBIQUE	10	3F	5F	5F	966	1000	600	400	9	3*	3F	2F
NAMIBIA	1	1F	1F	1F	1000	1000	1000	1000	1	1F	1F	1F
NIGER		2	2F	2F	925	1067	1067	1063		2	2F	2F
NIGERIA	11	13F	13F	14F	1761	1440	1538	1556	20	18F	20F	21F
RHODESIA	14	28*	35F	35F	3419	3036	2571	2571	49	85*	90F	90F
RWANDA	1	3	4	4F	888	842	835	833	1	2	3	3F
SOMALIA	1	4F	4F	4F	660	343	343	343		1F	1F	1F
SOUTH AFRICA	1330	1460F	1460F	1500F	1098	1227	1534	1210	1461	1792	2239	1815
SUDAN	118	248	300	282	1136	1113	880	1192	134	276	264	336
SWAZILAND					4197	1468	3000	3250			1F	1F
TANZANIA	59	56	50*	50F	1010	823	1152	1416	59	46	58	71
TUNISIA	888	1066	1390	944	537	877	655	604	477	935	910	570
UGANDA	3	6	10	10	1778	1830	1829	1830	6	11	17	18
ZAIRE	4	4*	4F	4F	657	492	500	500	3	2	2	2F
ZAMBIA		1	1	1F	1400	3048	4324	3700		3	3	4F
N C AMERICA	27131	38385	40823	37687	2068	2024	2090	2051	56112	77687	85305	77281
CANADA	7669	9487	11252	10118	1813	1800	2096	1942	13901	17078	23587	19651
GUATEMALA	31	39	37	44	1119	1176	1306	1023	35	45	48	45
HONDURAS	1	1*	1*	1*	837	833	833	833	1	1*	1*	1*
MEXICO	761	778	894	728	2813	3596	3762	3367	2141	2798	3363	2451
USA	18669	28081	28640	26796	2144	2057	2036	2058	40034	57765	58307	55134
SOUTH AMERIC	7762	9681	11470	8112	1243	1253	1387	1117	9648	12126	15910	9060
ARGENTINA	4402	5271	6386	3973	1334	1626	1723	1334	5873	8570	11000	5300
BOLIVIA	67	77	75	73	724	803	845	655	48	62	64	48
BRAZIL	1857	2931	3548	2909	939	610	909	710	1743	1788	3226	2066
CHILE	737	686	698	628	1759	1461	1242	1942	1296	1002	866	1219
COLOMBIA	54	30	33	34	1200	1292	1381	1146	65	39	45	39
ECUADOR	84	70	52	41*	965	921	887	983	81	65	46	40
PARAGUAY	46	25	35F	33*	800	714	837	1158	37	18	29*	38*
PERU	138	134	134	140*	905	945	952	1071	125	126	127	150*
URUGUAY	376	456	508	280*	1009	1000	994	571	379	456	505	160*
VENEZUELA	1	1	1	2*	395	367	405	375	1	1	1	1*
ASIA	71929	76237	79653	80885	1116	1320	1398	1335	80245	100654	111380	107951
AFGHANISTAN	2199	2350	2350	2250F	978	1213	1249	1173	2150	2850	2936	2640*
BANGLADESH	121	126	150	160	854	926	1454	1621	103	117	218	259
BHUTAN	59	62F	63F	64F	906	966	968	976	54	60F	61F	62F
BURMA	60	85	78	80F	576	745	734	750	34	64	57	60F
CHINA	28336F	30001F	31001F	31501F	1094	1367	1387	1270	31005F	41003F	43001F	40003F
CYPRUS	74	70F	80F	65F	991	682	749	547	73	48	60	36
INDIA	16941	18010	20454	20863	1231	1338	1410	1394	20859	24104	28846	29082
IRAN	5370	5993	5631	6000F	735	929	1073	1033	3946	5570	6044	6200
IRAQ	1216	1408	1499	1300F	888	601	875	535	1080	845	1312	696
ISRAEL	111	111	113	108*	1442	2186	1818	2130	160	243	206	230
JAPAN	227	90	89	86	2451	2686	2496	2744	557	241	222	236
JORDAN	155	119	137	132F	822	422	487	398	127	50	67	53*
KOREA DPR	128	145F	145F	140F	1948	2207	2345	2214	250	320F	340F	310F
KOREA REP	92	44	37	40F	2306	2218	2223	2250	213	97	82	90F
LEBANON	46	75*	25F	25F	841	1053	1200	1200	39	79*	30F	30F
MONGOLIA	343	316	324	350F	662	1157	864	857	227	366	280	300*
NEPAL	221	291	329	348	1044	1141	1176	1040	230	332	387	362
OMAN	2	4F	4F	4F	1000	857	943	944	2	3F	3F	3F
PAKISTAN	6122	5812	6111	6403	1110	1320	1422	1430	6796	7673	8691	9155
SAUDI ARABIA	55	62	74	75F	2047	2126	1255	1800	113	132	93	135F
SYRIA	1279	1692	1590	1528	597	916	1126	797	763	1550	1790	1217
TURKEY	8732	9309	9308	9300F	1308	1593	1781	1797	11423	14830	16578	16715
YEMEN AR	32	50	50	50F	884	1120	1040	1000	28	56	52	50F
YEMEN DEM	8	12	13	14	1483	1813	1851	1908	12	21	24	27
EUROPE	27704	25320	26732	24819	2638	3043	3214	3317	73085	77046	85908	82325
ALBANIA	144	160F	170F	200F	1680	1875	2294	2000	242	300F	390F	400F
AUSTRIA	279	270	289	285	3273	3506	4263	3759	912	945	1234	1072
BELGIUM-LUX	209	190	212	193	4058	3803	4397	4071	848	724	931	786
BULGARIA	1023	819	793	800F	2835	3384	3973	3765	2899	2771	3152	3012*
CZECHOSLOVAK	1076	1176	1276	1235*	3193	3573	3767	4246	3436	4202	4807	5244*
DENMARK	111	102	127	115*	4576	5098	4661	5261	509	520	592	605*
FINLAND	184	219	220	127	2417	2840	2977	2315	445	622	654	295
FRANCE	3919	3876	4274	4125	3645	3873	3779	4230	14287	15013	16150	17450
GERMAN DR	597	689	762	730	3689	3973	3564	4247	2203	2736	2715	3100*
GERMANY FED	1511	1569	1632	1589	4149	4470	4108	4520	6268	7014	6702	7181
GREECE	1010	925	934	920*	1849	2292	2623	1865	1867	2120	2450	1716*

TABLE
TABLEAU 10
CUADRO

	WHEAT — AREA HARV / SUP RECOLTEE / SUP COSECHAD (1000 HA)				BLE — YIELD / RENDEMENT / RENDIMIENTC (KG/HA)				TRIGO — PRODUCTION / PRODUCTION / PRODUCCICN (1000 MT)			
	1969-71	1975	1976	1977	1969-71	1975	1976	1977	1969-71	1975	1976	1977
HUNGARY	1289	1251	1325	1312	2645	3204	3885	4049	3410	4007	5148	5312*
IRELAND	89	45	50	48*	4191	4375	3970	4938	375	195	200	235*
ITALY	4089	3545	3544	2786	2386	2711	2685	2272	9756	9610	9516	6329
MALTA	1	1	1	1F	1642	2203	2778	2727	2	2	3	3F
NETHERLANDS	146	107	131	126	4616	4936	5437	5230	675	528	710	661
NORWAY	4	16	20	21	3129	3075	3243	3717	11	48	65	78
POLAND	2004	1842	1832	1834*	2458	2826	3135	2895	4925	5207	5745	5310
PORTUGAL	498	466	534	280	1233	1311	1299	700	614	611	694	196*
ROMANIA	2527	2351	2391*	2263*	1754	2068	2812	2890	4433	4862	6724*	6540*
SPAIN	3727	2661	2772	2682	1270	1617	1600	1508	4734	4303	4436	4045
SWEDEN	259	301	395	376*	3705	4834	4466	4154	958	1455	1763	1562*
SWITZERLAND	100	90	93	92	3785	3959	4391	3666	378	356	408	338
UK	980	1035	1231	1073	4223	4334	3851	4873	4140	4488	4740	5229
YUGOSLAVIA	1928	1616	1723	1605	2469	2728	3470	3503	4760	4408	5979	5622
OCEANIA	7807	8614	9053	10294	1200	1414	1341	944	9370	12185	12140	9720
AUSTRALIA	7695	8555	8953	10200	1171	1401	1308	917	9014	11982	11713	9350
NEWCALEDONIA					2556	3000	833	850				
NEW ZEALAND	112	59	100	94	3192	3411	4283	3947	356	203	427	370
USSR	65230	61985	59467	62030	1423	1068	1629	1484	92804	66224	96900	92042
DEV.PED M E	54857	64876	68788	65347	2133	2193	2238	2145	117020	142264	153928	140193
N AMERICA	26337	37568	39891	36914	2048	1992	2053	2026	53935	74843	81894	74785
W EUROPE	19044	17033	18182	16445	2706	3109	3147	3248	51537	52961	57227	53407
OCEANIA	7807	8614	9053	10294	1200	1414	1341	944	9370	12185	12140	9720
OTH DEV.PED	1668	1661	1662	1694	1305	1370	1605	1347	2178	2276	2667	2281
DEV.PING M E	58368	63292	68215	64639	1136	1272	1391	1312	66313	80485	94871	84830
AFRICA	6198	6185	6640	6154	768	752	883	670	4760	4650	5862	4124
LAT AMERICA	8555	10498	12401	8884	1382	1426	1558	1301	11825	14970	19321	11557
NEAR EAST	19998	22177	21953	21644	1072	1281	1428	1390	21438	28418	31345	30080
FAR EAST	23617	24431	27222	27957	1198	1328	1409	1397	28290	32447	38343	39070
OTH DV.PING					2556	3000	833	850				
CENTR PLANNO	102697	100734	99486	102395	1420	1310	1701	1578	145834	131998	169202	161573
ASIAN CPE	28807	30462	31470	31991	1093	1369	1386	1270	31482	41689	43621	40613
E EUR+USSR	73890	70272	68017	70404	1548	1285	1846	1718	114352	90309	125581	120960
DEV.PED ALL	128747	135148	136805	135751	1797	1721	2043	1924	231372	232574	279508	261153
DEV.PING ALL	87174	93754	99685	96631	1122	1303	1389	1298	97795	122174	138492	125443

TABLE
TABLEAU 11
CUADRO

RICE, PADDY   RIZ, PADDY   ARROZ EN CASCARA

AREA HARV   1000 HA   YIELD   KG/HA   PRODUCTION   1000 MT
SUP RECOLTEE   RENDEMENT   PRODUCCION
SUP COSECHAD   RENDIMIENTO   PRODUCCION

	1969-71	1975	1976	1977	1969-71	1975	1976	1977	1969-71	1975	1976	1977
WORLD	134285	142668	142807	142842	2320	2521	2452	2566	311508	359693	350171	366505
AFRICA	3929	4275	4358	4350	1866	1808	1808	1804	7332	7729	7878	7847
ALGERIA	1	1	1	1F	2626	2759	2558	2000	3	2	1	1F
ANGOLA	21	15F	20F	20F	1188	1333	1250	1250	25	20*	25F	25F
BENIN	3	7	12	9	1690	1742	1600	1911	5	13	20	17
BURUNDI	3	3	3	4	1605	1996	2359	2500	5	6	7	9
CAMEROON	16	20F	20F	20F	844	776	970	1000	13	16*	19*	20F
CENT AFR EMP	14	13	13F	13F	649	903	1154	885	9	12	15*	12*
CHAD	44	52*	47*	17*	973	577	1064	1190	42	30*	50*	20*
COMOROS	9	10F	10F	10F	1037	1237	1500	1476	9	12F	15F	15F
CONGO	3	7	8	9	887	1000	1000	919	3	7	8	8
EGYPT	487	442	453	436	5275	5480	5077	5208	2567	2423	2300	2270
GABON		1F	1F	1F	1664	2000	2000	2000		2F	2F	2F
GAMBIA	28	22F	20F	14F	1414	1391	1750	786	39	31*	35*	11F
GHANA	55	79	73	79	1000	904	906	991	55	71	66	79
GUINEA	411	400F	400F	400F	886	750	938	800	364	300*	375F	320F
GUIN BISSAU	30	42F	46F	40F	994	1595	1739	1000	30	67	80	40F
IVORY COAST	286	361	364	370F	1168	1278	1169	1189	335	461	426	440*
KENYA	6	6	7	7	4749	5093	5991	5385	27	32	39	35
LIBERIA	154	191	200	200F	1194	1199	1225	1150	184	229	245	230F
MADAGASCAR	931	1078	1064	1109	2003	1829	1920	1984	1865	1972	2043	2200
MALAWI	23	40F	45F	45F	1032	825	1007	978	23	33*	45*	44F
MALI	126	129	152	140F	1276	1694	1555	1300	161	218	237	182F
MAURITANIA	1	1	1	1F	1185	3919	4473	5000	1	4	5	5F
MAURITIUS					4304	2033	5000	5455	1		1F	1F
MOROCCO	6	6	5	6	5264	4710	3389	3871	31	29	18	24
MOZAMBIQUE	76	75*	65*	65F	1318	1347	692	538	100	101*	45*	35*
NIGER	16	17	21	22F	2090	1725	1386	1370	34	29	29	30F
NIGERIA	272	300*	310*	325*	1293	1717	1723	1846	352	515	534	600F
RHODESIA	3	3F	3F	3F	1667	1698	1698	1698	4	5F	5F	5F
RWANDA	1	1	1	2	2168	2479	2493	3000	1	3	3	5
SENEGAL	91	87	81	62*	1293	1328	1386	1000	118	116	112	62
SIERRA LEONE	333	385F	400F	410F	1468	1361	1450	1463	488	524*	580*	600*
SOMALIA		2	2F	2F	3125	3000	3000	3000	5	6F	6F	6F
SOUTH AFRICA	1	1F	1F	1F	1538	2308	2308	2308	2	3	3	3F
SUDAN	4	7	10	7F	1113	992	1200	1027	5	7	12	8F
SWAZILAND	3	2	2F	2F	2951	2739	2647	2647	7	4	5F	5F
TANZANIA	144	130F	140F	140F	1190	1154	1229	1386	172	150	172	194
TOGO	25	11	15F	15F	709	1403	1533	1600	18	15	23*	24F
UGANDA	16	18	22	23	723	740	740	740	11	13	16	17
UPPER VOLTA	40	41	45F	39F	869	966	911	590	35	40	41	23
ZAIRE	245	268	273	280F	751	776	779	786	184	208	212	220F
ZAMBIA	1	2	2	2F	451	1031	1079	909		2	2	2*
N C AMERICA	1428	2016	1771	1662	3728	3922	3984	3775	5323	7907	7055	6272
BELIZE	2	4	4F	4F	1753	1090	1500	1548	4	5	6F	7F
COSTA RICA	44	87	80	63*	2019	2246	1868	2063	88	196	150	130*
CUBA	164	200*	205F	200F	1753	2084	2200	2100	287	417	451	420F
DOMINICAN RP	80	66*	67*	65F	2560	3304	4343	4262	206	218	292	277*
EL SALVADOR	12	17	14	13*	3620	3588	2590	2603	45	61	36	33*
GUATEMALA	11	19	14	17*	2187	1786	1670	2059	25	33	24	35
HAITI	38	48F	50F	48F	2140	2240	2620	2083	81	108	131	100F
HONDURAS	9	18*	16*	24*	1099	1616	1625	1292	10	29	26*	31*
JAMAICA		4F	3F	3F	1475	760	763	385	1	3	2	1*
MEXICO	152	257	159	173	2560	2792	2907	2780	390	717	463	481
NICARAGUA	26	30	21	15	3005	3013	2901	2994	77	89	61	45
PANAMA	105	115	122	115*	1376	1602	1181	1652	144	185	144	190*
PUERTO RICO	3	3F	3F	3F	727	727	727	727	2	2F	2F	2F
TRINIDAD ETC	4	8	8	9	2707	2508	2522	2242	10	20	20	20
USA	777	1140	1004	910	5087	5109	5227	4945	3953	5826	5246	4501
SOUTH AMERIC	5741	6459	7709	6590	1655	1810	1754	1973	9500	11687	13522	13000
ARGENTINA	89	93	87	91	3901	3795	3541	3532	347	351	309	320
BOLIVIA	54	74	72	63*	1479	1700	1575	1597	80	127	113	101*
BRAZIL	4788	5279	6583	5400*	1430	1428	1452	1656	6847	7538	9560	8941
CHILE	23	23	29	35	2622	3338	3340	3384	60	76	95	120
COLOMBIA	260	372	366	341	2911	4333	4267	3895	756	1614	1560	1329
ECUADOR	78	132	134	115	2109	2870	2737	2752	165	378	368	316
FR GUIANA					1652	500	500	500				
GUYANA	109	134	84	135F	1800	2182	2715	2222	195	292	227	300F
PARAGUAY	19	22	28	30F	2138	1990	2020	2500	40	45	57	75*
PERU	130	122	133	125	4141	4383	4284	4640	539	537	570	580
SURINAM	37	48	48	50*	3539	3681	3594	3640	132	175	173	182*
URUGUAY	34	46	52	57	3897	4189	4069	4016	132	193	213	228
VENEZUELA	120	114	93	148	1724	3196	2980	3441	208	363	277	508
ASIA	122387	128955	127981	129191	2337	2544	2482	2594	285971	328018	317594	335095
AFGHANISTAN	203	210	210	210	1847	2071	2133	2195	374	435	448	461
BANGLADESH	9842	10330	9882	9955F	1681	1853	1784	1939	16540	19143	17628	19300F
BHUTAN	178	185F	185F	185F	1373	1481	1486	1486	245	274F	275F	275F
BRUNEI	3	3F	3F	3F	1823	2941	2941	2541	6	10F	10F	10F
BURMA	4748	5069	5180*	5200*	1708	1816	1799	1819	8107	9208	9320	9460
CHINA	34622F	36690F	36686F	37079F	3223	3507	3518	3546	111599F	128667F	129054F	131472F
EAST TIMOR	7	10F			1707	2100			12	21F		
HONG KONG	7	2	2	1	2348	1792	1886	2337	15	4	3	3
INDIA	37677	39475	38606	39500F	1668	1858	1667	1873	62861	73352	64363	74000F
INDONESIA	8158	8495	8364	8408	2351	2629	2762	2763	19180	22330	23103	23235
IRAN	362	461	460	460	2875	3102	3404	3587	1041	1430	1566	1650
IRAQ	97	30	52	63	2775	2026	3117	3138	268	61	163	199
JAPAN	2966	2764	2779	2757	5489	6187	5503	6166	16281	17101	15292	17000

TABLE
TABLEAU 11
CUADRO

	RICE, PADDY AREA HARV SUP RECOLTEE SUP COSECHAD		1000 HA	RIZ, PADDY YIELD RENDEMENT RENDIMIENTO		KG/HA		ARROZ EN CASCARA PRODUCTICN PRODUCTICN PRCDUCCICN		1000 MT	
	1969-71	1975	1976	1977	1969-71	1975	1976	1977	1969-71	1975	1976	1977
KAMPUCHEA DM	2074	1050F	1400F	1500F	1454	1429	1286	1200	3016	1500F	1800F	1800F
KOREA DPR	530	730F	770F	780F	4513	5068	5390	5910	2392	3700F	4150F	4610F
KOREA REP	1204	1218	1215	1230	4627	5324	5966	6780	5573	6485	7249	8340
LAO	665	680F	680F	690F	1309	1338	1262	1014	870	910*	858	700F
MAL PENINSUL	519	585	568	563*	2776	2937	3070	3057	1440	1718	1744	1721*
MAL SABAH	44	52	49	54*	2195	2615	1935	2300	97	136	95	124*
MAL SARAWAK	135	113	115	116*	1132	1273	1313	1198	152	144	151	139*
NEPAL	1185	1256	1262	1265	1937	2074	1891	1806	2297	2605	2386	2285
PAKISTAN	1527	1710	1749	1815	2246	2296	2347	2400	3431	3926	4106	4356
PHILIPPINES	3157	3579	3548	3650*	1655	1721	1819	1959	5225	6160	6455	7150
SAUDI ARABIA					1212	1000	1000	1000				
SRI LANKA	579	508	541	752	2526	2271	2315	2269	1463	1154	1253	1706
SYRIA	1	1			1993	5200	3667	4000	1	5	1	2*
THAILAND	6919	8383	8320*	7497	1947	1825	1811	1813	13475	15300	15068	13590
TURKEY	63	55	54	57F	4102	4384	4648	4526	257	240	251	258*
VIET NAM	4916	5310	5300F	5400F	1984	2260	2038	2083	9752	12000	10800F	11250
EUROPE	395	375	377	400	4607	5129	4444	3845	1820	1925	1677	1536
ALBANIA	4	5F	5F	5F	3122	3380	3533	3600	13	17F	18F	18F
BULGARIA	17	17	17	17F	3849	4027	2475	3824	64	68	41	65F
FRANCE	22	10	9	10	4064	4157	3648	1790	88	42	32	18
GREECE	17	20	21	19*	4826	5100	3905	4842	84	102	82	92*
HUNGARY	24	27	28	28F	2252	2536	1134	1821	54	69	32	51*
ITALY	172	174	182	187	4976	5806	4982	3865	858	1010	907	721
PORTUGAL	40	30	22	34	4380	4392	4363	3327	177	133	97	112
ROMANIA	28	22	21	28*	2371	3129	1789	1786	67	69	37	50F
SPAIN	63	62	64	64	6105	6098	6347	5896	384	379	406	379
YUGOSLAVIA	7	8	8	8*	4272	4625	2875	3750	32	37	23	30*
OCEANIA	49	88	86	104	5846	4741	5153	5351	289	418	444	555
AUSTRALIA	38	76	75	92	6999	5135	5575	5761	267	388	417	530
FIJI	10	10	9	9	1839	2255	2252	1918	18	23	21	18
PACIFIC IS					917	938	941	941				
PAPUA N GUIN	1	1F	1F	1F	3572	1740	2000	2000	2	2	2	2
SOLOMON IS	1	1F	1F	1F	2209	3715	3731	3769	2	5F	5F	5F
USSR	356	500	524	546	3573	4019	3819	4029	1272	2009	2001	2200
DEV.PED M E	4105	4286	4165	4082	5391	5838	5404	5729	22126	25021	22505	23386
N AMERICA	777	1140	1004	910	5087	5109	5227	4945	3953	5826	5246	4501
W EUROPE	322	305	306	322	5036	5593	5055	4205	1622	1703	1547	1352
OCEANIA	38	76	75	92	6999	5135	5575	5761	267	388	417	530
OTH DEV.PED	2967	2765	2780	2758	5488	6185	5501	6164	16283	17104	15295	17003
DEV.PING M E	87609	94031	93890	93377	1839	1984	1914	2052	161153	186574	179732	191604
AFRICA	3436	3824	3894	3905	1385	1385	1429	1425	4758	5295	5563	5567
LAT AMERICA	6392	7334	8476	7341	1701	1877	1809	2012	10870	13769	15332	14771
NEAR EAST	1216	1206	1240	1234	3712	3815	3824	3927	4514	4602	4742	4848
FAR EAST	76554	81654	80269	80885	1842	1995	1919	2057	140989	162878	154068	166394
OTH DV.PING	11	13	11	12	1951	2364	2396	2129	22	30	27	25
CENTR PLANND	42571	44351	44752	45383	3012	3339	3306	3339	128229	148098	147934	151516
ASIAN CPE	42142	43780	44156	44759	3008	3332	3302	3332	126759	145867	145804	149132
E EUR+USSR	429	571	595	624	3426	3909	3579	3821	1470	2231	2130	2384
DEV.PED ALL	4533	4857	4760	4706	5205	5611	5175	5476	23595	27252	24635	25770
DEV.PING ALL	129751	137811	138047	138136	2219	2412	2358	2467	287912	332442	325536	340735

98

TABLE
TABLEAU 12
CUADRO

	BARLEY — AREA HARV / SUP RECOLTEE / SUP COSECHAD (1000 HA)				ORGE — YIELD / RENDEMENT / RENDIMIENTO (KG/HA)				CEBADA — PRODUCTION / PRODUCTICN / PRCDUCCICN (1000 MT)			
	1969-71	1975	1976	1977	1969-71	1975	1976	1977	1969-71	1975	1976	1977
WORLD	75312	89034	89943	91368	1843	1685	2051	1894	138826	150003	184434	173094
AFRICA	4341	4546	4896	4885	858	790	1036	641	3724	3589	5072	3133
ALGERIA	773	855	932	800F	608	433	631	500	470	370F	589	400F
EGYPT	44	42	44	45F	2124	2825	2826	2778	93	118	123	125*
ETHIOPIA	913	970	849	940F	796	930	1116	883	727	902	948	830F
KENYA	13	20F	20F	20F	1135	1572	1548	1650	15	31	31	33F
LESOTHO	1	1F	1F	1F	481	350	350	400	1			
LIBYA	213	368	415	300F	326	520	468	667	70	192	196	200F
MAURITANIA					688	773	783	833				
MOROCCO	2003	1844	2142	2341	1093	861	1336	575	2190	1587	2862	1347
RHODESIA	1	5*	5*	5F	3714	4746	5000	5000	3	22*	25F	25F
RWANDA					865							
SOUTH AFRICA	42	50F	50F	50F	643	1060	1261	1480	27	53	63	74
TANZANIA		2	2	2F	1020	1000	1000	1000		2	2	2F
TUNISIA	333	387	429	379*	378	801	538	251	126	310	231	95*
WESTN SAHARA	1	2F	2F	2F	617	667	750	706	1	1F	1F	1F
ZAIRE	3		1F	1F	782	1199	600	600	2			
N C AMERICA	8676	8207	8076	8763	2275	2207	2374	2394	19740	18113	19173	20980
CANADA	4483	4468	4353	4649	2236	2131	2415	2477	10024	9520	10513	11515
GUATEMALA					1473	1519	1500	1478				
MEXICO	230	286	364	272	1044	1537	1508	1504	240	440	549	409
USA	3963	3453	3358	3841	2391	2361	2415	2357	9476	8153	8111	9056
SOUTH AMERIC	1004	994	1028	1051	1057	1126	1296	1069	1062	1119	1332	1123
ARGENTINA	431	439	476	459	1153	1191	1596	1077	497	523	760	494
BOLIVIA	94	112	116	112	668	713	793	534	63	80	92	60
BRAZIL	26	23	49	62	983	1120	1269	1298	26	25	62	80
CHILE	48	66	58	63	2019	1822	1538	2267	97	121	89	143
COLOMBIA	53	76	68	61	1733	1611	1050	1339	92	122	71	81
ECUADOR	127	72	61	62*	675	878	1013	806	85	63	62	50
PERU	184	163	163	188*	894	913	920	904	164	149	150	170*
URUGUAY	42	44	38	44*	896	844	1224	1000	38	38	47	44*
ASIA	20975	21004	20483	19831	1341	1352	1455	1363	28131	28397	29793	27032
AFGHANISTAN	316	320	310	310F	1151	1200	1290	1313	363	384	400	407
BANGLADESH	32	26	26	24	674	641	670	648	21	16	17	15
BHUTAN	8	9F	9F	9F	877	902	889	880	7	8F	8F	8F
CHINA	11101F	10601F	10001F	10301F	1468	1434	1540	1495	16301F	15200F	15401F	15401F
CYPRUS	73	75*	75*	70F	1237	921	1219	943	90	69	91	66
INDIA	2693	2885	2802	2218	981	1087	1139	1035	2642	3135	3192	2296
IRAN	1532	1532	1481	1500F	681	939	1004	1067	1042	1438	1487	1600*
IRAQ	582	567	576	700F	1190	770	1006	654	692	437	579	458
ISRAEL	20	28	27	26*	890	730	684	750	17	21	18	20
JAPAN	224	78	80	78	2807	2832	2619	2641	629	221	210	206
JORDAN	35	30F	30F	29F	706	393	441	397	25	12	13	12*
KOREA DPR	150	180F	185F	185F	2356	1833	1892	1838	353	330F	350F	340F
KOREA REP	723	711	711	516*	2199	2391	2474	1578	1589	1700	1759	814*
KUWAIT					2000	1923	2000	2077				
LEBANON	7	7F	5F	5F	859	714	1000	1000	6	5F	5F	5F
MONGOLIA	18	54	69	54F	632	1189	778	981	11	64	54	53*
NEPAL	27	28	26	25	900	928	962	840	24	26	25	21
PAKISTAN	151	194	186	180	642	708	700	707	97	137	130	127
SAUDI ARABIA	11	7	10	10F	1473	2368	1251	1500	16	17	12	15F
SYRIA	520	1011	1172	1021	631	590	904	330	328	596	1059	337
TURKEY	2611	2588	2635	2500F	1425	1739	1860	1900	3720	4500	4900	4750F
YEMEN AR	143	73	68	70F	1067	1096	1176	1143	153	80	80	80F
YEMEN DEM	1	1	1	1	3236	1648	1624	1534	4	2	2	2
EUROPE	16441	19289	18791	19347	2947	3083	2998	3375	48444	59463	56335	65297
ALBANIA	9	10F	10F	10F	915	980	1200	1200	8	10F	12F	12F
AUSTRIA	286	315	325	328	3334	3190	3965	3689	954	1006	1287	1212
BELGIUM-LUX	174	141	156	169	3488	3389	4117	4290	608	476	643	727
BULGARIA	416	575	524	550F	2662	2957	3400	3182	1108	1699	1781	1750*
CZECHOSLOVAK	810	973	854	900F	3141	3200	3399	3444	2543	3114	2901	3100*
DENMARK	1342	1443	1478	1513*	3856	3573	3248	4021	5175	5156	4801	6084*
FINLAND	395	464	507	577	2389	2677	3066	2508	943	1242	1553	1447
FRANCE	2828	2779	2781	2910	3123	3363	3067	3536	8829	9344	8530	10290
GERMAN DR	646	929	560	850F	3241	3963	3601	4000	2093	3681	3456	3400*
GERMANY FED	1456	1756	1735	1791	3586	3969	3738	4186	5219	6970	6487	7497
GREECE	335	395	391	387*	1954	2319	2448	1814	655	916	957	702*
HUNGARY	322	257	228	230F	2328	2729	3273	3261	749	701	747	750*
IRELAND	216	245	259	289*	3955	4157	3563	4702	854	1019	922	1355*
ITALY	180	249	274	290	1815	2603	2759	2334	326	648	755	677
MALTA	1	1	1	1F	1281	1812	2128	2200	2	2	2	2F
NETHERLANDS	101	83	62	66	3627	4041	4280	4351	366	336	263	287
NORWAY	183	180	173	179	2985	2475	2813	3516	545	445	486	630
POLAND	861	1335	1210	1235*	2536	2725	2990	2756	2182	3638	3617	3404
PORTUGAL	94	101	143	67	687	860	818	507	64	86	117	34*
ROMANIA	309	442	410	440*	1993	2153	3005	3695	615	952	1231	1626
SPAIN	2235	3262	3240	3198	1755	2063	1689	2097	3922	6728	5473	6707
SWEDEN	602	605	554	604*	3049	3145	3292	3298	1836	1903	1825	1992*
SWITZERLAND	39	45	43	45	3775	3871	4327	3872	147	173	186	175
UK	2317	2345	2182	2411	3564	3630	3505	4473	8257	8513	7648	10784
YUGOSLAVIA	287	360	293	306	1540	1953	2229	2124	442	703	653	650
OCEANIA	2091	2448	2407	2977	1240	1435	1325	966	2594	3513	3191	2876
AUSTRALIA	2024	2329	2320	2900	1171	1365	1227	883	2372	3179	2847	2560

TABLE
TABLEAU 12
CUADRO

	BARLEY / ORGE / CEBADA											
	AREA HARV / SUP RECOLTEE / SUP COSECHAD (1000 HA)				YIELD / RENDEMENT / RENDIMIENTC (KG/HA)				PRODUCTICN / PRODUCTICN / PRODUCCICN (1000 MT)			
	1969-71	1975	1976	1977	1969-71	1975	1976	1977	1969-71	1575	1576	1977
NEW ZEALAND	67	119	87	77	3310	2820	3932	4131	222	334	344	316
USSR	21782	32547	34261	34514	1613	1100	2030	1526	35132	35808	69539	52653
DEV.PED M E	23892	25292	24872	26753	2591	2655	2601	2803	61911	67146	64694	75002
N AMERICA	8446	7921	7712	8491	2309	2231	2415	2423	19500	17672	18624	20571
W EUROPE	13069	14768	14596	15132	2995	3092	2918	3387	39144	45666	42588	51255
OCEANIA	2091	2448	2407	2977	1240	1435	1325	966	2594	3513	3191	2876
OTH DEV.PED	286	156	157	154	2358	1885	1858	1945	674	295	292	300
DEV.PING M E	14997	15839	16360	15346	1055	1115	1262	1017	15819	17657	20650	15603
AFRICA	4042	4085	4383	4450	874	790	1070	609	3535	3226	4689	2734
LAT AMERICA	1235	1280	1392	1323	1055	1218	1351	1158	1302	1560	1881	1532
NEAR EAST	6088	6622	6826	6562	1084	1185	1311	1228	6602	7850	8949	8056
FAR EAST	3633	3851	3759	2971	1206	1304	1365	1104	4380	5021	5131	3281
CENTR PLANND	36422	47902	48711	49269	1677	1361	2034	1674	61096	65159	99090	82489
ASIAN CPE	11268	10834	10255	10540	1479	1439	1541	1499	16665	15595	15804	15794
E EUR+USSR	25154	37068	38456	38729	1766	1338	2166	1722	44431	49605	83285	66695
DEV.PED ALL	49046	62360	63328	65482	2168	1872	2337	2164	106342	116751	147979	141697
DEV.PING ALL	26266	26674	26615	25886	1237	1247	1370	1213	32484	33252	36455	31397

TABLE
TABLEAU **13**
CUADRO

	MAIZE — AREA HARV. SUP RECOLTEE SUP COSECHAD (1000 HA)				MAIS — YIELD RENDEMENT RENDIMIENTO (KG/HA)				MAIZ — PRODUCTION PRODUCTION PRODUCCION (1000 MT)			
	1969-71	1975	1976	1977	1969-71	1975	1976	1977	1969-71	1975	1976	1977
WORLD	108487	113797	116577	118453	2567	2849	2857	2952	278504	324257	333079	349676
AFRICA	17991	19652	19916	20134	1155	1277	1191	1300	20776	25091	23729	26172
ALGERIA	6	4	2	4F	953	1806	1279	1756	6	7	3	7F
ANGOLA	540	610*	600F	600F	864	738	750	750	467	450*	450F	450F
BENIN	359	310	350	323	559	699	630	727	201	217	221	235
BOTSWANA	34	40F	80F	65F	338	625	777	646	11	25	62	42*
BURUNDI	112	120F	122F	122F	1071	1143	1147	1148	120	137	140	140F
CAMEROON	238	340*	340F	350F	1190	1029	1044	1029	283	350*	355F	360F
CAPE VERDE	4	10F	8F	5F	431	500	500	400	2	5F	4	2
CENT AFR EMP	62	98	97F	100F	711	390	392	370	44	38	38	37*
CHAD	6	10	10F	10F	1532	968	980	950	9	10	10F	10F
COMOROS	3	4F	4F	4F	1061	1053	1050	1070	4	4F	4F	5F
CONGO	8	21	24	26	667	605	605	556	5	13	14	14
EGYPT	634	769	739*	777*	3741	3617	4121	3732	2370	2781	3047	2900F
ETHIOPIA	849	817	701	820F	1071	1800	1449	1402	909	1470	1015	1150F
GABON	2	2F	2F	2F	1059	1000	1000	1000	2	2F	2F	2F
GAMBIA	7	6	4	4*	400	1745	976	548	3	10	4	2*
GHANA	340	320	323	311	1227	1072	742	1179	417	343	240	366
GUINEA	387	410F	420F	450F	940	756	762	822	363	310F	320F	370F
GUIN BISSAU	3	5F	5F	3F	628	778	889	500	2	4F	4F	2F
IVORY COAST	166	240F	243F	250F	618	546	481	560	103	131F	117F	140F
KENYA	1167	1250*	1250F	1250F	1220	1280	1240	1360	1423	1600*	1550F	1700F
LESOTHO	134	85	80	140F	644	579	1563	643	86	49	125	90F
LIBYA	1	2F	2F	2F	945	1111	1000	1000	1	2F	2F	2F
MADAGASCAR	121	109	115	115F	1004	1096	1188	1174	122	120	136	135F
MALAWI	1039	1000F	1000*	1000*	1025	1000	1100	1250	1066	1000*	1100*	1250*
MALI	106	81	90	95F	770	868	892	895	81	71	81	85F
MAURITANIA	7	6F	6F	6F	580	508	516	524	4	3F	3F	3F
MAURITIUS		1	1F	1F	1984	2296	2264	2453		1	1F	1F
MOROCCO	474	492	433	425	801	754	1138	434	380	371	493	184
MOZAMBIQUE	363	680*	680*	600F	945	368	662	583	343	250*	450*	350F
NAMIBIA	32	34F	34F	34F	375	441	441	441	12	15F	15F	15F
NIGER	3	7	7F	7F	600	643	694	767	2	5	5F	6F
NIGERIA	1544	1675*	1725*	1800*	787	752	751	775	1215	1260F	1295F	1395F
REUNION	9	12F	13F	12F	953	1000	1038	1167	9	12	14	14F
RHODESIA	403	475*	475*	500F	2396	2947	2947	2600	966	1400*	1400*	1300F
RWANDA	50	63	65	66F	1084	1075	1080	1076	54	67	71	71F
SENEGAL	52	56	47	50F	814	866	997	911	42	49	47	46*
SIERRA LEONE	11	13F	13F	13F	980	1040	1038	1077	10	13F	14F	14F
SOMALIA	126	100F	120F	120F	912	1000	1000	1000	115	100F	120F	120F
SOUTH AFRICA	5290	5700F	5700F	5700F	1265	1604	1283	1704	6691	9140	7312	9714
SUDAN	39	89	85F	85F	780	615	588	529	31	55	50F	45F
SWAZILAND	89	66	75F	77F	846	1425	1467	1104	75	94	110	85
TANZANIA	1005	1100F	1300*	1300F	813	750	690	745	817	825	897	968
TOGO	144	110	110F	110F	1109	1234	1236	1273	160	135	136*	140F
UGANDA	343	476	525	551	1225	1200	1200	1200	420	571	629	661
UPPER VOLTA	92	144	90F	90F	659	587	511	556	60	84	46*	50F
ZAIRE	595	675	702	710F	716	734	726	725	426	495	510	515F
ZAMBIA	992	1020F	1100F	1050F	849	980	973	933	843	1000F	1070F	980F
N C AMERICA	33478	36407	38117	38374	4070	4457	4536	4612	136248	162265	172908	176984
ANTIGUA					2296	2000	1962	2000				
BARBADOS	1	1F	1F	1F	2412	2614	2614	2614	2	2F	2F	2F
BELIZE	9	11	12F	13F	1437	921	917	923	13	10	11F	12F
CANADA	490	635	709	729	5078	5740	5319	5904	2487	3645	3771	4303
COSTA RICA	50	65	53	43*	1123	1416	1681	1419	56	92	89	61*
CUBA	24	30*	30*	30*	917	690	667	667	22	21	20*	20*
DOMINICA					1235	1267	1333	1400				
DOMINICAN RP	27	25*	33*	20*	1712	1280	1394	1750	46	32	46*	35*
EL SALVADOR	203	246	234	245*	1671	1786	1464	1540	340	439	342	377*
GRENADA		1F	1F	1F	622	600	636	636				
GUADELOUPE		1			1778	2000	1500	1500	1	1		
GUATEMALA	671	514	514	590F	1118	1328	1335	1281	751	683	686	756
HAITI	231	238F	239F	239F	1058	756	754	1046	245	180*	180*	250*
HONDURAS	267	330	330*	412*	1117	1099	876	915	298	363	289	377*
JAMAICA	4	13*	13*	13*	1275	998	857	769	5	13	11	10*
MEXICO	7412	6694	6783	7387	1218	1264	1182	1217	9025	8459	8017	8991
MONTSERRAT					1000	1000	1000	1000				
NICARAGUA	260	210	228	240*	912	916	883	926	238	192	201	222
PANAMA	77	74	83	80*	859	877	769	963	66	65	64	77*
PUERTO RICO	2	1F			1043	908	908	908	2			
ST LUCIA					806	700	700	700				
ST VINCENT					1259	1429	1500	1571				
TRINIDAD ETC	1	1	1*	1*	4153	3921	4141	4092	2	5	5*	5*
USA	23749	27318	28854	28330	5164	5420	5517	5700	122649	148061	159172	161485
SOUTH AMERIC	16521	15962	16469	16796	1554	1719	1653	1849	25675	27438	27229	31051
ARGENTINA	3880	3070	2766	2532	2247	2508	2117	3278	8717	7700	5855	8300
BOLIVIA	223	230	233	244	1306	1325	1534	1228	291	305	357	299
BRAZIL	10021	10473	11176	11682	1365	1562	1597	1637	13680	16354	17845	19122
CHILE	70	92	96	116	3111	3594	2579	3075	217	329	248	355
COLOMBIA	684	573	648	604*	1251	1262	1365	1246	856	723	884	753
ECUADOR	312	277	239	240F	805	994	904	908	251	275	216	218
FR GUIANA					4420	4000	4000	4000		1F	1F	1F
GUYANA	1	3	4*	4*	1632	1913	1181	2024	2	6	4*	9*
PARAGUAY	158	223	257	275F	1249	1351	1366	1354	197	301	351	372*
PERU	373	363	385	410*	1621	1751	1883	1707	605	635	726	700*
SURINAM					1906	1253	1333	1667				
URUGUAY	194	153	177	138*	832	1028	1191	877	161	157	210	121*
VENEZUELA	606	506	489	551*	1152	1291	1089	1452	698	653	532	800*
ASIA	25293	26889	27026	27850	1854	2008	1993	1957	46887	54002	53853	54497

TABLE
TABLEAU 13
CUADRO

	AREA HARV / SUP RECOLTEE / SUP COSECHAD 1000 HA				YIELD / RENDEMENT / RENDIMIENTO KG/HA				PRODUCTION / PRODUCTION / PRODUCCION 1000 MT			
	1969-71	1975	1976	1977	1969-71	1975	1976	1977	1969-71	1975	1976	1977
AFGHANISTAN	456	484	482	500F	1551	1612	1660	1634	707	780	800	817
BANGLADESH	3	2	2	3F	870	902	897	923	3	2	2	2F
BHUTAN	49	52F	53F	54F	989	1054	1057	1065	49	55F	56F	57F
BURMA	112	175F	176F	170F	619	457	455	500	69	80	80F	85F
CHINA	10388F	10750F	11041F	11348F	2765	2990	2999	2962	28720F	32138F	33114F	33615F
EAST TIMOR	16	16F	16F		918	938	938		15	15F	15F	
INDIA	5794	6031	6054	6000F	1051	1203	1033	1133	6087	7256	6257	6800F
INDONESIA	2667	2445	2095	2550	965	1187	1228	1188	2575	2903	2572	3030
IRAN	25	30*	35*	40F	1400	2167	2286	2250	35	65	80*	90F
IRAQ	6	9	16F	30F	1495	2480	3436	2740	9	23	55	82
ISRAEL	1	3	3	3F	4806	5795	5669	5600	5	15	14	14
JAPAN	12	5	4	3	2649	2600	2750	2667	33	13	11	8
JORDAN	1	1	1		625	1186	1176	1200		1	1	1F
KAMPUCHEA DM	94	55F	55F	62F	1332	1182	1364	1290	125	65F	75F	80F
KOREA DPR	413	430F	420F	410F	3613	3953	4238	4439	1493	1700F	1780F	1820F
KOREA REP	44	35	35	50F	1450	1728	2394	1800	63	60	84	90F
KUWAIT					3000	3000	3000	3000				
LAO	15	16F	16F	16F	1711	1750	1875	2184	26	28F	30F	35
LEBANON	1	1F	1F	1F	904	1000	1000	1000	1	1F	1F	1F
MAL PENINSUL	3	3	3F	3F	2891	7610	7000	7000	10	20F	21F	21F
MAL SABAH	5	9	13	5F	686	468	397	800	3	4F	5F	4F
MALDIVES					4500	5000	5000	5000				
NEPAL	439	453	445	370F	1812	1651	1791	1759	796	748	797	651
PAKISTAN	640	620	624	640F	1088	1294	1224	1351	697	802	764	864*
PHILIPPINES	2415	3126	3257	3445	832	863	850	881	2009	2697	2767	3037
SAUDI ARABIA	1	2	2	2	5070	714	1568	1562	4	2	3	4
SRI LANKA	19	40	38	40F	763	683	703	875	14	27	27	35F
SYRIA	6	16	23	20*	1416	1696	2175	1700	8	27	51	34*
THAILAND	771	1180F	1200F	1150F	2567	2426	2229	1458	1979	2863	2675	1677
TURKEY	646	597	600	620F	1637	2010	2183	1935	1058	1200	1310	1200*
VIET NAM	239	250F	260F	260F	1136	1280	1231	1000	272	320F	320F	260F
YEMEN AR	8	50	50	50F	2016	1580	1440	1400	16	79	72	70F
YEMEN DEM	2	5*	6F	6F	2528	2511	2545	2333	5	13	14F	14F
EUROPE	11497	12155	11662	11845	3363	3932	3846	4185	38663	47789	44853	49572
ALBANIA	111	106F	125F	120F	1988	2075	2240	2500	220	220F	280F	300F
AUSTRIA	122	144	160	166	5547	6820	5860	6980	677	981	936	1159
BELGIUM-LUX	2	6	6	7	5510	5917	5278	5140	11	38	30	33
BULGARIA	623	652	731	670F	3913	4329	4146	3813	2436	2822	3031	2555*
CZECHOSLOVAK	127	153	143	178	4021	5490	3598	3371	511	843	514	600F
FRANCE	1454	1960	1394	1627	5104	4188	4020	5294	7422	8209	5603	8614
GERMAN DR	3			1F	2631	3978	2211	1000	9	2		1F
GERMANY FED	99	96	103	102	5052	5524	4679	5598	500	531	480	571
GREECE	162	127	126	129*	3073	3843	3976	4310	498	488	501	556*
HUNGARY	1272	1429	1351	1280*	3570	5018	3806	4805	4542	7172	5141	6150*
ITALY	986	897	889	983	4664	5938	5985	6568	4601	5326	5321	6456
NETHERLANDS	1	1*	1*	1*	4630	4615	5444	5714	5	6*	5*	4*
POLAND	5	15	52	90*	2449	5331	4408	2444	12	79	231	220*
PORTUGAL	432	392	368	343	1385	1290	1166	1172	599	506	429	402
ROMANIA	3170	3305	3378	3365*	2320	2796	3429	3002	7354	9241	11583	10103
SPAIN	526	485	442	445	3432	3699	3493	4235	1804	1794	1545	1885
SWITZERLAND	10	22	19	17	6168	6539	6189	6176	61	141	115	105
UK	1	1	1	1	4666	3000	2000	2000	2	3	2	2
YUGOSLAVIA	2392	2363	2374	2321	3093	3973	3836	4246	7399	9389	9106	9856
OCEANIA	89	80	84	91	2930	4319	4417	4465	261	344	369	406
AUSTRALIA	77	51	47	53	2387	2588	2799	2717	184	133	131	144
FIJI	2	2F	2F	2F	2438	2000	2000	2000	4	4F	4F	4F
GUAM					1130	1500	1500	1500				
NEWCALEDONIA			1F	1F	2222	3000	1000	1000	1	1	1F	1F
NEW HEBRIDES		1F	1F	1F	2208	1000	1000	1000	1	1F	1F	1F
NEW ZEALAND	9	25	33	34	7777	8271	7117	7574	70	205	232	256
PACIFIC IS					1125	1125	1125	1125				
PAPUA N GUIN					888	1151	480	480				
SAMOA					1250	1625	1750	1875				
USSR	3617	2652	3303	3362	2763	2763	3069	3270	9993	7328	10138	10993
DEV.PED M E	35815	40231	41230	40992	4347	4689	4723	5015	155700	188623	194716	205567
N AMERICA	24238	27953	29563	29059	5163	5427	5512	5705	125137	151706	162943	165788
W EUROPE	6187	6494	5882	6141	3811	4221	4093	4827	23580	27411	24073	29643
OCEANIA	86	76	79	87	2951	4435	4572	4608	255	338	363	400
OTH DEV.PED	5303	5708	5707	5706	1269	1606	1286	1706	6729	9168	7337	9736
DEV.PING M E	52609	53769	54487	56315	1276	1371	1324	1375	67118	73706	72155	77412
AFRICA	12027	13093	13390	13570	971	1002	995	996	11684	13114	13318	13511
LAT AMERICA	25761	24416	25024	26111	1428	1556	1486	1618	36787	37997	37194	42247
NEAR EAST	1826	2055	2042	2134	2325	2447	2686	2465	4247	5028	5486	5260
FAR EAST	12992	14201	14027	14495	1108	1237	1151	1131	14394	17561	16152	16388
OTH DV.PING	3	3	4	4	2262	1760	1472	1473	6	6	6	6
CENTR PLANND	20062	19797	20859	21146	2776	3128	3174	3154	55686	61929	66208	66697
ASIAN CPE	11135	11485	11776	12080	2749	2980	2997	2962	30610	34223	35289	35775
E EUR+USSR	8927	8313	9083	9066	2809	3333	3404	3411	25076	27706	30919	30922
DEV.PED ALL	44743	48543	50313	50058	4040	4456	4485	4724	180776	216329	225635	236489
DEV.PING ALL	63744	65254	66264	68395	1533	1654	1621	1655	97728	107929	107445	113187

TABLE
TABLEAU 14
CUADRO

RYE   SEIGLE   CENTENO

AREA HARV / SUP RECOLTEE / SUP COSECHAD — 1000 HA
YIELD / RENDEMENT / RENDIMIENTC — KG/HA
PRODUCTION / PRODUCTION / PRODUCCION — 1000 MT

	1969–71	1975	1976	1977	1969–71	1975	1976	1977	1969–71	1975	1976	1977
WORLD	18846	14965	16179	13904	1568	1586	1822	1709	29549	23734	29482	23767
AFRICA	24	22	22	23	418	250	256	258	10	6	6	6
MOROCCO	4	2F	2F	3F	800	750	739	720	3	2F	2F	2F
SOUTH AFRICA	20	20F	20F	20F	338	200	200	200	7	4	4	4
N C AMERICA	959	615	542	523	1521	1510	1513	1575	1458	928	820	824
CANADA	356	320	250	242	1332	1635	1758	1619	474	523	440	392
USA	603	295	292	281	1633	1374	1302	1537	984	405	380	432
SOUTH AMERIC	479	333	367	261	637	919	966	776	305	306	354	203
ARGENTINA	440	300	340	231	616	912	971	736	271	273	330	170
BOLIVIA	1	1	1	1	435	500	500	500		1		
BRAZIL	23	21	14	15	836	931	957	939	19	19	13	14
CHILE	9	9	10	11	1300	1273	974	1440	11	11	9	16
ECUADOR	5	2*	2F	2F	414	500	500	500	2	1*	1F	1F
PERU	1	1	1	1F	872	842	853	1000	1	1	1F	1F
ASIA	697	603	567	547	1194	1344	1404	1415	832	810	797	774
JAPAN	1	1F	1F	1F	1729	1000	1000	1000	1	1F	1F	1F
KOREA DPR	27	31F	32F	32F	1481	1613	1563	1625	40	50F	50F	52F
KOREA REP	6	6	4	4F	1681	1533	1274	1500	11	9	6	6F
TURKEY	663	565	530	510F	1177	1327	1396	1402	781	750	740	715
EUROPE	7060	5367	5622	5831	2081	2350	2402	2311	14689	12613	13504	13477
ALBANIA	9	9F	9F	10F	875	889	1111	1000	8	8F	10F	10F
AUSTRIA	143	119	120	119	2917	2923	3429	2959	417	347	410	351
BELGIUM-LUX	23	10	17	20	3551	3488	3165	3394	82	33	53	67
BULGARIA	22	17	13	18F	1227	1099	1205	1111	27	18	15	20F
CZECHOSLOVAK	243	190	187	210	2415	2794	3010	4143	586	530	561	870F
DENMARK	42	49	72	88*	3289	3327	2958	3636	137	163	213	320*
FINLAND	67	38	65	47	1929	2146	2733	1702	130	81	178	80
FRANCE	139	111	114	129	2132	2775	2603	2911	297	308	297	376
GERMAN DR	679	593	600	600F	2346	2636	2425	2500	1594	1563	1455	1500F
GERMANY FED	868	624	663	701	3298	3404	3165	3621	2862	2125	2100	2538
GREECE	7	5	4	4*	1201	1400	2250	1250	9	7	9	5*
HUNGARY	155	104	93	90*	1242	1410	1693	1633	193	147	157	147*
IRELAND					3086	2422	2469	2469	1	1F	1F	1F
ITALY	34	17	16	15	1889	2187	2206	2098	65	37	35	31
NETHERLANDS	60	18	21	21	3292	3453	3062	3483	196	63	65	74
NORWAY	1	1	2	2	3312	2979	3025	3398	5	4	7	8
POLAND	3766	2792	2934	3116*	1897	2246	2359	1990	7142	6270	6922	6200*
PORTUGAL	227	211	219	177	725	692	754	508	164	146	165	90
ROMANIA	45	41*	35F	40*	1153	1268	1400	1250	52	52*	49*	50*
SPAIN	320	228	224	221	912	1057	955	986	292	241	214	218
SWEDEN	78	95	123	116*	3058	3442	3480	3138	239	327	427	364*
SWITZERLAND	12	6	8	8	4013	3938	4520	4000	48	24	34	32
UK	5	6	8	11	2929	3167	2500	3455	14	19	20	38
YUGOSLAVIA	115	84	76	68	1147	1167	1382	1279	132	98	105	87
OCEANIA	39	15	23	22	489	494	431	563	19	7	10	13
AUSTRALIA	39	15	23	22*	476	467	413	545	18	7	10	12*
NEW ZEALAND						2500	2500	2500	1	1F	1F	1F
USSR	9588	8010	9035	6697	1276	1132	1549	1265	12235	9064	13991	8471
DEV.PED M E	3159	2272	2338	2314	2081	2185	2210	2386	6573	4965	5169	5522
N AMERICA	959	615	542	523	1521	1510	1513	1575	1458	928	820	824
W EUROPE	2141	1622	1752	1747	2376	2481	2474	2678	5088	4024	4334	4680
OCEANIA	39	15	23	22	489	494	431	563	19	7	10	13
OTH DEV.PED	21	21	21	21	387	238	238	238	8	5	5	5
DEV.PING M E	1153	906	903	777	954	1177	1219	1190	1099	1067	1102	925
AFRICA	4	2	2	3	800	750	739	720	3	2	2	2
LAT AMERICA	479	333	367	261	637	919	966	776	305	306	354	203
NEAR EAST	663	565	530	510	1177	1327	1396	1402	781	750	740	715
FAR EAST	6	6	4	4	1681	1533	1274	1500	11	9	6	6
CENTR PLANND	14534	11787	12937	10813	1505	1502	1794	1602	21876	17702	23211	17320
ASIAN CPE	27	31	32	32	1481	1613	1563	1625	40	50	50	52
E EUR+USSR	14507	11756	12905	10781	1505	1502	1795	1602	21836	17652	23161	17268
DEV.PED ALL	17666	14028	15244	13095	1608	1612	1858	1740	28409	22617	28330	22790
DEV.PING ALL	1180	937	935	809	966	1192	1231	1208	1139	1117	1152	977

TABLE
TABLEAU 15
CUADRO

	OATS — AREA HARV / SUP RECOLTEE / SUP COSECHAD (1000 HA)				AVOINE — YIELD / RENDEMENT / RENDIMIENTO (KG/HA)				AVENA — PRODUCTION / PRODUCTICN / PRODUCCICN (1000 MT)			
	1969-71	1975	1976	1977	1969-71	1975	1976	1977	1969-71	1975	1976	1977
WORLD	31336	30671	28864	30640	1768	1554	1719	1713	55403	47663	45606	52472
AFRICA	396	402	433	382	463	531	550	440	184	214	238	168
ALGERIA	67	88	119	70F	617	624	751	857	41	55F	89	60F
ETHIOPIA	10	10F	10F	10F	500	500	500	500	5	5F	5F	5F
KENYA	5	5F	5F	5F	760	900	1000	1000	4	5F	5*	5F
LESOTHO	1	1F	1F	1F	1429	1429	1429	1429	1	1F	1F	1F
MOROCCO	20	32	31	27	626	885	1180	285	13	29	36	8
SOUTH AFRICA	255	225F	225F	225F	404	458	378	320	103	103	85	72
TUNISIA	23	24F	25F	26F	306	208	200	212	7	5F	5F	6F
ZAIRE	16	17F	18F	18F	625	676	667	667	10	12F	12F	12F
N C AMERICA	9934	7769	7310	7637	1902	1786	1752	1991	18891	13873	12809	15205
CANADA	2834	2411	2409	2132	1943	1852	2005	2018	5508	4467	4831	4303
MEXICO	49	59	66	63	668	1473	727	730	33	87	48	46
USA	7051	5298	4834	5442	1893	1759	1640	1995	13350	9319	7930	10856
SOUTH AMERIC	518	567	562	530	1194	1185	1277	1435	618	672	717	760
ARGENTINA	328	338	383	370	1280	1282	1384	1541	420	433	530	570
BOLIVIA	7	4	4	3	659	687	690	671	5	3	2	2
BRAZIL	32	45	36	33	861	929	1076	1036	27	42	39	34
CHILE	76	94	79	75	1389	1388	1207	1649	106	131	96	124
COLOMBIA				2	1752	2000	2000	2000		1	1	3
ECUADOR	1	1*	1*	1*	967	1000	1000	1000	1	1*	1*	1*
PERU	1	1	1	1F	917	829	828	909	1	1	1	1F
URUGUAY	73	84	58	45*	804	733	830	556	59	61	48	25*
ASIA	2440	2315	2285	2280	985	1073	1093	1085	2404	2484	2496	2474
CHINA	1950F	1900F	1900F	1900F	906	1000	1000	1000	1767F	1900F	1900F	1900F
CYPRUS	2	2F	2F	2F	770	1000	1000	1000	1	2F	2F	2F
ISRAEL	1	1	1	1F	1339	915	1067	1067	1	1	1	1
JAPAN	30	13	10	8	2067	2120	2172	2250	63	28	22	18
KOREA DPR	74	76F	77F	77F	1306	1447	1494	1558	97	110F	115F	120F
LEBANON	2	1F	1F	1F	1929	1500	1500	1800	3	2F	1F	1F
MONGOLIA	53	59	49	65F	457	834	798	785	24	49	39	51*
SYRIA	3	3	2	2	775	897	818	889	2	3	2	2
TURKEY	326	260	243	225F	1366	1500	1708	1689	446	390	415	380
EUROPE	7243	6502	6002	5567	2471	2571	2353	2596	17898	16719	14120	14453
ALBANIA	23	21F	21F	21F	724	1315	1333	1333	16	28F	28F	28F
AUSTRIA	101	101	95	90	2798	3028	2984	3115	281	306	283	279
BELGIUM-LUX	89	83	59	44	3281	3109	2482	2872	293	257	147	127
BULGARIA	74	50	44	50F	1257	1119	1463	1200	93	56	65	60F
CZECHOSLOVAK	372	214	169	190F	2371	2765	2241	3158	882	591	379	600F
DENMARK	191	111	98	83*	3654	3306	2684	3470	699	367	263	288*
FINLAND	516	572	551	417	2516	2535	2854	2452	1297	1450	1573	1022
FRANCE	829	641	652	625	2795	2961	2195	3085	2317	1898	1431	1928
GERMAN DR	237	243	190	220F	3098	3205	2664	2955	735	780	506	650F
GERMANY FED	840	920	856	795	3371	3744	2919	3425	2832	3445	2497	2723
GREECE	81	70	65	63*	1326	1629	1656	1286	108	114	108	81*
HUNGARY	49	45	39	35*	1630	2053	2376	1829	79	92	92	64*
IRELAND	68	49	40	35*	3253	3368	3228	3681	222	165	130	127*
ITALY	297	239	236	226	1643	2119	1862	1572	488	506	439	355
NETHERLANDS	61	34	25	21	3995	4604	4065	4531	243	158	103	94
NORWAY	69	103	102	99	3134	2520	2800	3630	215	259	287	360
POLAND	1409	1291	1115	1096*	2239	2261	2417	2372	3156	2920	2695	2600*
PORTUGAL	164	207	215	147	561	584	589	374	92	121	127	55*
ROMANIA	130	70	45	45*	1062	811	1224	1333	138	57	55	60*
SPAIN	481	457	455	384	1049	1333	1161	1115	504	609	528	428
SWEDEN	505	464	450	446*	3092	2847	2782	3137	1561	1321	1251	1399*
SWITZERLAND	9	13	12	11	3787	4129	3935	3982	35	55	49	45
UK	375	233	235	194	3465	3412	3251	3974	1300	795	764	771
YUGOSLAVIA	273	270	232	231	1133	1363	1379	1338	310	368	320	309
OCEANIA	1409	1010	1005	1219	1018	1194	1108	848	1435	1206	1113	1033
AUSTRALIA	1390	988	991	1200	992	1155	1082	802	1378	1141	1072	962
NEW ZEALAND	19	22	14	19	2917	2968	3000	3822	57	65	41	71
USSR	9394	12107	11269	13026	1488	1032	1607	1411	13974	12495	18113	18379
DEV.PED M E	16530	13525	12863	12936	2012	2020	1888	2062	33258	27319	24281	26674
N AMERICA	9885	7709	7244	7574	1908	1788	1762	2002	18858	13786	12761	15159
W EUROPE	4950	4567	4379	3910	2586	2670	2352	2657	12798	12155	10300	10391
OCEANIA	1409	1010	1005	1219	1018	1194	1108	848	1435	1206	1113	1033
OTH DEV.PED	286	239	236	234	583	552	455	388	167	132	107	91
DEV.PING M E	1040	1069	1083	979	1138	1184	1235	1315	1184	1266	1338	1287
AFRICA	141	177	208	157	570	625	737	613	81	111	153	96
LAT AMERICA	567	626	628	593	1149	1213	1219	1360	651	759	765	806
NEAR EAST	332	265	248	229	1361	1491	1695	1678	452	396	420	384
CENTR PLANND	13765	16077	14918	16725	1523	1187	1608	1466	20961	19079	23987	24512
ASIAN CPE	2077	2035	2026	2042	909	1012	1014	1014	1888	2059	2054	2071
E EUR+USSR	11688	14042	12892	14683	1632	1212	1701	1528	19074	17020	21933	22441
DEV.PED ALL	28218	27567	25755	27619	1855	1608	1794	1778	52331	44338	46214	49115
DEV.PING ALL	3118	3104	3109	3021	985	1071	1091	1112	3072	3325	3392	3358

TABLE
TABLEAU 16
CUADRO

	MILLET AREA HARV 1000 HA SUP RECOLTEE SUP COSECHAD				MILLET YIELD KG/HA RENDEMENT RENDIMIENTO				MIJO PRODUCTION 1000 MT PRODUCTION PRODUCCION			
	1969-71	1975	1976	1977	1969-71	1975	1976	1977	1969-71	1975	1976	1977
WORLD	65123	63867	64329	65453	661	658	692	655	43050	42015	44489	42886
AFRICA	15853	15272	16048	16356	623	621	605	568	9873	9480	9708	9285
ANGOLA	92	93F	93F	93F	845	806	860	914	78	75F	80F	85F
BENIN	16	30F	15F	12	375	500	667	433	6	15*	10F	5
BOTSWANA	19.	10F	10F	10F	244	500	500	500	5	5F	5F	5F
BURUNDI	37	27	36	36F	700	895	821	833	26	25	29	30F
CAMEROON	477	430F	430F	430F	834	898	907	837	397	386	390F	360F
CENT AFR EMP	54	67	67F	60*	790	610	612	750	43	41	41*	45*
CHAD	925	1000F	1075*	1133*	682	523	472	507	631	523	507*	574*
EGYPT	206	205	199	210F	4120	3771	3805	3952	847	775	759	830*
ETHIOPIA	197	336	174	250F	597	980	1139	920	117	329	199	230F
GAMBIA	42	58	33	60*	992	802	667	533	42	47	22	32*
GHANA	197	200	210*	215*	513	610	333	326	101	122	70*	70*
GUIN BISSAU	11	13F	13F	9F	531	500	538	444	6	7F	7F	4F
IVORY COAST	63	75	75	75F	488	607	540	613	31	46	41	46*
KENYA	75	80F	80F	80F	1681	1750	1688	1750	127	140F	135F	140F
LIBYA	1	1F	1F	1F	698	1000	1000	1000	1	1F	1F	1F
MALI	1415	1121	1244	1250F	556	621	646	680	787	696	804	850F
MAURITANIA	263	160F	180F	180F	337	188	333	250	89	30*	60*	45*
MOROCCO	7	4F	4F	4F	603	725	744	744	4	3F	3F	3F
MOZAMBIQUE	19	20F	20F	20F	526	400	400	400	10	8F	8F	8F
NAMIBIA	51	55F	55F	55F	363	345	355	364	19	19F	20F	20F
NIGER	2313	1692	2532	2300F	421	344	472	435	974	581	1195	1000F
NIGERIA	5022	5000F	5000F	5200F	556	600	580	500	2792	3000F	2900*	2600F
RHODESIA	390	390F	390F	390F	590	564	564	564	230	220F	220F	220F
RWANDA	3	5	5	5F	641	566	575	580	2	3	3	3F
SENEGAL	1006	963	955	1000*	540	645	581	432	543	621	555	432*
SIERRA LEONE	8	8F	9F	9F	1089	1000	1000	1000	9	8F	9F	9F
SOUTH AFRICA	22	22F	22F	22F	682	682	682	682	15	15F	15F	15F
SUDAN	750	1111	1111	1200F	566	389	387	342	424	432	430	410F
TANZANIA	211	220*	200F	220F	861	727	650	682	181	160*	130F	150F
TOGO	190	199	200F	200F	638	595	600	600	121	119	120*	120F
UGANDA	739	567	498	524	997	901	900	900	737	511	448	471
UPPER VOLTA	843	911	911	900	418	420	384	367	352	383	350F	330F
ZAIRE	61	67	68	68F	793	780	801	809	48	52	54	55F
ZAMBIA	127	132F	134F	135F	619	644	672	637	79	85F	90F	86F
SOUTH AMERIC	160	194	231	255	1049	1029	1274	1333	168	200	294	340
ARGENTINA	160	194	231	255	1049	1029	1274	1333	168	200	294	340
ASIA	46207	45570	44997	45734	659	684	694	682	30439	31147	31227	31191
AFGHANISTAN	20	40	42	42F	1000	800	952	952	20	32	40	40F
BANGLADESH					829	770	789	786				
BHUTAN	4	4F	4F	4F	989	1054	1250	1275	4	4F	5F	5F
BURMA	220	209	210F	210F	294	295	281	286	65	62	59*	60F
CHINA	24738F	25004F	25604F	25504F	766	780	801	784	18940F	19506F	20504F	20007F
INDIA	19618	18873	17692	18500F	519	554	539	541	10182	10457	9544	10000F
IRAN	17	20F	20F	20F	962	1250	1250	1250	17	25F	25F	25F
IRAQ	3	1	1*	1*	933	893	917	900	3	1	1F	1F
JAPAN	7	3F	3F	3F	1724	1333	1333	1333	12	4F	4F	4F
KOREA DPR	462	460F	470F	470F	881	902	909	889	407	415F	427F	418F
KOREA REP	59	27	26	27F	798	950	1031	963	47	26	27	26F
MALDIVES					2810	1609	1596	1583				
NEPAL	109	126	122	125F	1172	1135	1131	1120	128	143	138	140*
PAKISTAN	713	624	648	682F	476	493	480	447	339	308	311	305*
SAUDI ARABIA	110	36	19	13	1075	292	349	717	118	11	7	10
SRI LANKA	23	48	44	41F	719	543	577	610	17	26	25	25F
SYRIA	25	23	20	26F	707	639	796	769	18	15	16	20F
TURKEY	39	25	24	25F	1387	1600	1417	1600	54	40	34	40F
YEMEN DEM	39	47	48	40F	1786	1600	1277	1625	70	74	61	65F
EUROPE	42	35	27	30	1331	1174	1283	1466	56	41	34	44
AUSTRIA	1	1	1F	1F	2523	2242	2308	3000	3	3F	3F	3F
BULGARIA	1	1	1	1F	1192	590	816	813	2		1	1F
CZECHOSLOVAK	1				1440	1637	2514	2130	1			
FRANCE	1	1	1	1*	1581	1722	1222	1667	1	2	1	2*
GREECE	1	1F	1F	1F	1026	1617	1500	1500	1	1F	1F	1F
HUNGARY	6	11F	5	6F	835	962	1401	1917	5	11	7	12F
POLAND	21	12	10*	12F	1222	742	800	917	26	9	8*	11F
PORTUGAL	6	6F	6F	6F	1724	1724	1552	1667	10	10F	9F	10F
SPAIN	1	1	1	1	2551	2700	2667	3000	2	3	2	3
YUGOSLAVIA	4	2	1	2	1512	1733	1429	1333	6	3	2	2F
OCEANIA	39	22	28	30	950	976	968	853	37	22	27	26
AUSTRALIA	39	22	28	30	950	976	968	853	37	22	27	26
USSR	2821	2774	2998	3048	878	406	1067	656	2477	1125	3198	2000
DEV.PED M E	81	58	64	66	1070	1054	1009	985	87	61	65	65
W EUROPE	13	11	11	11	1726	1869	1688	1855	23	21	18	20
OCEANIA	39	22	28	30	950	976	968	853	37	22	27	26
OTH DEV.PED	29	25	25	25	927	760	760	760	27	19	19	19
DEV.PING M E	36992	35547	35176	36346	571	588	576	561	21107	20888	20279	20372
AFRICA	14874	13932	14715	14923	577	593	578	538	8586	8257	8504	8029
LAT AMERICA	160	194	231	255	1049	1029	1274	1333	168	200	294	340
NEAR EAST	1211	1508	1486	1578	1298	931	924	913	1571	1405	1373	1442

TABLE
TABLEAU **16**
CUADRO

| | MILLET | | | | MILLET | | | | MIJO | | | |
	AREA HARV SUP RECOLTEE SUP COSECHAD	1000 HA			YIELD RENDEMENT RENDIMIENTO	KG/HA			PRODUCTION PRODUCTION PRODUCCICN	1000 MT		
	1969-71	1975	1976	1977	1969-71	1975	1976	1977	1969-71	1575	1976	1977
FAR EAST	20747	19912	18745	19589	520	554	539	539	10781	11026	10108	10561
CENTR PLANND	28050	28262	29088	29041	779	745	830	773	21857	21066	24145	22449
ASIAN CPE	25200	25464	26074	25974	768	782	803	786	19347	19921	20931	20425
E EUR+USSR	2850	2798	3014	3067	881	409	1066	660	2510	1145	3214	2024
DEV.PED ALL	2931	2856	3078	3133	886	422	1065	667	2597	1207	3279	2089
DEV.PING ALL	62192	61011	61251	62320	650	669	673	655	40453	40808	41210	40797

TABLE
TABLEAU **17**
CUADRO

| | SORGHUM | | | | SCRGHO | | | | SORGO | | | |
	AREA HARV SUP RECOLTEE SUP COSECHAD	1000 HA			YIELD RENDEMENT RENDIMIENTC	KG/HA			PRODUCTION PRODUCTION PRODUCCICN	1000 MT		
	1969-71	1975	1976	1977	1969-71	1975	1976	1977	1969-71	1975	1976	1977
WORLD	42375	43275	42739	43650	1108	1182	1214	1269	46954	51167	51886	55413
AFRICA	12791	13839	13842	13807	706	707	702	684	9036	9789	9714	9447
ALGERIA	3	1	1	1F	798	1460	1432	1455	2	2	2	2F
BENIN	91	58	84	98	565	890	670	771	52	52	57	76
BOTSWANA	133	100F	100F	50F	267	300	555	700	36	30*	56	35*
BURUNDI	93	116	100	100F	1127	1112	1203	1200	105	129	120	120F
ETHIOPIA	950	748	770	708	870	880	1120	948	827	658	863	671
GHANA	180	208	210*	218*	634	650	386	394	114	135	81*	86*
GUINEA	11	11F	11F	11F	727	455	455	455	8	5F	5F	5F
GUIN BISSAU	5	8F	8F	7F	573	600	625	571	3	5F	5F	4F
IVORY COAST	28	50	50	50F	507	637	640	700	14	32	32	35*
KENYA	201	206F	207F	208F	1070	1141	1087	1154	215	235F	225F	240F
LESOTHO	79	44	40	80F	696	555	1550	500	55	25	62	40F
MADAGASCAR	4	3	1	3F	541	809	199	667	2	3		2F
MALAWI	107	120F	120F	120F	730	833	875	875	78	100F	105F	105F
MOROCCO	67	58	49	41	1048	1282	397	114	70	75	19	5
MOZAMBIQUE	261	250F	250F	250F	774	720	1000	1000	202	180F	250*	250F
NAMIBIA	7	8F	8F	8F	329	400	400	400	2	3F	3F	3F
NIGER	589	791	633	610F	445	321	486	459	262	254	308	280F
NIGERIA	5572	5795*	5940F	6000*	652	619	620	625	3632	3590*	3680*	3750F
RHODESIA	70	70F	70F	70F	738	716	716	714	52	50F	50F	50F
RWANDA	132	133	138	139F	1068	1087	1121	1115	141	144	155	155F
SIERRA LEONE	7	7F	7F	7F	1436	1500	1529	1571	10	11F	11F	11F
SOMALIA	257	250F	250F	250F	493	400	480	480	126	100F	120F	120F
SOUTH AFRICA	358	333*	281*	308*	1050	1204	996	1253	376	401	280	386
SUDAN	1828	2596	2617	2500F	834	780	673	640	1525	2026	1762	1600F
SWAZILAND	7	3	4F	4F	432	672	714	771	3	2	3F	3F
TANZANIA	310	330F	335F	330F	654	848	776	727	203	280*	260F	240F
TUNISIA	13	14F	14F	14F	553	563	536	549	7	8F	8F	8F
UGANDA	299	311	327	344	1127	1500	1500	1500	337	467	491	516
UPPER VOLTA	1054	1138	1138	1200F	501	648	571	500	528	738	650F	600F
ZAMBIA	76	79F	80F	80F	649	658	675	638	49	52F	54F	51F
N C AMERICA	7243	7872	7735	7456	3096	2872	2852	3232	22427	22606	22061	24097
COSTA RICA	6	11	19	20	1639	1840	1743	1841	11	20	33	37
CUBA	3	1F	1F	1F	1100	1100	1100	1100	3	1F	1F	1F
DOMINICAN RP	4	5*	5*	7*	3496	3617	3422	3000	14	17	18	21*
EL SALVADOR	121	132	125	144*	1186	1322	1253	1226	144	175	156	176*
GUATEMALA	50	57	55	54F	915	1678	1752	1722	46	95	96	93
HAITI	214	220F	223F	224F	981	955	942	982	210	210F	210F	220F
HONDURAS	31	56	57*	71*	1378	954	821	867	43	53	47	62*
MEXICO	934	1116	1232	1190*	2766	2549	2565	2815	2584	2843	3160*	3350*
NETH ANTILLE	3	2F	2F	2F	602	625	706	706	2	1F	1F	1F
NICARAGUA	56	60	56	52	1015	1046	981	1018	57	63	55	53
USA	5820	6214	5960	5692	3318	3078	3068	3528	19314	19128	18284	20083
SOUTH AMERIC	2097	2488	2435	3257	1933	2374	2636	2484	4052	5906	6417	8091
ARGENTINA	1979	2010	1916	2630	1932	2456	2693	2559	3823	4938	5158	6730*
BRAZIL	1	230*	193	182	2222	1957	2536	2485	2	450	490	453
COLOMBIA	64	134	174	177	2407	2500	2464	2290	153	335	428	406
ECUADOR				1*			2000	2000				1*
PARAGUAY	4	6	7	6F	1272	1107	1317	1000	5	6	9	6F
PERU	4	10	15	18*	3072	3005	3132	2778	11	29	46	50*
URUGUAY	42	54	59	80*	1259	1430	2765	1500	53	77	162*	120*
VENEZUELA	4	44	72*	163*	1342	1600	1722	1994	5	70	124*	325*
ASIA	19682	18328	17866	18316	517	618	667	656	10172	11335	11914	12021
BANGLADESH	1	1	1	1F	683	762	737	778	1	1	1	1F
CHINA	4F	7F	7F	7F	1611	2649	1383	3054	7F	19F	10F	23F

TABLE
TABLEAU **17**
CUADRO

	SORGHUM AREA HARV / SUP RECOLTEE / SUP COSECHAD (1000 HA)				SORGHO YIELD / RENDEMENT / RENDIMIENTO (KG/HA)				SORGO PRODUCTION / PRODUCTION / PRODUCCION (1000 MT)			
	1969-71	1975	1976	1977	1969-71	1975	1976	1977	1969-71	1975	1976	1977
INDIA	17585	16092	15779	16000F	484	591	659	650	8516	9504	10396	10400F
IRAN	12	10*	10*	10F	944	1200	1000	1000	11	12*	10*	10F
IRAQ	4	5	10F	19F	1258	1054	1053	1058	5	6	10F	20F
ISRAEL	6	7	2	3F	2777	4351	5385	5440	16	32	13	14
JAPAN	1	1F	1F	1F	1143	700	1000	1000	1	1F	1F	1F
JORDAN	1	1		1F	1005	463	1477	750	1	1		1F
KOREA DPR	117	120F	120F	120F	986	1000	975	1000	115	120F	117F	120F
KOREA REP	9	6	6	6F	683	743	774	1000	6	4	5	6F
LEBANON	1	1F	1F	1F	587	538	833	833	1	1F	1F	1F
PAKISTAN	518	476	447	713F	594	591	585	561	308	281	261	400F
SAUDI ARABIA	135	237	180	122	1408	540	375	704	190	128	67	86
SRI LANKA	1	6	4	4F	648	1030	950	943	1	6	3	3F
THAILAND	47	130	143	93	2321	1538	1035	1201	108	200	148	112
VIET NAM	5	12F	12F	16F	1502	1000	1000	1563	8	12F	12F	25F
YEMEN AR	1235	1215	1145	1200F	711	830	750	667	878	1008	859	800F
EUROPE	143	159	159	188	3104	3279	3112	3432	444	521	495	646
ALBANIA	24	24F	26F	26F	1115	1167	1000	1154	27	28F	26F	30F
BULGARIA	5				2667				15			
CZECHOSLOVAK	1				3004	3682	4514	4505	3	1F	1F	1F
FRANCE	55	86	82	104	3711	3621	3283	3427	203	310	269	356
GREECE					1918	1250	1000	1000				
HUNGARY	1	6F	5	6F	1644	2545	2314	2182	1	14	12	12F
ITALY	5	3	4	9	3115	3787	4141	4992	15	13	17	47
SPAIN	46	35	36	38	3680	4156	4402	5005	168	144	159	190
YUGOSLAVIA	6	5	5	5	1856	2000	1981	1961	12	10	10	10F
OCEANIA	375	513	506	534	2030	1764	2232	1798	761	905	1128	960
AUSTRALIA	374	511	504	532	2031	1763	2230	1797	759	901	1124	956
FIJI				1F	2058	3333	3702	2000	1	2F	2*	1F
NEWCALEDONIA					1450	1400	1433	1450				
PACIFIC IS					2500	2500	2500	2500				
PAPUA N GUIN	1	1	1F	1F	1546	1487	2273	2364	1	2	3F	3F
USSR	43	77	196	90	1423	1377	801	1667	62	106	157	150F
DEV.PED M E	6670	7195	6876	6692	3128	2910	2932	3294	20863	20941	20157	22043
N AMERICA	5820	6214	5960	5692	3318	3078	3068	3528	19314	19128	18284	20083
W EUROPE	112	129	128	157	3566	3703	3571	3852	398	478	456	603
OCEANIA	374	511	504	532	2031	1763	2230	1797	759	901	1124	956
OTH DEV.PED	365	341	284	312	1078	1271	1033	1286	393	434	294	401
DEV.PING M E	35504	35834	35497	36693	728	835	884	900	25853	29926	31394	33009
AFRICA	10605	10910	10945	10999	673	675	701	678	7135	7362	7672	7461
LAT AMERICA	3520	4146	4209	5022	2036	2263	2422	2411	7165	9383	10194	12105
NEAR EAST	3216	4065	3962	3853	812	782	684	653	2611	3180	2709	2517
FAR EAST	18162	16711	16379	16817	492	598	660	649	8940	9996	10814	10922
OTH DV.PING	1	2	2	2	1732	1961	2666	2222	2	4	4	4
CENTR PLANND	201	246	367	265	1181	1220	914	1360	237	300	335	361
ASIAN CPE	126	139	139	143	1027	1085	998	1169	129	151	139	168
E EUR+USSR	75	107	227	122	1441	1395	862	1586	108	149	196	193
DEV.PED ALL	6745	7302	7103	6814	3109	2888	2865	3263	20971	21090	20354	22236
DEV.PING ALL	35630	35973	35636	36836	729	836	885	901	25983	30077	31532	33177

TABLE
TABLEAU **18**
CUADRO

ROOTS AND TUBERS,TOTAL   RACINES+TUBERCULES,TOTAL   RAICES+TUBERCULOS,TOTAL

AREA HARV / SUP RECOLTEE / SUP COSECHAD — 1000 HA
YIELD / RENDEMENT / RENDIMIENTO — KG/HA
PRODUCTION / PRODUCTION / PRODUCCION — 1000 MT

	AREA HARV 1969-71	1975	1976	1977	YIELD 1969-71	1975	1976	1977	PRODUCTION 1969-71	1975	1976	1977
WORLD	51195	51592	51400	51898	10601	10720	10958	10987	542720	553065	563261	570211
AFRICA	10511	11014	11362	11567	6684	6856	6879	6814	70259	75517	78157	78814
ALGERIA	43	70	69	72F	5906	8206	7094	6944	253	575	493	500F
ANGOLA	144	143F	143F	144F	12365	12497	12552	12962	1781	1792F	1800F	1860F
BENIN	185	133	130	156	7219	5188	7201	7809	1336	690	938	1216
BOTSWANA	1	1F	1F	1F	4306	4615	4846	5000	5	6F	6F	7F
BURUNDI	203	172	179F	186F	8850	11005	10814	10638	1799	1895	1937F	1979F
CAMEROON	496	594F	557F	576F	3464	3396	3607	3559	1717	2017	2009F	2049F
CAPE VERDE	3	3F	3	3F	5787	6178	9208	8327	14	16F	24	21F
CENT AFR EMP	279	306	315F	324F	2932	2973	2891	2967	818	910	912	963
CHAD	30	34F	36F	37F	3584	2979	2915	2857	108	101F	104F	106F
COMOROS	24	27F	27F	29F	3401	3464	3445	3392	81	93F	94F	97F
CONGO	152	133	149	163	4180	5712	5813	5359	635	761	868	877
EGYPT	35	47	60	66F	17438	18075	17296	16837	617	858	1039	1118
EQ GUINEA	23	29F	29F	30F	3058	2727	2740	2733	70	78F	80F	82F
ETHIOPIA	289	299F	303F	307F	3098	3144	3160	3175	894	940F	957F	974F
GABON	99	103F	103F	104F	2243	2246	2241	2244	221	231	231F	233F
GAMBIA	2	2F	2F	2F	3820	4000	4000	3500	8	9F	9F	7F
GHANA	714	607	641	645	6270	6930	6911	6868	4475	4207	4430	4430
GUINEA	87	75F	82F	83F	7271	8067	8049	8078	635	605F	660F	666F
GUIN BISSAU	7	7F	7F	7F	5641	4615	4615	6154	37	30F	30F	40F
IVORY COAST	557	549	659	679	4114	4961	4495	4514	2291	2721	2960	3065
KENYA	203	231F	247F	242F	6926	6765	6883	6880	1406	1563	1700F	1665F
LESOTHO	1	1F	1F	1F	11437	11644	11757	12000	7	9F	9F	9F
LIBERIA	83	79F	80F	80F	3655	3833	3875	3948	304	305F	310F	316F
LIBYA	2	17F	17F	17F	6450	4706	4706	4706	15	80*	80F	80F
MADAGASCAR	281	309	303F	304F	6398	5971	5916	5965	1800	1845	1795	1812
MALAWI	30	31F	31F	32F	5752	6677	6387	5594	174	207F	198F	179F
MALI	9	9F	9F	9F	8944	9141	9228	9315	80	84F	85F	86F
MAURITANIA	4	5F	5F	5F	1707	1093	1087	1120	6	5F	5F	6F
MAURITIUS	1	1	1F	1F	14721	15111	15133	16685	8	10	11F	12F
MOROCCO	28	19	16	17F	10119	10007	10359	10588	283	195	170	180F
MOZAMBIQUE	465	465F	466F	466F	5668	5111	5340	5444	2634	2378F	2486F	2534F
NAMIBIA	14	16F	16F	16F	8519	8750	8750	8768	115	140F	140F	144F
NIGER	27	23	33F	35F	7136	8542	7636	7143	191	192	252	250F
NIGERIA	2610	2737F	2828F	2848F	9608	10054	9982	9736	25073	27518F	28230F	27730F
REUNION	1	1	1	1F	11919	9943	10606	10727	9	11	11	12F
RHODESIA	2	2F	2F	2F	10967	11291	11291	11291	22	23F	23F	23F
RWANDA	121	148	147	148F	7285	8424	8918	8856	878	1245	1307	1310F
SAO TOME ETC					40558	39231	39808	40385	9	10F	10F	11F
SENEGAL	40	41	24	28F	4388	2902	5010	4813	174	118	123	135F
SEYCHELLES					6062	6118	6163	6207		1F	1F	1F
SIERRA LEONE	30	32F	33F	33F	3285	3281	3287	3308	97	105F	108F	110F
SOMALIA	3	3F	3F	3F	10480	10690	10667	10645	28	31F	32F	33F
SOUTH AFRICA	57	61F	61F	63F	11160	11885	11475	12413	635	725	700	782
SUDAN	68	85	85F	85F	3467	3693	3671	3600	237	312	312F	306F
SWAZILAND	4	2	4F	4F	3147	4307	3784	3784	13	11	14F	14F
TANZANIA	750	874	886F	908F	4920	4813	4877	4878	3690	4207	4322F	4431F
TOGO	218	106	110F	116F	7028	8280	8096	7960	1534	879	893	921F
TUNISIA	4	5F	5F	5F	19829	22222	22826	18367	69	100	105	90*
UGANDA	427	510	538	542	4449	3824	3936	3958	1898	1951	2116	2144
UPPER VOLTA	23	24F	28F	25F	4200	4917	4464	4400	98	118F	125F	110F
ZAIRE	1582	1788	1831	1863F	6822	6949	6949	6926	10793	12423	12722	12905F
ZAMBIA	54	56F	56F	57F	3288	3286	3286	3286	179	184F	184F	187F
N C AMERICA	1078	1028	1080	1071	18356	19663	20388	20410	19796	20207	22026	21849
ANTIGUA					4009	3571	3198	3318				
BARBADOS	2	1F	2F	2F	11415	8123	9170	9186	21	11	17	17F
BELIZE	1	1F	1F	1F	17308	17857	17816	17778	14	15F	16F	16F
BERMUDA					13827	15865	19760	15620	1	1	1	1
CANADA	121	106	107	112	19113	20735	21890	22393	2312	2205	2350	2498
COSTA RICA	6	4	4	4F	6686	8668	9432	9390	39	36	39	39F
CUBA	109	123F	128F	130F	5097	5235	5385	5404	556	646	688	700F
DOMINICA					26183	26409	26481	26540	8	9	9F	9F
DOMINICAN RP	31	28F	32F	30F	10990	11055	10878	11081	345	304	347	328
EL SALVADOR	2	2	2	2F	8672	12055	13314	14000	15	30	28	31F
GRENADA	1				5893	4333	4444	4444	5			
GUADELOUPE	4	5	4	4F	11353	14820	15577	15184	50	72	59	59F
GUATEMALA	9	10F	10F	11F	3917	3722	3676	3651	35	36	38F	39F
HAITI	52	55F	55F	56F	5059	5216	5297	5233	261	287F	292F	291F
HONDURAS	7	8	8F	8F	6303	2548	2654	2778	43	20	21F	22F
JAMAICA	15	21F	19F	19F	9846	10523	10324	10294	148	216	193	200F
MARTINIQUE	4	3	3	3	9890	10484	10484	10484	36	33	33	33
MEXICO	65	72	71	67	10312	12143	12227	12438	667	872	862	828
NICARAGUA	5	6	6	7	4104	4058	4059	4059	18	24	25	27
PANAMA	8	9F	9F	10F	8011	8141	8299	7917	60	73	76	76F
PUERTO RICO	10	8	8	8F	3961	5223	4776	4842	41	40	39	38
ST KITTS ETC					2774	6250	6818	7786				
ST LUCIA	2	2F	2F	2F	3975	4157	4152	4033	8	9F	10F	10F
ST VINCENT	3	3F	3F	3F	8248	8370	8298	8333	21	22	22	23F
TRINIDAD ETC	2	2	2F	2F	9445	11724	11620	11489	17	20	21F	22F
USA	621	559	604	592	24254	27225	27867	27945	15073	15226	16841	16544
SOUTH AMERIC	3888	3909	3927	4084	12001	10934	11049	10883	46659	42738	43392	44444
ARGENTINA	260	172	168	170	11407	12041	12537	13743	2964	2067	2101	2337
BOLIVIA	130	169	171	170	7312	7002	6999	6141	949	1185	1196	1041
BRAZIL	2439	2451	2464	2585	13789	11966	12202	11693	33634	29331	30062	30226
CHILE	77	73	69	87	9373	10284	7877	10768	721	746	547	937
COLOMBIA	275	374	373	387	8414	9129	9455	9856	2312	3414	3528	3815
ECUADOR	88	89	86	84	11000	12264	10671	10611	968	1086	920	891
FR GUIANA	1	1F	1F	1F	7378	6711	6711	6711	6	5F	5F	5F

TABLE
TABLEAU 18
CUADRO

ROOTS AND TUBERS,TOTAL   RACINES+TUBERCULES,TOTAL   RAICES+TUBERCULOS,TOTAL

	AREA HARV SUP RECOLTEE SUP COSECHAD		1000 HA	YIELD RENDEMENT RENDIMIENTO		KG/HA		PRODUCTION PRODUCTICN PRODUCCION		1000 MT	
	1969-71	1975	1976	1977	1969-71	1975	1976	1977	1969-71	1975	1976	1977
GUYANA	4	3F	3F	3F	6748	6871	6284	6294	25	23	21*	21F
PARAGUAY	112	121	124	130F	13889	13987	13850	14056	1559	1691	1724	1834F
PERU	387	345	351	349	6995	6920	6849	6528	2705	2387	2404	2419
SURINAM					6363	6474	6500	6492	2	2F	2F	2F
URUGUAY	35	40	41	41	6016	5025	6138	5244	213	200	251	215
VENEZUELA	80	71	76	76	7488	8414	8355	9235	600	600	631	700
ASIA	20253	21129	21439	21747	9019	10116	10238	10325	182669	213748	219485	224544
AFGHANISTAN	15	17	18	19F	10667	11441	11500	12000	160	195	207	228F
BANGLADESH	157	160	167	148	10637	9968	10112	10076	1672	1599	1694	1491
BHUTAN	5	6F	6F	6F	6664	6683	6715	6744	31	40F	42F	44F
BRUNEI					9379	8553	8974	9263	4	3F	3F	4F
BURMA	13	16	17	17F	5112	5073	5373	5335	65	81	90	91F
CHINA	14948F	15247F	15398F	15548F	8866	10150	10171	10236	132539F	154759F	156616F	159146F
CYPRUS	10	9	10	10F	18603	21050	23443	23311	189	185	225	240F
EAST TIMOR	14	15F			5699	5164			78	79F		
HONG KONG					10510	13115	7265	11538	4	2	1	1
INDIA	1078	1206	1266	1257	10888	11787	12448	12192	11735	14210	15755	15328
INDONESIA	2042	2002	1934	1958	6973	8205	8435	8221	14238	16429	16313	16097
IRAN	46	50F	68	65F	9072	10120	8088	8523	417	506	550	580F
IRAQ	1	5*	6F	6F	14256	8381	8364	8473	16	44*	46F	47F
ISRAEL	5	5	5	7F	25869	32929	32412	32554	131	163	175	212
JAPAN	348	313	306	307F	19531	16792	18329	18330	6795	5263	5614	5620F
JORDAN			1	1F	5717	10870	14773	14773	1	5	13	13F
KAMPUCHEA DM	6	5F	6F	6F	8839	8113	8389	8406	49	43F	51F	54F
KOREA DPR	117	139F	144F	155F	10571	11022	11417	11355	1233	1532F	1644F	1760F
KOREA REP	179	147	138	144F	14858	17838	17260	17292	2652	2629	2375	2495F
LAO	6	7F	7F	7F	6945	7015	7118	7217	42	47F	48F	50F
LEBANON	9	9F	9F	9F	10636	9438	9546	9651	98	81F	86F	91F
MACAU					10563	10000	10000	10000	1	1F	1F	1F
MAL PENINSUL	30	20	20F	20F	8462	15356	15385	15194	252	301	300F	305F
MAL SABAH	4	6	9	9F	6407	5332	3890	3947	28	34F	35F	37F
MAL SARAWAK	13	14F	14F	14F	10013	9191	10176	10210	132	130F	145F	146F
MALDIVES	1	1F	1F	1F	4979	5148	5118	5086	5	6F	6F	7F
MONGOLIA	3	4	5	4F	6056	9465	7660	8000	18	41	36	32*
NEPAL	60	66	66	65	5369	5700	5847	5240	324	374	383	338
OMAN					391	440	462	481				
PAKISTAN	36	45	46	44	9906	9792	10301	10516	354	442	477	477
PHILIPPINES	250	315F	326	294F	5127	5467	4971	5409	1280	1720	1622	1591F
SAUDI ARABIA					8527	2626	7636	7083	1	1F	1F	1F
SINGAPORE	1				9977	11791	11767	11111	7	2	2	2F
SRI LANKA	85	216	200	224F	5550	4630	4813	4474	471	1002	962	1000F
SYRIA	5	10	10	11F	11360	13158	12717	12273	62	125	126	135F
THAILAND	245	496	636	741	14264	13613	13364	14924	3497	6753	8500	11054
TURKEY	160	178	187	190F	12377	13989	15241	15263	1984	2490	2850	2900F
VIET NAM	356	391F	406F	451F	5808	6041	5946	6311	2068	2361F	2414F	2848F
YEMEN AR	5	7	7	7F	7242	10923	11176	11429	35	71	76	80F
YEMEN DEM					4752	5030	5030	5030	1	1	1	1F
EUROPE	7214	6362	6267	6123	17633	17248	17979	18729	127208	109735	112682	114680
ALBANIA	19	16F	16F	16F	5747	6710	7871	7871	109	104F	122F	122F
AUSTRIA	109	69	73	60	25506	22846	23816	22463	2787	1579	1746	1352
BELGIUM-LUX	46	37	38	46*	35363	34682	23175	28478	1631	1300	879	1310*
BULGARIA	30	30	29	29F	12781	10593	12050	12069	378	318	350	350F
CZECHOSLOVAK	331	250	240	236F	14683	14241	17552	16052	4864	3565	4214	3785
DENMARK	34	31	35	36*	23909	21484	16603	22222	815	666	576	800*
FAEROE IS					14611	13253	13253	13253	1	1F	1F	1F
FINLAND	56	49	53	46	16171	14012	17987	15873	906	680	948	737
FRANCE	402	311	281	298*	22342	21382	15217	27483	8986	6642	4279	8190*
GERMAN DR	643	574	599	564F	16225	13361	11375	17700	10432	7673	6816	9976*
GERMANY FED	580	415	415	396	27250	26127	23618	28404	15804	10853	9808	11251
GREECE	55	57	64	59	12728	15390	15477	15883	702	860	993	938
HUNGARY	169	130	116	100	11095	12531	12049	14142	1877	1634	1399	1419
ICELAND	1	1F	1F	1F	8051	8571	8750	8750	7	6	7	7F
IRELAND	55	41	47	53	26581	25086	24900	22464	1450	1018	1179	1200F
ITALY	278	180	175	180	13158	16491	17167	18555	3659	2966	3001	3335
LIECHTENSTEN	1	1F	1F	1F	18466	18679	18704	18727	9	10F	10F	10F
MALTA	3	2	2	2F	8099	8498	9375	9565	21	18	22	22F
NETHERLANDS	153	151	161	170	35169	33098	29729	33772	5367	5003	4783	5752
NORWAY	33	25	28	26	23480	17518	17339	23221	776	435	484	605
POLAND	2707	2581	2466	2400*	16630	17989	20256	17208	45012	46429	49951	41300*
PORTUGAL	126	123	131	133	10273	9489	8156	9086	1293	1168	1072	1211
ROMANIA	310	305	301	270*	8627	8905	15914	13844	2671	2716	4788	3738
SPAIN	393	389	395	383	12794	13849	14453	14623	5031	5382	5709	5602
SWEDEN	52	42	46	48	23495	19926	23251	28222	1221	837	1058	1346
SWITZERLAND	31	25	25	23*	32951	36585	31400	37826	1016	908	769	870*
UK	270	214	222	232	27223	21266	21572	28440	7359	4551	4789	6598
YUGOSLAVIA	328	314	308	315	9198	7624	9506	9060	3020	2394	2928	2854
OCEANIA	230	238	237	240	10393	10150	10194	10341	2390	2417	2417	2479
AMER SAMOA					15432	18291	18950	19582	5	9	9	10
AUSTRALIA	43	38	34	34	18310	19662	20523	21392	784	744	699	731
COOK ISLANDS					24867	39273	40925	41281	11	11	12	12
FIJI	14	15F	15F	15F	9399	9307	9306	9304	133	141F	142F	143F
FR POLYNESIA	1	1F	1F	1F	12245	12734	12713	12692	15	16F	16F	17F
GILBERT IS	1	1F	1F	1F	8368	8496	8547	8595	8	10F	10F	10F
GUAM					15532	15368	15022	14692	1	1	1F	1F
NEWCALEDONIA	3	2	2F	2F	6091	6365	6320	6292	19	14	14F	14F
NEW HEBRIDES	1	1F	1F	1F	14302	14286	14300	14600	12	14F	14F	15F
NEW ZEALAND	12	9	9F	9F	24452	25048	28619	29492	304	233	259*	267F
NIUE ISLAND	1	2	2	2F	1231	828	830	830	1	1F	1F	1F
PACIFIC IS	1	1F	1	1F	8317	8397	8382	8286	9	11F	12F	12F

TABLE
TABLEAU **18**
CUADRO

ROOTS AND TUBERS,TOTAL   RACINES+TUBERCULES,TOTAL   RAICES+TUBERCULOS,TOTAL

	AREA HARV SUP RECOLTEE SUP COSECHAD		1000 HA	YIELD RENDEMENT RENDIMIENTO		KG/HA		PRODUCTION PRODUCTION PRODUCCION		1000 MT	
	1969-71	1975	1976	1977	1969-71	1975	1976	1977	1969-71	1975	1976	1977
PAPUA N GUIN	135	148F	150F	152F	6755	6879	6879	6879	910	1019F	1034F	1048F
SAMOA	4	4F	5	5F	6843	6903	6149	6787	27	28	28F	32F
SOLOMON IS	6	6F	6F	6F	11624	12564	12479	12397	68	74F	75F	76F
TOKELAU					20500	20833	21667	22500				
TONGA	7	8F	8F	8F	11805	10824	10831	10904	83	89F	90F	91F
WALLIS ETC					5528	5684	5625	5567	1F	1F	1F	1F
USSR	8019	7912	7087	7067	11689	11211	12008	11801	93739	88703	85102	83400
DEV.PED M E	4213	3568	3628	3632	20861	20140	19757	22204	87898	71856	71680	80645
N AMERICA	742	666	712	704	23416	26189	26966	27065	17386	17431	19191	19042
W EUROPE	3006	2476	2500	2509	20581	19104	18013	21518	61863	47297	45041	53991
OCEANIA	55	47	43	43	19693	20725	22222	23087	1088	977	958	998
OTH DEV.PED	410	379	373	376	18447	16214	17411	17585	7561	6151	6489	6614
DEV.PING M E	19324	20439	20960	21420	8271	8383	8496	8480	159832	171333	178077	181637
AFRICA	10348	10804	11139	11335	6644	6807	6825	6751	68755	73542	76026	76528
LAT AMERICA	4224	4271	4296	4451	11617	10657	10760	10617	49069	45513	46225	47251
NEAR EAST	359	434	477	487	10681	11421	11757	11948	3832	4953	5612	5819
FAR EAST	4218	4740	4853	4950	8741	9681	10046	10213	36874	45884	48755	50557
OTH DV.PING	175	191	194	197	7454	7542	7523	7538	1303	1441	1460	1482
CENTR PLANND	27657	27585	26812	26846	10666	11234	11693	11470	294990	309876	313504	307929
ASIAN CPE	15430	15787	15958	16165	8808	10055	10074	10136	135907	158735	160761	163840
E EUR+USSR	12228	11798	10854	10681	13010	12810	14073	13490	159083	151141	152743	144089
DEV.PED ALL	16441	15366	14482	14313	15022	14512	15497	15701	246981	222996	224422	224734
DEV.PING ALL	34754	36226	36918	37585	8510	9111	9178	9192	295738	330069	338838	345477

TABLE
TABLEAU 19
CUADRO

| | POTATOES | | | | POMMES DE TERRE | | | | PATATAS | | | |
| | AREA HARV
SUP RECOLTEE
SUP COSECHAD | | 1000 HA | | YIELD
RENDEMENT
RENDIMIENTO | | KG/HA | | PRODUCTION
PRODUCTION
PRODUCCION | | 1000 MT | |
	1969-71	1975	1976	1977	1969-71	1975	1976	1977	1969-71	1975	1976	1977
WORLD	22383	21806	21037	20945	13280	13134	13853	13986	297239	286401	291422	292938
AFRICA	379	510	542	56C	7705	7874	7709	7810	2920	4014	4180	4377
ALGERIA	43	70	69	72F	5906	8206	7094	6944	253	575	493	500F
ANGOLA	5	5F	5F	6F	6296	5926	6481	7273	34	32F	35F	40F
BURUNDI	14	27	28F	29F	5435	5369	5330	5294	77	146	149F	151F
CAMEROON	11	15F	15F	16F	2809	2667	2667	2561	30	40F	40F	40F
CAPE VERDE					12186	14667	20833	13889	1	1F	2	1F
CENT AFR EMP					1968	2000	2000	2000		1F	1F	1F
CHAD					4243	3333	3333	3333	1	1F	1F	1F
CONGO					7083	8056	8194	8108	2	3F	3F	3F
EGYPT	30	41	54	60F	16570	17489	16608	16167	496	720	893	970*
ETHIOPIA	30	32F	33F	33F	5303	5313	5345	5376	161	170F	174F	177F
KENYA	55	80F	90F	90F	3745	4160	4111	4167	206	333	370F	375F
LIBYA	2	17F	17F	17F	6450	4706	4706	4706	15	80*	80F	80F
MADAGASCAR	16	24	24F	24F	6428	5024	5000	5000	106	123	120	120F
MALAWI	25	25F	25F	26F	3387	3480	3520	3423	85	87F	88F	89F
MAURITANIA					20670	15000	13846	13571	2	2F	2F	2F
MAURITIUS		1	1F	1F	15027	15204	15385	16923	7	10	10F	11F
MOROCCO	28	19	16	17F	10119	10007	10359	10588	283	195	170	180F
MOZAMBIQUE	6	6F	6F	6F	6917	6667	7000	6667	40	40F	42F	40F
NIGERIA	2	2F	2F	2F	12514	14000	13636	13636	25	28F	30F	30F
REUNION					11121	3521	5164	5333	1	2	2	2F
RHODESIA	2	2F	2F	2F	11167	11500	11500	11500	22	23F	23F	23F
RWANDA	19	23	26	26F	7153	6604	6604	6538	134	150	170	170F
SENEGAL	1	1F	1F	1F	6543	6000	5823	5663	4	5	5F	5F
SOUTH AFRICA	44	48F	48F	50F	13688	14292	13729	14800	596	686	659	740
SUDAN	1	1F	1F	1F	17446	20692	19231	16923	25	27	25F	22F
SWAZILAND	2	1F	2F	2F	2350	2218	2500	2500	5	2	5F	5F
TANZANIA	16	21	22F	23F	3918	3762	3818	3826	61	79	84F	88F
TUNISIA	4	5F	5F	5F	19829	22222	22826	18367	69	100	105	90*
UGANDA	17	35	43	45	8800	9099	8600	8601	147	321	366	384
ZAIRE	5	6*	6*	6F	5483	5333	5433	5583	28	32*	33*	34F
ZAMBIA					8823	9000	9000	9000	3	3F	3F	3F
N C AMERICA	762	703	750	740	22957	25274	26036	26203	17482	17757	19537	19394
BERMUDA					15565	16020	20000	16213	1	1	1	1
CANADA	121	106	107	112	19113	20735	21890	22393	2312	2205	2350	2498
COSTA RICA	3	2	2	2F	9902	11523	11645	11667	27	24	25	25F
CUBA	9	13F	15F	15F	9359	9342	9673	9820	83	117	145	147F
DOMINICA					4764	5714	5857	6000	1	1F	1F	1F
DOMINICAN RP	2	1F	2F	2F	12703	12143	12222	12000	23	17	22	18*
EL SALVADOR		1	1	1F	7367	18500	19700	19750	3	13	16	16F
GUADELOUPE					7000	7000	7000	7000				
GUATEMALA	7	7F	8F	8F	4209	4061	4000	3974	28	28	30F	31F
HAITI		1F	1F	1F	15217	14400	13455	14545	7	7F	7F	8F
HONDURAS	1				6966	11750	12000	11905	4	5F	5F	5F
JAMAICA	1	1F	1F	1F	9731	9723	9554	10125	10	14	8	8F
MEXICO	46	57	56	52	10631	12130	12268	12558	489	693	687	653
NICARAGUA					4000	4109	4115	4124	1	2	2	2
PANAMA	1	1F	1F	1F	10057	10886	11036	11111	9	9	11	10F
ST KITTS ETC					800	7500	9000	9077				
USA	571	512	556	546	25386	28574	29164	29246	14483	14622	16228	15972
SOUTH AMERIC	1038	944	969	977	8389	8760	8988	9536	8707	8267	8710	9321
ARGENTINA	190	111	108	112	11617	12182	14177	15802	2212	1349	1528	1777
BOLIVIA	95	128	128	126	6968	6532	6421	5406	660	834	824	679
BRAZIL	214	193	202	194	7260	8661	8969	9803	1557	1669	1816	1900
CHILE	76	72	68	86	9312	10316	7875	10813	707	738	539	928
COLOMBIA	84	110	125	128	10407	12000	12126	12567	871	1320	1516	1609
ECUADOR	47	39	41	40*	11818	12643	12924	12766	560	499	533	504*
PARAGUAY	2	1			4126	6286	8028	7955	6	4	4	4F
PERU	293	251*	256*	250*	6413	6299	6445	6400	1877	1581	1650	1600*
URUGUAY	22	26	26	26*	6040	4689	6412	5000	135	121	166	130*
VENEZUELA	15	14	14	16	8258	10751	9982	12025	121	152	135	190
ASIA	4931	5343	5393	5448	9380	10685	11203	11183	46249	57096	60412	60922
AFGHANISTAN	15	17	18	19F	10667	11441	11500	12000	160	195	207	228F
BANGLADESH	85	94	96	77	9863	9379	9413	9503	842	880	903	735
BHUTAN	4	5F	6F	6F	6667	6679	6709	6737	27	35F	37F	38F
BURMA	9	11	11	11F	4576	4342	4843	4832	40	46	55	55F
CHINA	3663F	3870F	3853F	3904F	8745	10348	10704	10669	32029F	40040F	41244F	41646F
CYPRUS	10	9F	9F	10F	18641	21176	23656	23500	183	180F	220F	235F
EAST TIMOR					9519	8000			1	1F		
INDIA	501	587	622	634	8950	10598	11738	11494	4482	6225	7306	7287
INDONESIA	16	21F	18	19F	7058	7042	7056	7058	114	150F	127	133F
IRAN	46	50F	68	65F	9072	10120	8088	8923	417	506	550	580F
IRAQ	1	5*	6F	6F	14256	8381	8364	8473	16	44*	46F	47F
ISRAEL	5	5	5	7F	25869	32929	32412	32554	131	163	175	212
JAPAN	164	201	197	197F	21257	16224	18995	18995	3490	3261	3742	3742F
JORDAN			1	1F	5717	10870	14773	14773	1	5	13	13F
KOREA DPR	96	115F	120F	130F	10000	10435	10833	10769	960	1200F	1300F	1400F
KOREA REP	54	53	50	50F	11100	12807	11790	13000	598	675	591	650F
LAO	3	3F	3F	3F	5467	6240	6320	6400	14	16F	16F	16F
LEBANON	9	9F	9F	9F	10694	9412	9528	9634	96	80F	85F	90F
MONGOLIA	3	4	5	4F	6056	9465	7660	8000	18	41	36	32*
NEPAL	49	54	53	52	5425	5726	5925	5173	265	307	314	269
PAKISTAN	19	28	29	26	10955	10455	11221	12345	213	289	321	318
PHILIPPINES	3	4F	5	5F	7006	5112	4312	4396	22	20	20	21F
SAUDI ARABIA					8527	2626	7636	7083	1		1F	1F
SRI LANKA	3	3	3	4F	9533	8944	8955	8571	30	28	28	30F

TABLE
TABLEAU **19**
CUADRO

POTATOES				POMMES DE TERRE				PATATAS			
AREA HARV SUP RECOLTEE SUP COSECHAD	1000 HA			YIELD RENDEMENT RENDIMIENTO	KG/HA			PRODUCTION PRODUCTICN PRODUCCICN	1000 MT		
	1969-71	1975	1976	1977	1969-71	1975	1976	1977	1969-71	1975	1976	1977

	1969-71	1975	1976	1977	1969-71	1975	1976	1977	1969-71	1975	1976	1977
SYRIA	5	10	10	11F	11360	13158	12717	12273	62	125	126	135F
THAILAND	1	2F	2F	2F	7650	6667	6250	5882	9	10F	10F	10F
TURKEY	160	178	187	190F	12377	13989	15241	15263	1984	2490	2850	2900F
VIET NAM	1	1F	1F	1F	17793	13750	14000	13846	10	11F	14F	18F
YEMEN AR	5	7	7	7F	7242	10923	11176	11429	35	71	76	80F
YEMEN DEM					4752	5030	5030	5030	1	1	1	1F
EUROPE	7199	6348	6253	6109	17649	17264	17996	18747	127059	109592	112531	114535
ALBANIA	19	16F	16F	16F	5747	6710	7871	7871	109	104F	122F	122F
AUSTRIA	109	69	73	60	25506	22846	23816	22463	2787	1579	1746	1352
BELGIUM-LUX	46	37	38	46*	35363	34682	23175	28478	1631	1300	879	1310*
BULGARIA	30	30	29	29F	12781	10593	12050	12069	378	318	350	350F
CZECHOSLOVAK	331	250	240	236F	14683	14241	17552	16052	4864	3565	4214	3785
DENMARK	34	31	35	36*	23909	21484	16603	22222	815	666	576	800*
FAEROE IS					14611	13253	13253	13253	1	1F	1F	1F
FINLAND	56	49	53	46	16171	14012	17987	15873	906	680	948	737
FRANCE	402	311	281	298*	22342	21382	15217	27483	8986	6642	4279	8190*
GERMAN DR	643	574	599	564*	16225	13361	11375	17700	10432	7673	6816	9976*
GERMANY FED	580	415	415	356	27250	26127	23618	28404	15804	10853	9808	11251
GREECE	55	57	64	59F	12727	15404	15484	15893	700	878	991	936*
HUNGARY	169	130	116	100*	11097	12538	12059	14160	1874	1630	1396	1416*
ICELAND	1	1F	1F	1F	8051	8571	8750	8750	7	6	7	7F
IRELAND	55	41	47	53	26581	25086	24900	22464	1450	1018	1179	1200F
ITALY	277	179	174	179	13133	16472	17133	18535	3632	2943	2975	3310
LIECHTENSTEN	1	1F	1F	1F	18466	18679	18704	18727	9	10F	10F	10F
MALTA	3	2	2	2F	8099	8498	9375	9565	21	18	22	22F
NETHERLANDS	153	151	161	170	35169	33098	29729	33772	5367	5003	4783	5752
NORWAY	33	25	28	26	23480	17518	17339	23221	776	435	484	605
POLAND	2707	2581	2466	2400*	16630	17989	20256	17208	45012	46429	49951	41300*
PORTUGAL	117	115	123	125	10446	9585	8153	9146	1222	1098	1003	1144
ROMANIA	310	305	301	270*	8627	8905	15914	13844	2671	2716	4788	3738
SPAIN	389	385	391	379	12811	13871	14480	14652	4985	5338	5659	5553
SWEDEN	52	42	46	48	23495	19926	23251	28222	1221	837	1058	1346
SWITZERLAND	31	25	25	23*	32951	36585	31400	37826	1016	908	769	870*
UK	270	214	222	232	27223	21266	21572	28440	7359	4551	4789	6598
YUGOSLAVIA	328	314	308	315	9198	7624	9506	9060	3020	2394	2928	2854
OCEANIA	55	47	43	43	19694	20682	22156	23013	1082	971	950	990
AUSTRALIA	43	38	34	34	18384	19719	20607	21490	782	742	697	729
FIJI					5569	5833	5600	5385	1	1F	1F	1F
FR POLYNESIA					10000	10000	10000	10000	1	1F	1F	1F
NEWCALEDONIA					7683	10000	10000	10000	1	1	1F	1F
NEW ZEALAND	12	9	9F	9F	24790	25470	29070	29919	297	226*	250*	257F
PAPUA N GUIN					5889	6500	6500	6500				
USSR	8019	7912	7087	7067	11689	11211	12008	11801	93739	88703	85102	83400
DEV.PED M E	3950	3381	3443	3449	21218	20429	20039	22614	83810	69062	68993	77997
N AMERICA	692	618	664	658	24289	27225	27987	28084	16796	16827	18578	18470
W EUROPE	2991	2462	2486	2495	20633	19154	18055	21578	61718	47158	44893	53848
OCEANIA	55	46	42	43	19790	20814	22323	23195	1079	967	947	986
OTH DEV.PED	213	254	250	254	19817	16184	18274	18515	4217	4110	4576	4694
DEV.PING M E	2443	2638	2762	2776	8732	9441	9811	9998	21332	24910	27095	27758
AFRICA	302	402	422	432	5926	6217	5976	5936	1788	2502	2523	2565
LAT AMERICA	1108	1028	1056	1060	8478	8945	9159	9666	9392	9196	9668	10245
NEAR EAST	286	344	387	396	12226	13168	13378	13578	3491	4524	5173	5381
FAR EAST	747	864	897	888	8906	10054	10847	10775	6657	8684	9727	9563
OTH DV.PING		1	1	1	8402	8626	8803	8276	4	5	5	5
CENTR PLANND	15989	15788	14833	14720	12014	12189	13169	12716	192097	192429	195334	187183
ASIAN CPE	3762	3990	3979	4039	8776	10350	10705	10670	33016	41292	42594	43096
E EUR+USSR	12227	11798	10854	10681	13010	12810	14073	13490	159080	151137	152740	144087
DEV.PED ALL	16177	15178	14297	14130	15014	14507	15510	15717	242890	220159	221733	222084
DEV.PING ALL	6205	6628	6741	6815	8759	9988	10339	10397	54349	66202	69689	70854

TABLE
TABLEAU 20
CUADRO

	SWEET POTATOES				PATATES DOUCES				BATATAS (CAMOTES)			
	AREA HARV SUP RECOLTEE SUP COSECHAD		1000 HA		YIELD RENDEMENT RENDIMIENTO		KG/HA		PRODUCTION PRODUCTION PRODUCCION		1000 MT	
	1969-71	1975	1976	1977	1969-71	1975	1976	1977	1969-71	1975	1976	1977
WORLD	13941	14014	14230	14334	8770	9662	9562	9638	122270	135400	136074	138148
AFRICA	755	799	819	829	6081	6236	6316	6336	4589	4985	5175	5253
ANGOLA	18	18F	18F	18F	8339	8889	9167	9444	147	160F	165F	170F
BENIN	17	11	10	11F	3878	3882	3902	3786	66	44	39	42F
BURUNDI	101	76F	81F	85F	8231	10658	10440	10246	831	810F	842F	873F
CAMEROON	38	49F	49F	49F	2883	3265	3265	3265	111	160F	160F	160F
CAPE VERDE					11117	12000	25000	20000	3	3F	7	5F
CENT AFR EMP	17	20F	20F	20F	2980	3000	3040	3080	51	60F	61F	62F
CHAD	13	14F	15F	15F	4131	3203	3158	3114	52	46F	46F	46F
COMOROS	1	1F	1F	2F	10159	9630	9571	8750	11	13F	13F	14F
CONGO	14	16F	17F	17F	6005	6038	6118	6176	83	96F	104F	105F
EGYPT	4	4	3	3F	21309	19836	20530	20294	87	75	69	69F
EQ GUINEA	7	9F	9F	9F	3727	3488	3523	3556	27	30F	31F	32F
GABON	2	2F	2F	2F	1500	495	455	435	3	1*	1F	1F
GUINEA	12	15F	18F	18F	7297	6667	6667	6626	90	100F	120F	121F
IVORY COAST	11	13F	14F	14F	1927	1846	1852	1857	21	24	25F	26F
KENYA	55	56F	57F	57F	9217	9464	9298	9474	510	530F	530F	540F
LIBERIA	1	2F	2F	2F	9998	10000	10000	10333	13	15F	15F	16F
MADAGASCAR	59	63	63F	63F	5864	4389	4444	4546	348	276	280F	287F
MALI	3	3F	3F	3F	12625	11667	11867	12067	35	35F	36F	36F
MAURITANIA	3	4F	5F	5F	629	386	378	391	2	2F	2F	2F
MAURITIUS					12667	13000	12000	13000				
MOZAMBIQUE	8	8F	9F	9F	4813	4217	4706	4706	40	35F	40F	40F
NIGER	1	3	3	3F	6729	6640	18333	16667	9	17	55	50F
NIGERIA	15	15F	16F	16F	10718	12667	12500	12500	164	190F	200F	200F
REUNION					18000	22727	22727	22083	5	5	5	5F
RHODESIA					3157	3269	3269	3269				
RWANDA	65	87	90	90F	5788	7755	7743	7722	379	675	694	695F
SENEGAL	3	3	3	3F	4186	1467	1571	1714	11	4	4F	5F
SIERRA LEONE	5	5F	5F	5F	2190	2000	2019	2000	11	10F	11F	11F
SOMALIA					9772	10000	10000	10000	3	3F	3F	3F
SOUTH AFRICA	13	13F	13F	13F	2900	3000	3154	3231	39	39	41	42
SUDAN	1	2F	2F	2F	26456	23529	24118	23529	21	40F	41F	40F
SWAZILAND	2	1	2F	2F	4013	5706	5294	5294	8	9	9F	9F
TANZANIA	39	52F	53F	54F	6163	6154	6226	6204	238	320F	330F	335F
TOGO	1	1F	1F	1F	2500	2333	2308	2143	3	3F	3F	3F
UGANDA	133	135F	135F	137F	5200	4667	4815	4818	693	630F	650F	660F
UPPER VOLTA	12	12F	15F	12F	2611	2917	2667	2917	31	35F	40F	35F
ZAIRE	74	83	84	86F	5729	5715	5742	5756	426	474	485	495F
ZAMBIA	2	3F	3F	3F	7041	6923	6923	6815	17	18F	18F	18F
N C AMERICA	167	157	163	160	7407	7798	7839	7625	1240	1228	1276	1219
ANTIGUA					4032	3825	2969	3030				
BARBADOS	1		1F	1F	9620	8334	8330	8400	6	3	4	4F
BERMUDA					5538	8000	8000	6333				
CUBA	63	65F	66F	67F	3777	3846	3864	3830	237	250F	255F	257F
DOMINICA					10000	10074	10000	10000	1	1*	1F	1F
DOMINICAN RP	9	5F	9F	7F	9322	9800	9333	10000	87	49	84	65*
GRENADA					4878	4000	4125	4125				
GUADELOUPE	2	3	2	2F	12322	14615	18750	18750	29	38	38	38F
HAITI	17	19F	19F	19F	5188	5027	5081	5000	88	93F	94F	95F
HONDURAS					2632	2750	2750	2750	1	1F	1F	1F
JAMAICA	2	2F	2F	2F	8097	7541	7784	7818	16	15	16	17F
MARTINIQUE	3	3	3	3	9575	10000	10000	10000	30	27	27	27
MEXICO	14	10	10F	10F	9843	13167	13000	13000	142	134	130F	130F
PUERTO RICO	3	1	2	2F	3496	5654	4132	4108	10	8	8	7
ST KITTS ETC					2886	5000	5000	5688				
ST LUCIA					2368	2500	2667	2626	1	1F	1F	1F
ST VINCENT	1	1F	1F	1F	4674	3629	3674	3700	4	4*	4*	4F
TRINIDAD ETC					3366	11189	10400	9643	2	3	3F	3F
USA	51	47	48	46	11538	12684	12784	12429	586	600	610	568
SOUTH AMERIC	271	251	247	248	11177	10665	10462	10443	3033	2677	2579	2592
ARGENTINA	44	41	38	36	10410	10208	9211	9116	457	418	348	330
BOLIVIA	2	2	3F	3F	7471	7392	6800	6800	12	16	17F	17F
BRAZIL	183	160F	156F	156F	11775	11563	11538	11489	2155	1850F	1800F	1815
CHILE	1	1*	1*	1*	14000	8000	8000	7500	14	8*	8*	9*
ECUADOR	3	3	3	3F	3574	4963	5973	5893	10	14	17	17F
FR GUIANA					6570	6875	6875	6875	1	1F	1F	1F
PARAGUAY	10	14	14F	15F	10710	8280	8571	8667	110	114	120F	130F
PERU	13	14*	15*	15*	12502	10714	10333	10667	167	150*	155	160*
SURINAM					4679	5000	5000	5000				
URUGUAY	13	14F	15F	15F	5975	5643	5667	5667	78	79	85F	85F
VENEZUELA	2	2	2F	2F	12470	12362	12435	12174	29	27	29F	28F
ASIA	12639	12689	12883	12978	8923	9916	9807	9892	112780	125824	126339	128377
BANGLADESH	72	67	72	71	11557	10798	11050	10704	830	719	791	756
BRUNEI					6490	8333	8333	8333		1F	1F	1F
BURMA	3	4F	4F	4F	4951	5405	5405	5410	15	20F	20F	21F
CHINA	11262F	11357F	11524F	11624F	8895	10074	9984	10080	100178F	114403F	115051F	117164F
EAST TIMOR	3	3F			5933	5962			15	16F		
HONG KONG					10510	13115	7265	11538	4	2	1	1
INDIA	225	231	251	231	10052	7195	7206	6755	2260	1658	1811	1561
INDONESIA	361	311	300	315	6131	7823	8060	7787	2215	2433	2418	2453
JAPAN	134	69	66	66F	19356	20640	19497	19497	2590	1418	1279	1279F
KAMPUCHEA DM	2	2F	3F	3F	8731	8500	8800	8846	21	17F	22F	23F
KOREA DPR	21	24F	24F	25F	13226	13833	14333	14400	273	332F	344F	360F
KOREA REP	125	95	87	94F	16489	20651	20406	19580	2053	1953	1783	1844F
LAO	2	3F	3F	3F	7016	6429	6552	6667	16	18F	19F	20F
MAL PENINSUL	6	2	2F	2F	4472	16432	15000	15000	25	32	30F	30F
MAL SABAH	1	1	1	1F	5112	3693	3827	4000	5	5F	5F	6F

TABLE
TABLEAU **20**
CUADRO

	SWEET POTATOES				PATATES DOUCES				BATATAS (CAMOTES)			
	AREA HARV SUP RECOLTEE SUP COSECHAD		1000 HA		YIELD RENDEMENT RENDIMIENTO		KG/HA		PRODUCTION PRODUCTION PRODUCCION		1000 MT	
	1969-71	1975	1976	1977	1969-71	1975	1976	1977	1969-71	1975	1976	1977
MALDIVES					6000	10000	10000	10000				
PAKISTAN	16	18F	18F	18F	8660	8743	8814	8860	141	153F	156F	159F
PHILIPPINES	134	180F	192	160F	5069	4801	4063	4688	680	864	781	750F
SINGAPORE					9380	12375	12337	11957	4	1	1	1F
SRI LANKA	17	49	46	50F	3926	3992	4093	4400	65	194	186	220F
THAILAND	32	41	41F	42F	8660	9436	9512	9639	280	385	390F	400F
VIET NAM	223	235F	250F	270F	4977	5106	5000	4926	1108	1200F	1250F	1330F
EUROPE	14	13	13	13	9807	9976	10412	10385	140	130	139	134
GREECE					12935	11111	12778	12778	2	2	2F	2F
ITALY	2	1	1	1	17870	19360	22018	21601	27	23	26	25
PORTUGAL	9	9F	9F	8F	7782	7890	7918	7927	69	67	67	65F
SPAIN	4	3	4	3F	11229	11969	12371	12353	42	38	43	42F
OCEANIA	95	105	106	106	5168	5328	5358	5376	489	557	565	572
AMER SAMOA					4500	5000	5000	5000				
AUSTRALIA					5719	10045	10318	10000	1	2	3	3F
COOK ISLANDS					19231	22500	23333	23333	2	1*	1F	1F
FIJI	1	1F	1F	1F	9776	10278	10556	10833	7	7F	8F	8F
FR POLYNESIA					4377	5000	5000	5000	1	2F	2F	2F
GUAM					3667	2313	2313	2313				
NEWCALEDONIA	1				4250	5000	4878	4762	3	2F	2F	2F
NEW ZEALAND					15883	16710	20000	21333	7	7	9*	10F
NIUE ISLAND					16333	21875	23333	23333				
PACIFIC IS					7237	7333	7036	7021	3	3F	3F	3F
PAPUA N GUIN	82	91F	92F	92F	4237	4464	4483	4501	346	405F	411F	416F
SOLOMON IS	5	5F	5F	5F	9518	10000	9878	9760	46	48F	48F	49F
TONGA	5	6F	6F	6F	13797	12381	12429	12476	73	78F	78F	79F
DEV.PED M E	213	143	140	138	15797	15391	14821	14754	3363	2197	2081	2036
N AMERICA	51	47	48	46	11538	12684	12784	12429	586	600	610	568
W EUROPE	14	13	13	13	9807	9976	10412	10385	140	130	139	134
OCEANIA	1	1	1	1	12373	14508	16311	16892	9	10	12	13
OTH DEV.PED	147	82	79	79	17865	17834	16794	16807	2629	1457	1320	1321
DEV.PING M E	2220	2253	2290	2275	7804	7655	7567	7576	17327	17250	17326	17235
AFRICA	736	781	801	811	6034	6187	6270	6291	4443	4831	5024	5102
LAT AMERICA	388	361	362	362	9503	9150	8975	8949	3687	3304	3246	3243
NEAR EAST	5	5	5	5	22149	20985	21739	21376	107	115	110	109
FAR EAST	997	1002	1017	991	8634	8435	8253	8298	8609	8453	8393	8221
OTH DV.PING	94	104	105	106	5113	5269	5282	5295	480	547	554	560
CENTR PLANND	11508	11618	11800	11921	8827	9981	9887	9972	101580	115952	116667	118877
ASIAN CPE	11508	11618	11800	11921	8827	9981	9887	9972	101580	115952	116667	118877
DEV.PED ALL	213	143	140	138	15797	15391	14821	14754	3363	2197	2081	2036
DEV.PING ALL	13728	13871	14090	14196	8661	9603	9510	9588	118907	133203	133993	136112

TABLE
TABLEAU **21**
CUADRO

	CASSAVA — AREA HARV / SUP RECOLTEE / SUP COSECHAD (1000 HA)				MANIOC — YIELD / RENDEMENT / RENDIMIENTO (KG/HA)				YUCA (MANDIOCA) — PRODUCTION / PRODUCCION (1000 MT)			
	1969-71	1975	1976	1977	1969-71	1975	1976	1977	1969-71	1975	1976	1977
WORLD	11020	11952	12158	12575	8746	8641	8785	8761	96378	103284	106807	110167
AFRICA	6085	6469	6611	6731	6536	6572	6624	6576	39772	42511	43794	44263
ANGOLA	121	120F	120F	120F	13223	13333	13333	13750	1600	1600F	1600F	1650F
BENIN	110	75	65	83	6633	4519	5860	7323	733	340	381	610
BURUNDI	80	60F	61F	63F	10542	14833	14593	14363	843	890F	896F	902F
CAMEROON	186	190F	190F	191F	3952	4284	4211	4241	734	814*	800F	810F
CAPE VERDE					10285	15000	30000	30000	2	3F	6	6F
CENT AFR EMP	262	286	295F	304F	2930	2972	2882	2962	767	850	850*	900
CHAD	17	19F	21F	22F	3173	2808	2738	2676	55	54F	56F	59F
COMOROS	23	26F	26F	27F	3088	3137	3115	3074	70	80F	81F	83F
CONGO	138	117	132	145	3993	5660	5767	5301	551	662	761	769
EQ GUINEA	16	20F	20F	21F	2745	2400	2402	2381	43	48F	49F	50F
GABON	63	65F	65F	65F	2723	2769	2769	2779	171	180F	180F	182F
GAMBIA	2	2F	2F	2F	3820	4000	4000	3500	8	9F	9F	7F
GHANA	343	285	300F	300F	6771	8413	8333	8333	2325	2398	2500F	2500F
GUINEA	60	45F	48F	48F	8028	10000	10000	10092	482	450F	480F	484F
IVORY COAST	166	168	185F	185F	3300	3871	3676	3723	546	650F	680F	689F
KENYA	93	95F	100F	95F	7446	7368	8000	7895	690	700F	800F	750F
LIBERIA	79	75F	75F	75F	3356	3490	3533	3600	264	260F	265F	270F
MADAGASCAR	188	205	200F	200F	6514	6548	6460	6500	1227	1344	1292	1300
MALAWI	5	6F	6F	6F	16841	20000	18333	15000	90	120F	110F	90F
MALI	5	5F	5F	5F	7213	8000	8000	8000	36	40F	40F	40F
MAURITIUS					16467	17583	15000	17500				
MOZAMBIQUE	450	450F	450F	450F	5669	5111	5333	5444	2549	2300F	2400F	2450F
NIGER	25	20	30F	32F	7157	8780	6567	6250	182	176	197	200F
NIGERIA	894	1050F	1080F	1100F	10592	10000	10000	9636	9473	10500F	10800F	10600F
REUNION					8333	9756	9756	10000	3	4	4	4F
RWANDA	29	32	25	25F	11358	12314	16831	16640	333	394	415	416F
SAO TOME ETC					10515	10385	10577	10769	2	3F	3F	3F
SENEGAL	36	37	21	24F	4365	2943	5441	5139	159	109	114	125F
SEYCHELLES					6062	6118	6163	6207		1F	1F	1F
SIERRA LEONE	23	26F	26F	26F	3307	3333	3346	3385	75	85F	87F	89F
SOMALIA	2	3F	3F	3F	10573	10769	10741	10714	25	28F	29F	30F
SUDAN	32	43	43F	43F	2866	3062	3047	3047	93	130	131F	131F
TANZANIA	695	800F	810F	830F	4870	4750	4815	4819	3384	3800F	3900F	4000F
TOGO	114	57	58F	60F	6466	7860	7759	7856	735	448	450*	468F
UGANDA	277	340F	360F	360F	3825	2941	3056	3056	1058	1000F	1100F	1100F
UPPER VOLTA	5	6F	6F	6F	6000	5833	5833	5000	30	35F	35F	30F
ZAIRE	1493	1688	1729	1760F	6880	7016	7014	6989	10275	11844	12130	12300F
ZAMBIA	52	53F	53F	54F	3084	3075	3075	3078	159	163F	163F	166F
N C AMERICA	104	112	113	115	6381	6353	6435	6437	667	710	725	739
ANTIGUA					3625	2533	2588	2684				
BARBADOS					26355	25714	25278	25278	1	1F	1F	1F
COSTA RICA	3	2	2	2F	3782	5714	7124	7000	12	12	14	14F
CUBA	33	36F	37F	38F	6566	6877	6929	6579	217	246F	254F	262F
DOMINICA					9958	10037	10000	10000	1	1*	1F	1F
DOMINICAN RP	15	15F	15F	16F	11490	11333	11400	11290	173	170	171	175*
EL SALVADOR	1	2	1	1F	9102	9474	9470	10714	12	17	13	15F
GUADELOUPE					11250	20000	20000	20000	5	6	4	4F
GUATEMALA	2	3F	3F	3F	3000	2808	2778	2750	6	7F	8F	8F
HAITI	32	34F	34F	34F	3971	4265	4353	4265	128	145F	148F	145F
HONDURAS	5	6	6F	6F	7276	1820	1930	2069	35	11	11F	12F
JAMAICA	2	2F	2F	2F	6733	8472	9100	8833	13	19	21	21F
MARTINIQUE					14057	20000	20000	20000	3	3	3	3
NICARAGUA	4	5	6	6	4111	4055	4055	4055	17	22	23	25
PANAMA	4	5F	5F	5F	8547	8866	8956	8889	32	40	40	40F
PUERTO RICO	1	1F	1F	1F	3474	3462	3462	3462	4	5F	5F	5F
ST LUCIA					4000	4111	4130	4083	1	1F	1F	1F
ST VINCENT					13516	11765	12222	12632	3	2F	2F	2F
TRINIDAD ETC					8966	12342	12368	12368	3	5	5F	5F
SOUTH AMERIC	2483	2623	2615	2759	13873	11947	12090	11605	34445	31338	31612	32022
ARGENTINA	26	20	22	21	11555	15018	10213	10748	296	300	225	230
BOLIVIA	18	22	22	23	12671	13101	13850	12800	223	285	305	294
BRAZIL	2042	2098	2105	2233	14655	12300	12562	11871	29922	25812	26446	26511
COLOMBIA	185	257	240	251	7476	7873	8028	8427	1380	2021	1927	2113
ECUADOR	35	43	38	38F	10753	12767	9052	9158	382	554	348	348F
FR GUIANA	1	1F	1F	1F	7564	6667	6667	6667	5	4F	4F	4F
PARAGUAY	100	107	110F	115F	14368	14773	14545	14783	1442	1573	1600F	1700F
PERU	37	38*	36*	38*	12716	12369	11167	11842	477	470	402*	450*
SURINAM					6846	6667	6667	6667	2	2F	2F	2F
VENEZUELA	39	37	40	39*	8080	8483	8825	9487	317	317	353	370*
ASIA	2331	2730	2801	2951	9142	10446	10878	11160	21307	28519	30468	32932
BRUNEI					7759	7833	8333	8667	2	2F	3F	3F
BURMA	1	2	2	2F	10378	8878	8824	8333	10	15	15F	15F
CHINA	21F	19F	18F	19F	14740	15036	16292	16665	305F	279F	294F	308F
EAST TIMOR	7	8F			2816	2073			20	17F		
INDIA	352	388	392	392	14176	16321	16934	16523	4993	6326	6638	6480
INDONESIA	1424	1410	1356	1356	7512	8898	9195	8974	10695	12546	12468	12169
KAMPUCHEA DM	2	2F	3F	3F	10244	8333	8462	8571	24	20F	22F	24F
LAO	1	1F	1F	1F	9734	9571	9714	9857	12	13F	14F	14F
MAL PENINSUL	21	15	15F	16F	9240	15772	16000	15731	191	238	240F	245F
MAL SABAH	3	5	8	8F	6775	5812	3902	3538	23	29F	30F	32F
MAL SARAWAK	6	6F	7F	7F	10030	10156	10231	10303	58	65F	67F	68F
PHILIPPINES	82	100F	100F	100F	5297	6793	6793	6793	436	679	679F	679F
SINGAPORE					10744	11192	11143	10309	4	1	1	1F
SRI LANKA	65	165	151	170F	5773	4736	4945	4412	376	780	748	750F
THAILAND	212	454	593	697	15156	14012	13649	15261	3208	6358	8100	10644

TABLE
TABLEAU 21
CUADRO

	CASSAVA AREA HARV SUP RECOLTEE SUP COSECHAD		1000 HA		MANIOC YIELD RENDEMENT RENDIMIENTO		KG/HA		YUCA (MANDIOCA) PRODUCTION PRODUCTION PRODUCCION		1000 MT	
	1969-71	1975	1976	1977	1969-71	1975	1976	1977	1969-71	1975	1976	1977
VIET NAM	133	155F	155F	180F	7153	7419	7419	8333	950	1150F	1150F	1500F
OCEANIA	17	19	19	19	10975	11022	11033	11066	187	205	209	212
AMER SAMOA					26800	15000	15000	15000				
COOK ISLANDS					27445	32258	32258	32258	4	4F	4	4*
FIJI	7	8F	8F	8F	12305	12080	12160	12240	85	91F	91F	92F
FR POLYNESIA					17877	18788	18788	18788	6	6F	6F	6F
NEWCALEDONIA					22699	30000	29268	29268	4	1	1F	1F
NIUE ISLAND					10000	4545	4545	4545				
PACIFIC IS	1	1F	1	1F	9032	8953	9071	8925	6	8F	8F	8F
PAPUA N GUIN	7	8F	8F	8F	10188	10782	10775	10768	72	84F	86F	88F
SAMOA					18750	18750	18750	18750				
TONGA	2	2F	2F	2F	5742	5795	5800	5950	10	11F	12F	12F
DEV.PING M E	10864	11776	11983	12374	8753	8647	8791	8755	95099	101836	105341	108335
AFRICA	6053	6426	6568	6688	6556	6595	6648	6599	39680	42381	43663	44132
LAT AMERICA	2587	2735	2727	2874	13570	11718	11856	11399	35112	32049	32336	32760
NEAR EAST	32	43	43	43	2866	3062	3047	3047	93	130	131	131
FAR EAST	2175	2554	2625	2749	9209	10598	11047	11311	20029	27070	29002	31099
OTH DV.PING	17	19	19	19	10975	11022	11033	11066	187	205	209	212
CENTR PLANND	156	176	176	201	8206	8234	8346	9102	1278	1449	1466	1832
ASIAN CPE	156	176	176	201	8206	8234	8346	9102	1278	1449	1466	1832
DEV.PING ALL	11020	11952	12158	12575	8746	8641	8785	8761	96378	103284	106807	110167

TABLE
TABLEAU 22
CUADRO

PULSES,TOTAL   LEGUMINEUSES SECHES,TCT.   LEGUMBRES SECAS,TCTAL

AREA HARV   1000 HA   YIELD   KG/HA   PRCDUCTICN   1000 MT
SUP RECOLTEE   RENDEMENT   PRCDUCTICN
SUP COSECHAD   RENDIMIENTC   PRCDUCCICN

	1969-71	1975	1976	1977	1969-71	1975	1976	1977	1969-71	1S75	1976	1977
WORLD	66099	67949	68929	7C197	685	654	721	683	45295	44451	45726	47959
AFRICA	10916	11274	11573	11667	422	448	456	422	4610	5055	5280	4919
ALGERIA	87	96	97	100F	499	773	774	593	44	75	75	59F
ANGOLA	120	120F	120F	120F	579	583	583	600	70	70F	70F	72F
BENIN	85	43	52	74	354	336	373	392	30	14	19	29
BOTSWANA	22	30F	30F	30F	569	500	567	600	12	15F	17F	18F
BURUNDI	253	269	281F	295F	643	611	606	602	162	164	170F	178F
CAMEROON	120	144F	147F	152F	526	583	599	605	63	84F	88F	92F
CAPE VERDE			1		2655	3333	1667	1667	1	1F	1	
CENT AFR EMP	10	11F	11F	11F	500	500	505	509	5	5F	6F	6F
CHAD	137	141F	141F	142F	455	398	402	405	62	56F	57F	57F
COMOROS	3	3F	3F	3F	556	547	548	544	2	2F	2F	2F
CONGO	3	3F	3F	3F	613	720	731	741	2	2F	2F	2F
EGYPT	185	162	172	178F	1985	2068	2075	2054	368	335	357	365
ETHIOPIA	833	880	826	795	784	745	833	883	653	656	688	702
GABON					498	200	500	500				
GAMBIA	11	12F	12F	12F	250	233	233	208	3	3F	3F	3F
GHANA	111	147	126F	125F	120	116	95	64	13	17	12F	8F
GUINEA	51	50F	50F	51F	520	540	540	543	27	27F	27F	28F
GUIN BISSAU	3	3F	3F	3F	500	500	600	533	2	2F	2F	2F
IVORY COAST	12	12F	12F	12F	600	625	625	625	7	8F	8F	8F
KENYA	620	620F	600F	620F	473	500	483	508	293	310F	290F	315F
LESOTHO	26	40	41F	42F	334	362	383	405	9	14	16	17F
LIBERIA	5	6F	6F	6F	486	467	500	500	2	3F	3F	3F
LIBYA	5	7F	7F	7F	520	985	989	989	3	6F	7F	7F
MADAGASCAR	77	72	78F	77F	843	854	855	882	65	61	66	68
MALAWI	59	61	61F	62F	591	573	590	605	35	35	36F	38F
MALI	30	30F	30F	30F	1100	1100	1123	1147	33	33F	34F	34F
MAURITANIA	30	34F	36F	38F	364	288	278	268	11	10F	10F	10F
MAURITIUS	1	2F	2F	2F	405	508	526	600	1	1	1F	1F
MOROCCO	425	564	557	434	813	802	844	374	345	453	470	162
MOZAMBIQUE	120	120F	120F	120F	545	500	542	567	65	60F	65F	68F
NAMIBIA	3	3F	3F	4F	1871	2059	2059	2057	6	7F	7F	7F
NIGER	1040	865	881	925F	110	266	284	277	115	231	250	256F
NIGERIA	3973	4110F	4160F	4210F	214	219	224	190	849	902F	932F	800F
REUNION	1	1	1	1	1713	1588	1588	1613	1	1	1	1
RHODESIA	51	51F	51F	51F	483	508	519	531	25	26F	27F	27F
RWANDA	237	262	274	275F	884	801	805	807	209	210	220	222F
SENEGAL	68	62	50	55F	323	335	326	382	22	21	16	21F
SIERRA LEONE	47	48F	49F	49F	527	532	537	540	25	26F	26F	26F
SOMALIA	12	13F	13F	13F	304	346	346	346	4	5F	5F	5F
SOUTH AFRICA	247	154	152	183	398	541	561	423	98	83	85	77
SUDAN	58	71	72	74F	1108	1041	1074	1071	64	74	77	80F
SWAZILAND	2	2F	2F	2F	373	524	545	545	1	1F	1F	1F
TANZANIA	419	405	482	498F	415	448	435	439	174	181	210	219F
TOGO	94	71	75F	78F	450	293	300	315	42	21	23F	24F
TUNISIA	112	128F	132F	134F	345	784	724	713	38	100	96	96F
UGANDA	443	579	732	755	592	532	498	499	262	308	365	377
UPPER VOLTA	437	510F	560F	550F	350	353	321	300	153	180F	180F	165F
ZAIRE	214	237	241	245F	587	612	614	615	125	145	148	151F
ZAMBIA	17	18F	20F	20F	620	611	600	596	10	11F	12F	12F
N C AMERICA	3388	3435	2979	3222	786	824	824	761	2663	2832	2454	2452
BARBADOS	1	1F	1F	1F	1193	1183	1200	1213	1	1F	1F	1F
BELIZE	2	2	2F	2F	810	561	567	600	2	1	1F	1F
CANADA	71	96	91	105	1479	1455	1466	1578	105	139	134	165
COSTA RICA	22	36	36	35F	403	456	516	429	9	16	19	15*
CUBA	35	35F	35F	35F	648	671	671	671	23	24F	24F	24F
DOMINICA	1	1F	1F	1F	545	545	545	545	1	1F	1F	1F
DOMINICAN RP	48	56	60	61	1201	914	1000	886	57	51	60	54
EL SALVADOR	36	56	53	55*	836	710	758	727	30	40	40	40*
GRENADA	1	1F	1F	1F	615	558	554	554				
GUATEMALA	109	110	156	144	614	611	517	485	67	67	81	70
HAITI	93	95F	95F	96F	830	905	917	923	77	86F	87F	88F
HONDURAS	74	80*	90*	91*	585	536	534	549	43	43	48	50*
JAMAICA	7	7F	12F	13F	681	767	851	836	5	5	10	11F
MEXICO	2056	2003	1492	1780	542	635	585	546	1113	1271	873	973
NICARAGUA	63	56	68	82	887	785	785	785	56	44	53	64
PANAMA	18	18	19	18	332	307	307	275	6	6	6	5
PUERTO RICO	8	11	10	10F	809	630	660	663	6	7	6	7
ST LUCIA					1947	2000	2000	2000				
ST VINCENT					722	775	850	900				
TRINIDAD ETC	2	2F	2F	2F	1807	1455	1460	1485	3	3*	3*	3F
USA	743	770	754	691	1426	1334	1334	1275	1059	1027	1006	881
SOUTH AMERIC	4700	5264	5172	5804	638	578	516	568	3000	3040	2667	3297
ARGENTINA	99	169	178	219	825	899	1287	1268	82	152	230	277
BOLIVIA	17	17	18F	19F	630	962	881	959	10	17	16	18
BRAZIL	3868	4331	4238	4759	634	544	455	508	2453	2358	1930	2417
CHILE	96	105	120	152	886	926	783	1021	85	97	94	155
COLOMBIA	170	221	197	212	601	638	582	565	102	141	115	120
ECUADOR	131	103	102	102F	524	524	552	553	69	54	57	56F
GUYANA		1	1F	2F	862	685	660	690		1	1*	1*
PARAGUAY	59	80	83	86F	621	800	803	798	37	64	67	68F
PERU	140	132	128	128F	816	840	837	842	114	111	107	108F
SURINAM					939	1163	1154	1231				
URUGUAY	5	5F	6F	6F	845	839	888	898	4	5F	5F	5F
VENEZUELA	115	101	101*	120*	381	416	465	592	44	42	47*	71*
ASIA	37033	38422	40322	40585	651	660	699	668	24120	25341	28167	27096
AFGHANISTAN	32	34F	35F	35F	1474	1632	1623	1629	47	56F	56F	57F

TABLE
TABLEAU 22
CUADRO

	PULSES,TOTAL				LEGUMINEUSES SECHES,TOT.				LEGUMBRES SECAS,TOTAL			
	AREA HARV / SUP RECOLTEE / SUP COSECHAD		1000 HA		YIELD / RENDEMENT / RENDIMIENTO		KG/HA		PRODUCTION / PRODUCTION / PRODUCCION		1000 MT	
	1969-71	1975	1976	1977	1969-71	1975	1976	1977	1969-71	1975	1976	1977
BANGLADESH	367	312	308	327	803	729	722	725	295	228	222	237
BHUTAN	4	5F	5F	5F	385	400	404	407	2	2F	2F	2F
BURMA	396	442	453	454F	586	624	619	619	232	276	280	281F
CHINA	10066F	10960F	11173F	11423F	921	1020	1023	1023	9273F	11175F	11430F	11681F
CYPRUS	17	20	21	21F	802	661	671	683	13	13	14	14F
EAST TIMOR	6	7F			633	881			4	6F		
INDIA	21978	22380	23721	23587	517	473	541	485	11365	10592	12833	11440
INDONESIA	500	589F	607F	625F	500	501	502	502	250	295F	305F	314F
IRAN	216	202	207	218F	938	941	966	966	202	190*	200*	211F
IRAQ	51	51	58	63	872	753	787	835	44	39	46	53
ISRAEL	8	5	5	5F	995	1931	1584	1535	8	10	8	8
JAPAN	177	129	118	118F	1221	1283	1298	1298	216	166	153	153F
JORDAN	31	23	30	28F	710	360	396	354	22	8	12	10
KAMPUCHEA DM	40	28F	31F	33F	681	536	581	606	27	15F	18F	20F
KOREA DPR	315	330F	335F	340F	654	758	776	794	206	250F	260F	270F
KOREA REP	61	61	62	65F	615	767	873	862	37	47	54	56F
LAO	9	9F	9F	9F	1400	1489	1495	1500	12	13F	14F	14F
LEBANON	13	21F	21F	22F	814	914	907	900	10	19F	19F	20F
MALDIVES					600	600	600	600				
MONGOLIA	1	4	4	4F	222	658	528	658		3	2	3F
NEPAL	112	108F	108F	109F	432	407	407	413	48	44F	44F	45F
OMAN					868	1500	1520	1560				
PAKISTAN	1354	1388	1465	1497	509	521	531	555	689	723	777	831
PHILIPPINES	51	58F	70	70F	591	512	800	803	30	29	56	56F
SAUDI ARABIA	2	3F	3F	3F	1563	1643	1621	1600	3	5F	5F	5F
SRI LANKA	8	25	27	27F	670	542	474	500	5	14	13	13F
SYRIA	230	223	312	345	715	662	880	746	164	147	275	257
THAILAND	212	173	231	228F	866	836	651	665	184	145	150	152F
TURKEY	538	567	630	644F	1098	1190	1192	1122	591	675	752	722F
VIET NAM	194	194F	195F	206F	457	448	462	471	89	87F	92F	97F
YEMEN AR	47	71	76	75F	1071	1000	1000	1000	50	71	76	75F
EUROPE	4817	3672	3437	3493	691	729	668	701	3329	2676	2296	2449
ALBANIA	58	52F	54F	56F	278	324	320	316	16	17F	17F	18F
AUSTRIA	3	2	2	2F	2268	2351	2474	2406	7	5	5	5F
BELGIUM-LUX	7	5	4	3	2898	3304	2358	3065	19	15	8	8
BULGARIA	148	111	103	102F	811	768	698	830	120	85	72	85F
CZECHOSLOVAK	82	92	65	63F	1177	1523	1102	1596	96	140	72	100F
DENMARK	27	7	7	7F	3034	3008	2681	2665	83	21	18	19F
FINLAND	1	5	5	10	1829	2019	3021	1553	3	11	15	16
FRANCE	77	79	73	74	1753	2175	1215	2185	135	172	89	161
GERMAN DR	59	49*	49*	45F	1464	1580	1163	1170	87	77*	57	57F
GERMANY FED	32	28	22	16	2569	3029	2111	2606	95	86	47	43
GREECE	128	104	101	104	1007	1147	1131	1147	129	120	115	120
HUNGARY	211	143	142	146F	653	975	982	1019	138	140	140	149F
IRELAND	4	3	2*	2F	4612	1059	1096	625	19	3F	3F	2F
ITALY	613	318	307	265	1000	1262	1265	1273	613	401	389	337
MALTA	1	1	1	1F	3001	2226	2406	2197	3	3	3	3F
NETHERLANDS	15	12	10	8	3060	3014	2323	2290	46	36	22	17
NORWAY					2500	2500	2500	2500				
POLAND	213	205	208F	211F	1174	1049	1065	1077	250	215	222F	227F
PORTUGAL	457	391	366	342	247	238	235	211	113	93	86	72
ROMANIA	1261	904	838	906	169	123	106	127	214	112	89	115
SPAIN	818	675	624	632	727	730	643	656	594	492	401	414
SWEDEN	8	10	11	11F	1759	1470	2047	2047	14	15	22	22F
SWITZERLAND		1	1			3462	2066	3571		4	2	1
UK	101	68	71	73	2491	2631	2704	3000	252	179	192	219
YUGOSLAVIA	493	408	371	409	578	578	573	585	285	236	213	239
OCEANIA	103	211	232	212	1020	880	914	689	105	186	212	146
AUSTRALIA	42	147	166	144	799	727	778	413	33	107	129	59
FIJI	1	2F	2F	2F	709	941	1024	1000	1	2F	2*	2F
NEWCALEDONIA					825	667	667	667				
NEW ZEALAND	23	21F	22F	23F	2235	2684	2716	2744	52	56F	59F	62F
PACIFIC IS					600	600	600	600				
PAPUA N GUIN	35	40F	40F	41F	500	500	500	500	18	20F	20F	21F
SOLOMON IS	2	2F	2F	2F	731	784	789	789	1	1F	2F	2F
USSR	5143	5670	5213	5215F	1452	938	1659	1457	7468	5321	8651	7600F
DEV.PED M E	4095	3439	3285	3228	972	1012	975	962	3981	3479	3202	3105
N AMERICA	814	865	846	796	1431	1347	1348	1315	1164	1166	1140	1046
W EUROPE	2785	2117	1978	1960	865	893	823	867	2409	1890	1628	1698
OCEANIA	65	168	188	167	1314	970	1003	731	85	163	188	122
OTH DEV.PED	432	288	274	306	745	898	896	779	322	259	246	238
DEV.PING M E	44214	45769	47229	48215	528	510	538	507	23331	23336	25403	24434
AFRICA	10420	10880	11172	11225	391	419	426	391	4076	4555	4755	4391
LAT AMERICA	7274	7834	7305	8230	618	601	545	572	4498	4706	3980	4703
NEAR EAST	1424	1454	1643	1712	1112	1127	1153	1095	1583	1638	1894	1875
FAR EAST	25058	25557	27065	27003	525	486	545	498	13154	12413	14750	13441
OTH DV.PING	38	43	44	45	518	530	534	534	20	23	24	24
CENTR PLANND	17791	18742	18414	18754	1011	941	1147	1089	17983	17637	21121	20421
ASIAN CPE	10615	11516	11742	12006	904	1001	1005	1005	9595	11530	11802	12070
E EUR+USSR	7175	7226	6673	6748	1169	845	1397	1237	8388	6107	9319	8350
DEV.PED ALL	11271	10664	9958	9976	1097	899	1257	1148	12368	9585	12521	11455
DEV.PING ALL	54829	57284	58971	60221	601	609	631	606	32926	34866	37205	36504

TABLE
TABLEAU **23**
CUADRC

	BEANS, DRY			HARICOTS SECS			FRIJCLES SECOS				
	AREA HARV / SUP RECOLTEE / SUP COSECHAO 1000 HA			YIELD / RENDEMENT / RENDIMIENTC KG/HA			PRODUCTICN / PRODUCTION / PRODUCCICN 1000 MT				
	1969-71	1975	1976	1977	1969-71	1975	1976	1977	1969-71	1975	1976	1977
WORLD	22597	24161	23400	25082	509	527	516	515	11512	12737	12085	12912
AFRICA	1838	2034	2220	2291	595	576	571	570	1C93	1170	1267	1307
ALGERIA	6	4	2	4F	312	969	860	818	2	3	2	4F
ANGOLA	120	120F	120F	120F	579	583	583	600	70	70F	70F	72F
BENIN	79	36	44	67	354	296	346	377	28	11	15	25
BURUNDI	215	245F	252F	259F	616	604	596	589	132	148F	150F	153F
CAMEROON	101	123F	125F	13CF	508	593	608	615	51	73F	76F	80F
CHAD	97	100F	10CF	100F	438	440	447	454	42	44F	45F	45F
EGYPT	5	7*	7*	7F	1942	2264	2515	2458	10	16*	17*	18F
ETHICPIA	69	47	45	46F	684	760	649	696	47	35	29	32F
LESOTHO	15	30	30F	3CF	238	292	290	317	4	9	9	10F
MADAGASCAR	69	65F	71F	70F	847	842	841	871	58	55	60	61
MALAWI	13	12F	12F	12F	539	494	500	500	7	6	6F	6F
MOROCCO	5	4	2	9	786	767	800	659	4	3	1F	6F
REUNION		1F	1F	1F	1713	1639	1639	1613	1	1*	1*	1F
RHODESIA	50	50F	5CF	50F	487	512	524	536	24	26F	26F	27F
RWANDA	161	191	203	203F	900	801	805	808	145	153	163	164F
SOMALIA	12	13F	13F	13F	304	346	346	346	4	5F	5F	5F
SOUTH AFRICA	75	70	68*	69*	729	829	889	742	55	58	60	51*
SUDAN	3	5	4	4F	1035	807	768	750	3	4*	3*	3F
SWAZILAND	2	2F	2F	2F	373	524	545	545	1	1F	1F	1F
TANZANIA	243	268	294	30CF	512	500	497	509	124	134	146	153F
TOGO	78	61	62F	65F	451	259	258	278	35	16	16F	18F
UGANDA	263	407	537	55CF	589	480	480	480	155	195	258	264F
ZAIRE	157	174*	176*	180F	583	605	609	608	91	1C5*	107*	109F
N C AMERICA	2798	2873	2520	2620	731	774	790	740	2C45	2222	1991	1938
BELIZE	2	2	2F	2F	810	561	567	600	2	1	1F	1F
CANADA	38	66	67	70	1584	1379	1358	1567	61	90	91	110
COSTA RICA	22	36	36	35F	403	456	516	429	9	16	19	15*
CUBA	35	35F	35F	35F	648	671	671	671	23	24F	24F	24F
DOMINICAN RP	31	42*	45*	47*	844	719	836	687	26	30	38	32*
EL SALVACOR	36	56	53	55*	836	710	758	727	30	40	40	40*
GUATEMALA	94	93	138	126*	684	692	565	532	64	65	78	67
HAITI	41	41F	42F	42F	996	1098	1108	1108	41	45F	46F	46F
HONDURAS	74	80*	90*	91*	585	536	534	549	43	43	48	50*
MEXICO	1778	1753	1316	1506	503	586	562	495	894	1027	740	745
NICARAGUA	63	56	68	82	887	785	785	785	56	44	53	64
PANAMA	16	16	17	16	290	252	253	212	5	4	4	3
PUERTO RICO	4	4F	4F	4F	592	600	600	600	2	2F	2F	2F
USA	563	593	607	511	1403	1333	1330	1446	790	791	807	739
SOUTH AMERIC	4172	4770	4686	5270	643	572	501	553	2682	2728	2348	2915
ARGENTINA	53	137	147	167	822	791	1160	1080	43	109	171	180
BOLIVIA	4	3	3F	3F	452	810	833	833	2	2	3F	3F
BRAZIL	3685	4143	4047	4564	642	548	455	510	2366	2271	1842	2327
CHILE	57	68	82	97	1078	1089	862	1155	62	74	70	112
COLOMBIA	67	121	101	112	641	745	669	645	43	90	68	72
ECUADOR	78	63	68	68F	468	417	471	471	36	26	32	32F
PARAGUAY	50	67	70F	72F	617	783	786	778	31	52	55F	56F
PERU	77	75F	75F	75F	804	853	853	867	62	64F	64F	65F
URUGUAY	4	4F	4F	4F	550	512	568	568	2	2F	3F	3F
VENEZUELA	97	89	89*	108*	353	416	461	602	34	37	41*	65*
ASIA	11044	12361	11966	12835	432	471	481	468	4769	5827	5759	6006
BANGLADESH	75	67	68	68F	801	756	717	765	60	51	49	52F
BURMA	153	194F	194F	194F	741	825	825	825	113	160F	160F	160F
CHINA	2308F	2520F	2623F	2673F	715	843	850	872	1650F	2125F	2230F	2331F
CYPRUS	1	1F			1780	1203	1467	1556	1	1	1	1F
EAST TIMOR	5	5F			674	1077			3	6F		
INDIA	7593	8750	8200	9000F	283	316	315	300	2152	2767	2581	2700F
IRAN	114	94*	96*	1C5F	985	1032	1063	1055	112	97*	102*	111F
IRAQ	12	13	11	12F	928	528	674	711	11	7	8	9F
JAPAN	160	120	109	109F	1236	1292	1319	1319	198	156	144	144F
KAMPUCHEA DM	40	28F	31F	33F	681	536	581	606	27	15F	18F	20F
KOREA REP	49	45	46	49F	624	793	919	909	31	35	42	44F
LEBANON	1	1F	1F	1F	1299	1354	1354	1354	2	1F	1F	1F
PAKISTAN	105	126	114	12CF	471	491	478	500	50	62	55	60F
PHILIPPINES	47	54F	63	63F	482	425	659	659	23	23	41	41F
SRI LANKA	4	15F	15F	15F	576	534	357	400	2	8	5	6F
SYRIA	4	6	7	7F	1250	1404	1606	1846	5	8	11	12F
THAILAND	192	153	210	206F	824	788	595	607	158	121	125	125F
TURKEY	106	96	104	102F	1366	1641	1547	1569	145	158	161	160F
VIET NAM	77	74F	74F	78F	350	365	365	385	27	27F	27F	30F
EUROPE	2706	2084	1955	2018	313	345	321	327	846	718	629	660
ALBANIA	50	43F	46F	48F	260	312	307	303	13	14F	14F	15F
AUSTRIA					2010	2099	2083	2083		1F	1F	1F
BELGIUM-LUX	1	1	1	1*	2674	2610	1340	2006	1	4	2	2*
BULGARIA	110	94	91	90F	692	754	628	778	76	71	57	70F
CZECHOSLOVAK	2	4	2	3F	1074	1004	354	400	2	4	1	1F
FRANCE	37	25	24	24*	1252	1169	880	1574	46	29	21	38*
GERMAN DR					1514	1862	1239	1277				
GREECE	51	39	38	38*	1031	1154	1184	1184	52	45	45	45*
HUNGARY	107	70	67F	70F	115	259	117	214	12	18	8	15F
ITALY	177	71	70	64	875	1478	1555	1552	155	106	109	100
NETHERLANDS	4	6	5	4	2542	2374	1247	1009	9	14	6	4
POLAND	22	8	9F	10F	1508	1125	1289	1420	34	9	12F	14F
PORTUGAL	345	301	271	258	165	154	136	128	57	46	37	33

TABLE
TABLEAU 23
CUADRO

BEANS, DRY   HARICOTS SECS   FRIJOLES SECOS

AREA HARV   PRODUCTION
SUP RECOLTEE   1000 HA   YIELD   KG/HA   PRODUCTION   1000 MT
SUP COSECHAD   RENDEMENT   PRODUCCICN
                              RENDIMIENTO

	1969-71	1975	1976	1977	1969-71	1975	1976	1977	1969-71	1975	1976	1977
ROMANIA	1159	879	831	880F	72	93	92	93	83	82	76	82F
SPAIN	196	167	162	152	600	647	611	439	117	108	99	67
SWEDEN	1	1	1	1F	1721	1200	1600	1600	2	1	2	2F
YUGOSLAVIA	445	373F	337F	375F	414	448	415	461	184	167	140	173*
OCEANIA	5	4	8	8	322	758	799	775	1	3	6	6
AUSTRALIA	5	4	8	8F	322	758	799	775	1	3	6	6F
USSR	35	36	46	40F	2132	1917	1870	2000	75	69	86	80F
DEV.PED M E	2098	1838	1767	1684	825	881	887	898	1731	1619	1568	1513
N AMERICA	602	659	674	581	1415	1338	1333	1460	851	882	897	849
W EUROPE	1256	985	909	917	498	529	507	505	625	521	461	463
OCEANIA	5	4	8	8	322	758	799	775	1	3	6	6
OTH DEV.PED	235	190	177	178	1075	1122	1155	1095	253	214	204	195
DEV.PING M E	16589	18566	17812	19473	469	468	448	449	7782	8684	7989	8742
AFRICA	1754	1951	2141	2211	584	560	554	559	1025	1092	1187	1235
LAT AMERICA	6368	6984	6532	7310	609	583	527	548	3876	4068	3441	4004
NEAR EAST	246	223	230	238	1177	1311	1318	1319	289	292	303	314
FAR EAST	8222	9409	8909	9714	315	344	343	328	2591	3232	3058	3189
CENTR PLANND	3910	3757	3820	3925	511	648	662	677	2000	2434	2529	2658
ASIAN CPE	2425	2622	2728	2784	703	827	834	855	1704	2167	2275	2381
E EUR+USSR	1485	1135	1092	1141	199	235	233	243	296	266	254	277
DEV.PED ALL	3583	2973	2855	2825	566	634	637	634	2027	1885	1822	1790
DEV.PING ALL	19014	21188	20541	22258	499	512	500	500	9486	10852	10263	11122

TABLE
TABLEAU 24
CUADRO

		BROAD BEANS, DRY		FEVES SECHES			HABAS SECAS				
	AREA HARV SUP RECOLTEE SUP COSECHAO		1000 HA	YIELO RENDEMENT RENDIMIENTO		KG/HA		PRODUCTION PRODUCTION PRCDUCCICN			1000 MT	
	1969-71	1975	1976	1977	1969-71	1975	1976	1977	1969-71	1975	1976	1977
WORLD	5396	5478	5492	5524	1056	1131	1133	1103	5699	6156	6225	6101
AFRICA	723	755	725	730	1183	1154	1299	1089	855	871	941	795
ALGERIA	28	35	36	35F	687	970	944	708	19	34	34	25F
EGYPT	127	96*	102*	102F	2196	2262	2321	2321	278	217*	237*	237F
ETHIOPIA	280	274	265	271F	1077	1110	1247	1243	302	304	330	337F
LIBYA	3	5F	5F	5F	502	1000	1000	1000	2	5F	5F	5F
MOROCCO	179	220	191	190	1093	966	1203	493	196	213	230	94
SIERRA LEONE	45	47F	47F	47F	522	527	532	535	24	25F	25F	25F
SUDAN	10	16	15	16F	1497	1128	1261	1290	15	18*	19	20F
TUNISIA	51	62F	63F	64F	395	887	952	813	20	55	60	52F
N C AMERICA	78	69	79	75	534	674	665	658	42	47	53	49
DOMINICAN RP	6	6F	6F	7F	860	938	938	954	5	6F	6F	6F
GUATEMALA	15	17F	18F	18F	177	165	156	156	3	3F	3F	3F
MEXICO	56	46	55	50	594	826	800	800	34	38	44	40
SOUTH AMERIC	242	250	246	251	545	569	568	575	132	142	140	145
ARGENTINA	1	1	1	1F	3289	2470	3511	3500	4	2	3	4F
BOLIVIA	8	10F	10F	11F	607	842	888	993	5	8*	9*	11*
BRAZIL	183	188F	191F	195F	470	463	461	462	86	87F	88F	90F
ECUADOR	22	19	14	14F	605	828	898	929	14	16	13	13F
PARAGUAY	4	9	10	10F	647	886	890	900	3	8	9	9F
PERU	23	23*	20*	20F	901	889	880	880	20	20*	18*	18F
URUGUAY					2715	2650	2650	2571	1	1F	1F	1F
ASIA	3665	3902	3971	4074	1024	1136	1131	1115	3753	4433	4492	4544
CHINA	3597F	3840F	3900F	4000F	1020	1133	1128	1113	3670F	4350F	4400F	4450F
CYPRUS	3	3F	3F	3F	877	1000	1000	1071	2	3F	3F	3F
IRAQ	18	20	26*	29*	1036	895	942	965	19	18	25F	28F
ISRAEL					600	357	227	227				
JAPAN	7	2	2*	2F	1127	1300	1043	1043	7	3*	2*	2F
JORDAN	1				1837	465	606	606	2			
LEBANON	1	1F	1F	1F	976	1207	1207	1207	1	1F	1F	1F
SYRIA	8	6	8	9F	1235	1603	1671	1333	9	9	14	12F
TURKEY	32	31	30	30F	1326	1613	1583	1600	42	50	48	48F
EUROPE	687	501	471	404	1336	1402	1272	1408	917	703	599	569
AUSTRIA	1				3048	2957	3055	2973	2	1F	1F	1F
BELGIUM-LUX	1				2924				4			
BULGARIA					955	967	778	778				
CZECHOSLOVAK	6	38	27	23F	1704	2011	932	2212	10	77	25	50F
DENMARK	17				3144	2688	2655	2583	52	1		
FRANCE	19	23	21	19*	2047	2278	1333	2399	39	52	28	46*
GERMAN DR	7	6*	6*	6F	2031	2046	1572	1586	15	12*	9*	9F
GERMANY FED	22	17	13	8	3083	3210	2216	3161	67	53	30	24
GREECE	13	11F	11F	11F	1181	1429	1429	1429	15	15F	15F	15F
ITALY	347	206	200	163	1075	1222	1194	1218	373	252	239	198
MALTA	1	1	1	1F	1876	1796	2009	1625	1	1	1	1F
NETHERLANDS					2173	2000	2000	2000				
PORTUGAL	53	44	47	38*	615	658	660	658	33	29	31	25F
SPAIN	124	117	105	96	1041	992	1032	909	129	116	108	87
SWITZERLAND		1	1			3462	2066	3571		4	2	1
UK	76	38	40F	40F	2332	2395	2750	2750	176	91	110F	110F
DEV.PED M E	680	460	441	378	1323	1343	1287	1354	900	617	567	512
W EUROPE	673	457	438	375	1325	1343	1289	1357	893	614	565	509
OTH DEV.PED	7	2	3	3	1112	1238	972	972	8	3	2	2
DEV.PING M E	1105	1134	1119	1127	999	1005	1094	958	1104	1141	1224	1080
AFRICA	583	638	603	608	961	989	1128	877	560	631	680	533
LAT AMERICA	320	319	326	326	543	592	592	594	174	189	193	194
NEAR EAST	202	177	190	194	1833	1810	1843	1824	370	321	351	354
CENTR PLANNO	3610	3884	3933	4028	1023	1143	1128	1119	3695	4438	4434	4509
ASIAN CPE	3597	3840	3900	4000	1020	1133	1128	1113	3670	4350	4400	4450
E EUR+USSR	13	44	33	28	1878	2014	1046	2082	25	88	34	59
DEV.PED ALL	693	503	473	406	1334	1401	1270	1405	925	705	601	571
DEV.PING ALL	4702	4974	5019	5127	1015	1104	1121	1079	4774	5491	5624	5530

TABLE
TABLEAU 25
CUADRO

	PEAS, DRY				POIS SECS				GUISANTES SECOS			
	AREA HARV / SUP RECOLTEE / SUP COSECHAD 1000 HA				YIELD / RENDEMENT / RENDIMIENTO KG/HA				PRODUCTION / PRODUCTION / PRODUCCION 1000 MT			
	1969-71	1975	1976	1977	1969-71	1975	1976	1977	1969-71	1975	1976	1977
WORLD	9666	10450	10364	10239	1156	1008	1306	1214	11179	10534	13534	12431
AFRICA	397	502	510	511	653	601	748	616	259	302	382	315
ALGERIA	5	6	5	6F	396	631	646	667	2	4	4	4F
BURUNDI	38	24	29F	36F	799	679	690	694	30	16	20F	25F
EGYPT	3	3	3	4F	1474	1667	1700	1649	4	4	6	6F
ETHIOPIA	105	140	142	145F	646	370	762	758	68	52	108	110F
LESOTHO	11	10	11F	12F	464	569	636	625	5	6	7F	8F
LIBYA					1483	2313	2294	2294		1F	1F	1F
MOROCCO	61	137	133	117	635	717	842	284	39	99	112	33
RWANDA	76	72	71	72F	849	801	802	806	64	57	57	58F
SIERRA LEONE	2	2F	2F	2F	659	688	688	688	1	1F	1F	1F
SOUTH AFRICA	7	8F	8F	8F	559	1650	1750	1875	4	13	14F	15F
TANZANIA	21	23F	24F	26F	276	233	238	242	6	5F	6F	6F
UGANDA	21	26F	28F	30F	381	462	500	500	8	12*	14F	15F
ZAIRE	46	52*	53*	54F	585	612	610	613	27	32*	32*	33F
N C AMERICA	156	143	97	137	1661	1587	1774	921	260	227	173	126
CANADA	32	30	25	35	1354	1619	1759	1600	44	49	43	56
DOMINICAN RP	1	1F	1F	1F	759	821	832	853	1	1F	1F	1F
MEXICO	5	5	7*	5*	613	630	571	800	3	3	4*	4*
USA	118	107*	65*	96*	1795	1625	1924	683	212	175	125	66
SOUTH AMERIC	138	128	131	140	733	734	837	915	101	94	110	128
ARGENTINA	20	15	18	20	965	1322	2003	2513	19	19	36	50
BOLIVIA	4	4F	4F	4F	909	1439	976	1048	3	6*	4F	4F
CHILE	11	9	9	16	713	723	838	875	8	6	7	14
COLOMBIA	44	53*	55*	53*	680	547	545	500	30	29*	30*	27*
ECUADOR	25	17	16	16F	609	575	578	563	15	10	9	9F
GUYANA		1	1F	2F	862	685	660	690		1	1*	1*
PARAGUAY	5	3	4F	4F	632	902	914	917	3	3	3F	3F
PERU	24	22*	20*	20F	777	781	770	762	18	17*	15*	15F
URUGUAY	1	1F	1F	1F	2018	2000	2000	2000	1	2F	2F	2F
VENEZUELA	5	4	4*	4*	521	350	500	500	3	1	2*	2*
ASIA	5199	5338	5408	5487	931	977	998	982	4842	5218	5398	5388
BURMA	24	21	21	22F	656	566	813	837	16	12	17	18F
CHINA	4161F	4600F	4650F	4750F	950	1022	1032	1032	3953F	4700F	4800F	4900F
INDIA	982	688	705	683	850	687	769	629	835	472	542	430
IRAN	21	20*	22*	23F	1232	1150	1273	1270	26	23*	28*	30F
ISRAEL					1319	5000	5200	5200	1	1F	1F	1F
JAPAN	5	3	3	3F	1116	1233	1067	1067	6	4*	3*	3F
KOREA REP	1	2	2	2F	612	697	733	750		2	2	2F
LEBANON					955	875	875	875				
SYRIA		1	1	1F	814	700	750	875		1	1	1F
TURKEY	4	3	3	3F	1181	1273	1379	1400	4	4	4	4F
EUROPE	409	324	291	323	1585	1794	1743	1838	648	581	507	593
AUSTRIA	1	1*	1F	1F	2262	2614	2632	2625	2	2*	2*	2F
BELGIUM-LUX	4	3	2	1*	3029	3688	2696	3435	13	9	5	4*
BULGARIA	24	12	10	10F	1433	969	1312	1316	34	11	13	13F
CZECHOSLOVAK	14	34	26	27F	1618	1305	1445	1481	23	45	37	40F
DENMARK	7	5	5	5F	2853	3126	3004	2960	20	17	14	15F
FINLAND	1	5	5	10	1829	2019	3021	1553	3	11	15	16
FRANCE	12	20	17	17*	3305	3632	1817	3556	38	73	31	60*
GERMAN DR	25	21*	21*	21F	1786	2020	1441	1456	44	42*	30*	30F
GREECE	2	2F	2F	2F	1164	1250	1250	1250	3	2F	2F	2F
HUNGARY	80	61	61	62F	1395	1904	2076	2078	111	116	126	128F
ITALY	8	4	4	5	1225	1172	1235	1123	10	5	5	5
NETHERLANDS	11	6	5	3	3234	3646	3532	3782	37	22	16	13
NORWAY					2500	2500	2500	2500				
POLAND	50	67F	69F	71F	1242	1239	1246	1254	62	83F	86F	89F
ROMANIA	100	23	6	25F	1282	1258	1919	1280	128	29	12	32F
SPAIN	24	10	9	9*	846	802	744	778	20	8	6	7*
SWEDEN	5	5	5	5F	1758	1370	2000	2000	8	6	11	11F
UK	25	30	31	33	2963	2930	2645	3303	75	88	82	109
YUGOSLAVIA	16	16	16	17	1067	794	1052	1024	17	12	16	17F
OCEANIA	52	39	40	41	1533	1940	1952	1980	80	76	78	81
AUSTRALIA	29	18	18	18F	966	1094	1023	1017	28	20	18	18F
NEW ZEALAND	23	21F	22F	23F	2235	2684	2716	2744	52	56F	59F	62F
USSR	3316	3975	3886	3600F	1505	1015	1772	1611	4989	4036	6886	5800F
DEV.PED M E	333	294	239	291	1781	1947	1954	1660	593	572	468	483
N AMERICA	151	138	89	131	1700	1624	1879	926	256	223	168	121
W EUROPE	117	106	99	108	2103	2407	2058	2424	246	255	204	262
OCEANIA	52	39	40	41	1533	1940	1952	1980	80	76	78	81
OTH DEV.PED	13	11	11	11	806	1598	1644	1733	10	18	19	20
DEV.PING M E	1565	1363	1396	1383	793	660	771	663	1242	900	1076	916
AFRICA	386	492	499	499	648	577	724	587	250	284	361	293
LAT AMERICA	144	134	139	146	729	731	824	910	105	98	115	133
NEAR EAST	29	27	30	31	1246	1207	1329	1327	36	32	39	42
FAR EAST	1007	711	729	707	846	683	770	636	852	486	561	449
CENTR PLANNO	7768	8793	8728	8565	1203	1031	1374	1288	9344	9063	11990	11032
ASIAN CPE	4161	4600	4650	4750	950	1022	1032	1032	3953	4700	4800	4900
E EUR+USSR	3608	4193	4078	3815	1494	1040	1763	1607	5391	4363	7190	6132
DEV.PED ALL	3940	4487	4317	4106	1519	1100	1774	1611	5984	4935	7657	6615
DEV.PING ALL	5726	5963	6046	6133	907	939	972	948	5195	5600	5876	5816

TABLE
TABLEAU 26
CUADRO

	CHICK-PEAS				POIS CHICHES				GARBANZOS			
	AREA HARV SUP RECOLTEE SUP COSECHAD	1000 HA			YIELD RENDEMENT RENDIMIENTO	KG/HA			PRODUCTION PRODUCTION PRODUCCION	1000 MT		
	1969-71	1975	1976	1977	1969-71	1975	1976	1977	1969-71	1975	1976	1977
WORLD	9903	9344	10633	10245	656	596	688	676	6494	5565	7311	6924
AFRICA	431	386	403	318	567	736	555	600	244	284	224	191
ALGERIA	25	34	33	35F	452	729	764	511	11	25	25	18F
EGYPT	3	3	4	4F	1791	1587	1732	1722	5	4	6	6F
ETHIOPIA	181	187	210	173	723	790	520	661	131	148	109	114
LIBYA		1F	1F	1F	409	533	541	532				
MOROCCO	117	99	100	42	603	618	513	259	71	61	51	11
SUDAN	3	2	3	3F	907	895	933	929	3	1	3	3F
TANZANIA	71	30	19	27F	171	300	263	245	12	9	5	7F
TUNISIA	27	28F	30F	30F	354	1214	733	1000	10	34	22	30F
UGANDA	4	4F	4F	4F	500	500	500	500	2	2F	2F	2F
N C AMERICA	210	191	105	210	852	1023	725	834	179	195	76	175
DOMINICAN RP					1016	1095	1091	1087				
MEXICO	210	191	105*	210*	851	1023	724	833	179	195	76*	175*
SOUTH AMERIC	58	45	36	40	521	633	489	558	30	28	18	22
ARGENTINA	5	7	4	2	927	1101	1042	936	4	8	4	2
BOLIVIA	1	1	1F	1F	476	544	550	600	1		1F	1F
CHILE	12	8	7	8	445	630	386	603	5	5	3	5
COLOMBIA	34	25*	21*	25*	471	480	381	480	16	12*	8*	12*
PERU	6	4*	4*	4F	658	761	761	750	4	3*	3*	3F
ASIA	8948	8538	9916	9502	657	580	695	677	5882	4954	6894	6437
BANGLADESH	71	56	54	56	821	704	718	736	58	39	39	41
BURMA	129	130	147	147F	542	517	484	485	70	67	71	71F
CYPRUS		1	1	1F	493	570	759	748				
INDIA	7565	7042	8320	7856	663	570	707	683	5018	4015	5880	5366
IRAN	35	35*	36*	37F	1106	1143	1111	1093	38	40*	40*	40F
IRAQ	5	12	14	15F	669	652	525	628	4	8	7	9
ISRAEL	2	1	1	1F	1245	2479	1652	1667	2	3	2	2
JORDAN	1	4	2	2F	432	251	221	225	1	1		
LEBANON	2	3F	3F	3F	665	880	960	1040	2	2F	2F	3F
NEPAL	65	65F	65F	66F	495	462	462	470	32	30F	30F	31F
PAKISTAN	934	996	1068	1094	546	552	563	593	509	550	601	649
SYRIA	38	55	68	75F	778	484	751	733	30	27	51	55F
TURKEY	100	140	138	151F	1177	1229	1232	1113	118	172	170	168F
EUROPE	256	184	173	174	622	580	573	564	159	107	99	98
BULGARIA	1	2	1	1F	1320	851	993	927	1	2	1	1F
GREECE	15	14	12	15*	933	1071	917	1000	14	15	11	15*
ITALY	39	16	16	17	887	1085	1098	1004	34	18	17	17
PORTUGAL	50	40	41	39	386	377	357	282	19	15	15	11
SPAIN	149	111	103	102	594	509	534	530	88	56	55	54
YUGOSLAVIA	3	1	1	1F	648	1000	833	833	2	1	1	1F
DEV.PED M E	257	183	173	175	623	590	578	570	160	108	100	100
W EUROPE	255	182	172	173	619	577	571	563	158	105	98	98
OTH DEV.PED	2	1	1	1	1245	2479	1652	1667	2	3	2	2
DEV.PING M E	9644	9159	10459	10070	657	596	689	678	6333	5459	7210	6823
AFRICA	424	382	396	311	556	730	542	585	236	279	215	182
LAT AMERICA	268	236	141	250	780	949	664	789	209	224	94	198
NEAR EAST	188	253	267	289	1060	1009	1050	985	200	255	280	285
FAR EAST	8764	8289	9655	9219	649	567	686	668	5688	4701	6621	6159
CENTR PLANND	1	2	1	1	1320	851	993	927	1	2	1	1
E EUR+USSR	1	2	1	1	1320	851	993	927	1	2	1	1
DEV.PED ALL	258	185	174	175	626	593	580	572	162	110	101	100
DEV.PING ALL	9644	9159	10459	10070	657	596	689	678	6333	5459	7210	6823

TABLE
TABLEAU 27
CUADRO

	AREA HARV 1000 HA SUP RECOLTEE SUP COSECHAD				YIELD KG/HA RENDEMENT RENDIMIENTC				PRODUCTION 1000 MT PRODUCTICN PRODUCCION			
	1969-71	1975	1976	1977	1969-71	1975	1976	1977	1969-71	1975	1976	1977
WORLD	1632	1879	1926	1970	609	588	641	644	995	1104	1236	1269
AFRICA	179	231	183	154	602	619	793	737	108	143	145	113
ALGERIA	21	15	18	17F	392	468	495	446	8	7	9	7F
EGYPT	22	24	27	32F	1607	1610	1423	1424	36	39	38	45*
ETHIOPIA	101	142	72	67	460	430	760	769	46	61	55	51
MOROCCO	33	46	63	36	512	739	660	223	17	34	41	8
TUNISIA	3	3F	3F	3F	329	467	516	516	1	1F	2F	2F
N C AMERICA	36	45	59	60	1185	1093	1076	1085	42	49	64	65
MEXICO	6	9	9F	9F	755	913	944	934	5	8	9F	9F
USA	29	36	50F	51F	1276	1139	1100	1112	37	41	55F	56F
SOUTH AMERIC	66	57	57	87	578	677	730	893	38	38	41	78
ARGENTINA	20	8	9	29	537	1588	1801	1443	11	13	16	41
CHILE	16	21	23	30	645	587	602	782	10	12	14	24
COLOMBIA	25	22*	20*	22*	527	455	450	409	13	10*	9*	9*
ECUADOR	2	4	3	3F	403	534	579	594	1	2	2	2F
PERU	3	2*	2*	3F	1030	612	612	577	3	2*	2*	2F
ASIA	1211	1401	1498	1513	565	554	620	563	684	776	928	852
BANGLADESH	74	70	66	66F	742	677	687	739	55	47	45	49
BURMA	5	4	5	5F	330	285	249	245	2	1	1	1F
CYPRUS	1				742	757	864	938				
INDIA	761	953	937	926	505	480	492	465	385	457	461	431
IRAN	46	53F	53F	53F	563	566	566	568	26	30*	30*	30F
IRAQ	9	5	6	6F	561	938	902	952	5	5	5	6
ISRAEL					353	556	556	556				
JORDAN	21	15	23	21F	710	350	409	333	15	5	9	7*
LEBANON	3	3F	4F	4F	795	647	611	579	2	2F	2F	2F
PAKISTAN	65	75	73	73F	334	351	389	390	22	26	28	28F
SYRIA	121	98	147	178	606	681	930	658	73	67	136	117
TURKEY	105	125	186	180F	949	1084	1129	1000	100	135	210	180F
EUROPE	84	94	94	97	774	894	558	935	65	84	52	90
BULGARIA	5	1			506	121	508	510	2			
CZECHOSLOVAK	1	1	1	1F	712	600	547	515	1	1	1	1F
FRANCE	9	11	11	13*	1201	1621	792	1227	11	18	9	16*
GERMAN DR					286	1294	1125	1000				
GREECE	8	5	5	5*	833	1200	1000	1200	7	6	5	6*
HUNGARY	2				587	178	546	571	1			
ITALY	7	3	2	2	818	947	870	821	5	3	2	2
SPAIN	51	71	73	73	715	779	478	880	37	55	35	65
YUGOSLAVIA	1	1	1	1	816	857	857	900	1	1	1	1F
USSR	56	51	36	60F	1012	255	139	1167	57	13	5	70F
DEV.PED M E	106	127	142	145	928	976	749	1002	99	124	107	146
N AMERICA	29	36	50	51	1276	1139	1100	1112	37	41	55	56
W EUROPE	77	91	92	94	797	912	558	944	61	83	51	89
OTH DEV.PED					353	556	556	556				
DEV.PING M E	1462	1698	1746	1763	571	569	643	597	835	966	1123	1052
AFRICA	157	207	156	122	460	503	685	560	72	104	107	68
LAT AMERICA	73	66	66	96	593	710	759	897	43	47	50	86
NEAR EAST	328	323	445	474	785	876	969	818	257	283	431	388
FAR EAST	905	1102	1080	1070	511	483	496	476	463	532	535	509
CENTR PLANND	64	54	38	62	956	258	161	1144	61	14	6	71
E EUR+USSR	64	54	38	62	956	258	161	1144	61	14	6	71
DEV.PED ALL	170	181	180	208	938	762	624	1045	160	138	113	217
DEV.PING ALL	1462	1698	1746	1763	571	569	643	597	835	966	1123	1052

TABLE
TABLEAU 28
CUADRO

SOYBEANS GRAINES DE SOJA SOJA

	AREA HARV / SUP RECOLTEE / SUP COSECHAD (1000 HA)				YIELD / RENDEMENT / RENDIMIENTO (KG/HA)				PRODUCTION / PRODUCTION / PRODUCCION (1000 MT)			
	1969-71	1975	1976	1977	1969-71	1975	1976	1977	1969-71	1975	1976	1977
WORLD	35314	45968	44742	49426	1324	1516	1409	1568	46747	69670	63025	77502
AFRICA	188	207	223	237	396	496	507	533	75	103	113	126
EGYPT		4*	7*	8F	1346	1601	1875			5*	11*	15F
LIBERIA	4	4F	5F	5F	336	350	333	333	1	1F	2F	2F
NIGERIA	162	170F	180F	190F	376	382	389	368	61	65F	70F	70F
RHODESIA					600	600	600	600				
RWANDA	1	3	4	5F	781	819	837	844	1	3	4	4F
SOUTH AFRICA	12	13*	12*	15*	396	1538	1583	1873	5	20	19	28
TANZANIA	2	5F	4F	4F	256	178	150	150	1	1	1	1F
UGANDA	4	5	7	7	1000	1000	716	714	4	5	5	5
ZAIRE	2	3F	3F	3F	800	600	400	400	2	2F	1*	1F
ZAMBIA		1	1	1F		1533	632	1000		1	1	1F
N C AMERICA	17308	22185	20300	23928	1831	1946	1754	1994	31683	43181	35596	47720
CANADA	138	158	153	202	1860	2324	1637	2556	257	367	250	517
EL SALVADOR					1800	2400						
MEXICO	134	344	172	290*	1876	2029	1756	1690	252	699	302	490*
NICARAGUA		1*	1*	1F	1618	1618	1667			1*	1*	1F
USA	17036	21682	19974	23435	1830	1942	1754	1993	31174	42114	35042	46712
SOUTH AMERIC	1433	6448	7108	8036	1222	1676	1736	1746	1751	10810	12338	14028
ARGENTINA	30	356	434	660	1299	1363	1603	2121	39	485	695	1400
BOLIVIA	1	9	12	7	1499	1262	1270	1206	1	12	15	9
BRAZIL	1314	5824	6416	7059	1178	1699	1750	1714	1547	9892	11227	12100F
CHILE	1	1*	2*	1*	1251	1000	1111	1100	2	1*	2*	1*
COLOMBIA	58	88	38	53	1954	1924	1997	2047	114	169	75	109
ECUADOR	1	9	11	15*	1043	1372	1309	1284	1	12	15	19*
GUYANA					657	371	371	371				
PARAGUAY	27	150	173	230F	1690	1465	1635	1631	45	220	284	375*
PERU		1F	2F	3F	1194	1143	1300	1111		2	3	3F
SURINAM					971	1355	1125	1125				
URUGUAY	1	10*	20	8*	1000	1719	1150	1500	1	17*	23	12*
ASIA	15412	16067	16038	16106	817	893	876	909	12594	14347	14046	14639
BURMA	19	20F	21F	22F	644	622	476	472	12	12	10*	10F
CHINA	13859F	14141F	14236F	14236F	822	895	875	910	11398F	12662F	12453F	12955F
INDIA	4	160F	160F	160F	545	750	750	749	2	120*	120F	120F
INDONESIA	643	752	636	663	728	785	758	795	468	590	482	527
IRAN	6	54*	75*	75F	875	1296	1360	1368	5	70*	102*	103F
IRAQ				1F		1333	1429	1538		1F	1F	1F
JAPAN	100	87	83	83F	1279	1445	1321	1337	128	126	110	111*
KAMPUCHEA DM	7	4*	4*	4*	733	1000	1000	1000	5	4*	4*	4*
KOREA DPR	370	395F	400F	400F	689	734	750	775	255	290F	300F	310F
KOREA REP	292	274	247	280F	780	1134	1192	1214	228	311	295	340*
LAO	4	4F	4F	4F	1000	975	976	976	4	4F	4F	4F
MAL PENINSUL					1504	3817	3923	4000		1F	1F	1F
PHILIPPINES	1	8	11*	15*	846	722	763	700	1	6	9*	11*
SRI LANKA		1F	1F	1F		1000	1000	1000		1F	1F	1F
THAILAND	53	110F	100F	102F	965	1036	1136	959	51	114	114	98
TURKEY	9	6	6	6*	1300	1144	1328	1333	11	7	9	8*
VIET NAM	45	50F	52F	54F	578	600	635	667	26	30F	33F	36F
EUROPE	108	203	285	300	1081	1850	1430	1444	117	376	408	433
BULGARIA	9	36	56	55F	876	2210	1755	1802	8	80	99	99F
HUNGARY		24*	39	35*	714	1600	1086	1714		38*	42	60*
ITALY					2012	3000	2967	2967			1	
ROMANIA	93	121	155	171*	1095	1762	1372	1129	102	213	213	193*
SPAIN	1	8	4	8F	1373	1801	1483	1875	2	14	6	15*
YUGOSLAVIA	4	14	31	31	1126	2113	1548	2129	5	30	48	66*
OCEANIA	5	46	26	33	1111	1609	1699	1678	5	74	45	55
AUSTRALIA	5	46	26	33	1111	1606	1696	1678	5	74	45	55
NEW ZEALAND						3000	3000					
USSR	860	811	762	786	606	962	630	636	521	780	480	500*
DEV.PED M E	17297	22008	20283	23808	1826	1942	1751	1995	31576	42745	35520	47504
N AMERICA	17174	21840	20127	23638	1830	1945	1753	1998	31431	42480	35293	47229
W EUROPE	6	22	35	39	1194	2011	1545	2079	7	44	54	81
OCEANIA	5	46	26	33	1111	1609	1699	1678	5	74	45	55
OTH DEV.PED	112	100	95	98	1183	1457	1354	1419	133	146	129	139
DEV.PING M E	2774	8377	8755	9878	1029	1531	1586	1604	2855	12828	13882	15840
AFRICA	176	191	203	214	396	409	405	389	70	78	82	83
LAT AMERICA	1568	6793	7281	8327	1278	1694	1736	1744	2003	11510	12642	14519
NEAR EAST	15	64	89	90	1124	1285	1377	1412	17	82	122	127
FAR EAST	1015	1329	1181	1247	754	872	876	891	766	1158	1035	1111
CENTR PLANND	15243	15583	15704	15741	808	905	868	899	12315	14097	13624	14157
ASIAN CPE	14281	14590	14692	14694	818	890	871	906	11684	12986	12790	13305
E EUR+USSR	962	992	1012	1047	656	1120	824	814	631	1111	834	852
DEV.PED ALL	18259	23000	21296	24855	1764	1907	1707	1946	32208	43856	36354	48357
DEV.PING ALL	17055	22967	23447	24571	852	1124	1138	1186	14539	25814	26672	29145

TABLE
TABLEAU **29**
CUADRO

GROUNDNUTS IN SHELL · ARACHIDES NON DECORTIQ · MANI CON CASCARA

	AREA HARV / SUP RECOLTEE / SUP COSECHAD (1000 HA)				YIELD / RENDEMENT / RENDIMIENTO (KG/HA)				PRODUCTION / PRODUCCION (1000 MT)			
	1969-71	1975	1976	1977	1969-71	1975	1976	1977	1969-71	1975	1976	1977
WORLD	19700	19084	18703	18531	924	1018	956	942	18200	19428	17871	17459
AFRICA	7047	6468	6412	6059	783	819	799	738	5520	5299	5123	4473
ANGOLA	37	40*	40F	40F	536	500	500	625	20	20*	20F	25F
BENIN	87	58	73	73F	533	598	624	651	46	35	46	48F
BOTSWANA	5	5F	7F	7F	994	1280	1000	1077	5	6F	7F	7F
BURUNDI	16	20	20F	20F	1323	1205	1203	1203	21	24	24F	24F
CAMEROON	201	202*	205F	205F	885	817	829	732	178	165*	170F	150F
CAPE VERDE					1967							
CENT AFR EMP	106	114	113F	114F	647	1164	1150	1158	68	133	130F	132*
CHAD	143	101*	101F	101F	668	693	693	693	95	70*	70F	70F
CONGO	21	29	32	33	819	720	720	720	17	21	23	24
EGYPT	23	13	13	16F	1760	2083	2108	2188	40	28	28	35F
ETHIOPIA	41	45F	46F	47F	592	600	605	610	24	27F	28F	29F
GABON	2	2F	2F	2F	1000	1000	1000	1000	2	2F	2F	2F
GAMBIA	80	105F	110F	110F	1563	1429	1409	1318	125	150F	155F	145
GHANA	90	102	105*	105*	827	1086	571	714	74	111	60*	75F
GUINEA	30	32F	32F	33F	843	875	938	951	25	28F	30F	31F
GUIN BISSAU	87	90F	90F	80F	412	411	500	369	36	37	45	30F
IVORY COAST	52	57	55F	55F	800	842	818	847	42	48	45F	47F
KENYA	4	5F	5F	5F	582	600	600	600	3	3F	3F	3F
LIBERIA	3	4F	4F	4F	610	671	665	675	2	3F	3F	3F
LIBYA	5	6F	6F	7F	2089	2258	2258	2308	11	14*	14F	15F
MADAGASCAR	41	37	36F	40F	1021	1059	1076	1000	42	39	39	40
MALAWI	306	239*	239*	200F	595	690	690	500	182	165*	165*	100F
MALI	257	200	199	205F	561	1137	1294	1124	144	227	258	230*
MAURITANIA	4	2F	2F	2F	504	344	337	325	2	1F	1F	1F
MAURITIUS					2519	3499	3750	4000				
MOROCCO	4	16	11	19	688	1176	1193	433	3	19	13	8
MOZAMBIQUE	220	200F	200F	200F	636	400	500	500	140	80F	100F	100F
NIGER	357	318	178	160F	623	132	445	375	223	42	79	60F
NIGERIA	1846	920F	920F	920F	900	304	543	359	1660	280F	500F	330F
REUNION					1000	806	806	839				
RHODESIA	170	170*	170F	170F	655	735	706	706	111	125*	120*	120F
RWANDA	8	15	15	15F	869	963	917	933	7	14	13	14F
SENEGAL	1006	1201	1331	1000F	789	1229	895	700	794	1476	1192	700*
SIERRA LEONE	27	25F	27F	27F	734	720	704	704	20	18F	19*	19F
SOMALIA	10	10F	10F	10F	933	970	970	980	9	10F	10F	10F
SOUTH AFRICA	370	220	156	175	984	1227	981	1378	364	270	153	241
SUDAN	490	868	769	798F	756	1073	1076	1065	370	931	827	850*
TANZANIA	49	77	124	125F	662	597	597	591	32	46	74	74F
TOGO	42	36	36F	37F	466	554	639	620	20	20	23*	23F
UGANDA	263	243	293	250F	786	750	675	830	207	182	198	208
UPPER VOLTA	140	180	164*	168F	487	500	533	531	68	90	87*	89F
ZAIRE	376	432	443	450F	688	712	722	733	258	308	319	330F
ZAMBIA	30	30F	30F	32F	955	1033	1000	1019	28	31F	30F	32F
N C AMERICA	740	741	720	717	1981	2558	2535	2502	1465	1896	1825	1794
CUBA	15	15F	15F	15F	1000	1000	1000	1000	15	15F	15F	15F
DOMINICAN RP	68	45F	35F	35F	1113	1111	1143	1143	76	50	40	40*
EL SALVADOR		1		1F	1125	1440	935	1100		1		1F
GUATEMALA					1825	1931	1943	2000	1	1	1F	1F
HAITI	4	4F	4F	4F	506	512	512	512	2	2F	2F	2F
HONDURAS					1100	1250	1250	1250				
JAMAICA	1	1F	1F	1F	1061	1053	975	960	1	1	1	1F
MEXICO	59	62	43	43	1366	1118	1302	1302	81	69	56	56
NICARAGUA	1	4	4	5	1666	1700	1700	1701	1	7	8	8
ST KITTS ETC					2214	1368	1338	1346				
ST VINCENT					973	926	963	1000				
USA	591	609	616	613	2182	2875	2763	2725	1289	1750	1701	1670
SOUTH AMERIC	979	784	759	675	1230	1144	1219	1492	1204	896	925	1008
ARGENTINA	255	357	309	363	1099	1052	1094	1653	280	375	338	600
BOLIVIA	6	10	11	12	1362	1500	1299	1315	8	15	14	16
BRAZIL	670	348	375	231	1307	1267	1369	1404	876	441	514	324
COLOMBIA		2	2	1	1381	1611	1158	1154	1	3	2	2
ECUADOR	8	16	12	15*	897	725	662	733	7	11	8	11F
GUYANA					600	603	609	622				
PARAGUAY	22	20	22F	22F	788	891	818	818	17	18	18*	18F
PERU	4	5F	5F	5F	1359	1417	1440	1462	5	7F	7F	8F
SURINAM					1046	1012	1061	1061				
URUGUAY	3	3	3	4*	707	813	800	500	2	3	2	2*
VENEZUELA	11	22	19	21*	764	1017	1105	1286	8	22	21	27*
ASIA	10886	11054	10772	11036	915	1020	923	918	9959	11279	9939	10127
BANGLADESH	28	22	20F	20F	1601	1461	1170	1250	45	33	23	25F
BURMA	655	690	690F	695F	751	596	754	763	492	411	520F	530F
CHINA	2165F	2214F	2259F	2355F	1216	1306	1323	1181	2634F	2891F	2989F	2781F
CYPRUS					2099	1625	1669	1786		1*	1	1F
EAST TIMOR	1	1F			1004	1083			1	1F		
INDIA	7287	7222	6948	7000F	797	935	757	786	5807	6755	5262	5500*
INDONESIA	376	475	411	506	1230	1326	1338	1330	462	630	550	673
IRAN		2*	3F		1500	1520				3*	4F	
IRAQ					1175	1518	1400	1636				
ISRAEL	5	5	6	6F	3684	3643	3865	3889	17	19	24	25
JAPAN	59	41	38	38F	2038	1720	1730	1667	120	71	65	63*
JORDAN							1000					
KAMPUCHEA DM	20	14F	14F	15F	1070	1037	1071	1067	21	14	15F	16F
KOREA REP	7	6	6	7F	1177	1148	1198	1154	8	7	8	8F
LAO	2	2F	2F	2F	867	933	933	933	1	1F	1F	1F
LEBANON	3	4F	4F	4F	1151	1056	1027	1000	4	4F	4F	4F

TABLE
TABLEAU **29**
CUADRO

| GROUNDNUTS IN SHELL | | | | ARACHIDES NON DECORTIC | | | | MANI CON CASCARA | | | |
| AREA HARV SUP RECOLTEE SUP COSECHAO | | 1000 HA | | YIELD RENDEMENT RENDIMIENTO | | KG/HA | | PRODUCTION PRODUCTION PRODUCCION | | | 1000 MT |
	1969-71	1975	1976	1977	1969-71	1975	1976	1977	1969-71	1975	1976	1977
MAL PENINSUL	3	7	7F	7F	1830	3621	3571	3549	5	25F	25F	25F
PAKISTAN	38	44	45	45F	1433	1415	1421	1421	55	62	64	64F
PHILIPPINES	32	55	61*	67*	534	660	673	657	17	36	41*	44*
SRI LANKA	5	11	11	12F	983	1648	1708	1652	5	18	19	19F
SYRIA	10	13	13F	14F	1836	1659	1769	1755	18	21	23F	24F
THAILAND	97	120F	125F	130F	1317	1185	1215	1344	128	142	152	175*
TURKEY	16	18	20	22*	2528	2222	2738	2364	40	40	55	52*
VIET NAM	78	90F	90F	91F	1006	1056	1056	1066	78	95F	95F	97F
EUROPE	10	9	10	10	1890	2524	2062	2177	19	24	20	23
BULGARIA	1	2	2	2F	1157	2099	1094	1167	2	5	3	3F
GREECE	4	4	4	4F	2199	2750	2250	2250	9	11	9	9F
ITALY	1	1	1	1F	2307	2268	2462	2436	3	1	2	2F
SPAIN	3	2	2	3F	1810	2696	2727	2900	5	6	6	9*
YUGOSLAVIA	1				1149	1500	1000	1000	1			
OCEANIA	38	27	31	33	855	1256	1241	1003	33	34	38	33
AUSTRALIA	35	24	27	30	873	1325	1300	1030	30	32	35	31
FIJI	1	1F	1F	1F	432	768	780	800		1F	1F	1F
NEW HEBRIDES					500	585	592	600				
PAPUA N GUIN	2	2F	2F	2F	764	750	750	750	2	2F	2F	2F
TONGA					667	667	667	667				
USSR			1	1F	428	500	1125	1125			1	1F
DEV.PED M E	1068	906	850	870	1721	2384	2348	2356	1838	2160	1996	2049
N AMERICA	591	609	616	613	2182	2875	2763	2725	1289	1750	1701	1670
W EUROPE	9	7	7	8	2015	2653	2384	2481	17	19	17	20
OCEANIA	35	24	27	30	873	1325	1300	1030	30	32	35	31
OTH DEV.PED	434	266	200	219	1157	1350	1210	1500	502	359	242	329
DEV.PING M E	16368	15858	15487	15197	833	899	825	823	13628	14263	12773	12512
AFRICA	6160	5361	5467	5064	769	757	750	658	4736	4056	4101	3332
LAT AMERICA	1127	916	863	779	1225	1138	1215	1452	1381	1043	1049	1131
NEAR EAST	546	924	828	861	883	1127	1154	1140	483	1042	956	981
FAR EAST	8531	8654	8325	8490	824	938	801	832	7026	8121	6665	7065
OTH DV.PING	4	3	3	3	673	745	749	755	2	2	2	2
CENTR PLANND	2264	2320	2366	2464	1208	1295	1311	1176	2735	3005	3102	2898
ASIAN CPE	2262	2318	2363	2461	1208	1295	1312	1176	2733	3000	3099	2894
E EUR+USSR	2	2	3	3	980	2085	1102	1156	2	5	4	4
DEV.PED ALL	1070	908	853	873	1720	2383	2343	2351	1840	2164	1999	2053
DEV.PING ALL	18631	18176	17850	17658	878	950	889	872	16360	17264	15872	15406

TABLE / TABLEAU / CUADRO **30**

	CASTOR BEANS — AREA HARV / SUP RECOLTEE / SUP COSECHAD — 1000 HA				GRAINES DE RICIN — YIELD / RENDEMENT / RENDIMIENTO — KG/HA				RICINO — PRODUCTION / PRODUCTION / PRODUCCION — 1000 MT			
	1969-71	1975	1976	1977	1969-71	1975	1976	1977	1969-71	1975	1976	1977
WORLD	1435	1592	1227	1342	588	531	513	514	844	846	630	690
AFRICA	104	84	79	78	545	572	541	592	57	48	43	46
ANGOLA	11	12F	12F	12F	253	250	250	292	3	3F	3F	4F
BENIN	1	1F	1F	1F	550	600	600	600		1F	1F	1F
BURUNDI					200	200	200	200				
CAPE VERDE					4077							
ETHIOPIA	22	13F	10F	10F	581	1062	1000	1010	13	14	10F	10F
KENYA	12	12F	12F	12F	257	208	208	250	3	3F	3F	3F
LIBYA					1060	1833	1833	1833		1F	1F	1F
MADAGASCAR	6	3F	3F	3F	178	135	133	133	1			
MOROCCO					983	1111	1050	1050				
MOZAMBIQUE	5	6F	6F	6F	393	333	333	333	2	2F	2F	2F
SOUTH AFRICA	8	8F	8F	8F	609	625	625	625	5	5F	5F	5F
SUDAN	16	19	18F	21F	1020	833	833	927	16	16	15*	19*
TANZANIA	19	6F	6F	2F	597	500	500	500	11	3*	3*	1*
TOGO	1				437	477	500	500				
UGANDA	4	2F	2F	2F	469	208	208	208	2	1F	1F	1F
N C AMERICA	23	15	14	14	694	417	518	504	16	6	7	7
DOMINICAN RP					382	409	422	435				
HAITI	2	3F	3F	3F	347	520	520	520	1	1F	1F	1F
MEXICO	9	10	9*	9*	500	421	578	556	5	4	5*	5*
USA	11	2F	2F	2F	935	265	265	265	10	1F	1F	1F
SOUTH AMERIC	415	420	300	277	984	894	832	893	408	376	250	247
ARGENTINA	7	1			815	1331	1082	643	5	1		
BRAZIL	373	395	262	246	973	892	814	906	363	353	213	223
ECUADOR	23	3	17	23*	1078	1049	789	711	25	3	13	16*
PARAGUAY	12	21	22	8F	1249	876	1078	1000	15	18	24	8*
PERU					2000	2025	2025	2025				
ASIA	681	843	633	753	414	402	448	423	282	339	284	318
BANGLADESH					634	589	633	645				
CHINA	180F	170F	165F	165F	485	400	424	424	87F	68F	70F	70F
INDIA	411	590	375	487	304	356	381	353	125	210	143	172
INDONESIA	10	25F	25F	25F	353	171	180	185	4	4F	5F	5F
IRAN	17	7*	4*	4F	529	571	1000	1000	9	4*	4*	4F
KAMPUCHEA DM					1353	1000	947	1000				
KOREA REP	2	2F	2F	2F	769	700	700	700	2	1F	1F	1F
PAKISTAN	17	11	18F	19F	760	750	750	755	13	8	14F	15F
PHILIPPINES		4F	4F	4F	1927	1950	1625	1500		8*	7*	6*
SYRIA					290	570	570	570				
THAILAND	39	30F	36F	42F	1021	1073	1069	1029	40	32	39	43
VIET NAM	4	4F	4F	4F	500	500	500	500	2	2F	2F	2F
EUROPE	21	20	17	20	726	598	305	575	15	12	5	12
BULGARIA	1				1308	7000			1			
ROMANIA	19	20	17	20F	663	596	304	575	13	12	5	12F
YUGOSLAVIA	1				1675	1000	1250	1250	1			
USSR	191	210	183	200F	349	314	224	300	67	66	41	60F
DEV.PED M E	19	10	10	10	838	557	554	554	16	6	6	6
N AMERICA	11	2	2	2	935	265	265	265	10	1	1	1
W EURCPE	1				1675	1000	1250	1250	1			
OTH DEV.PED	8	8	8	8	609	625	625	625	5	5	5	5
DEV.PING M E	1020	1178	848	943	645	588	597	574	658	693	506	541
AFRICA	81	56	53	49	444	467	422	440	36	26	22	22
LAT AMERICA	427	433	312	289	970	880	821	878	414	381	257	254
NEAR EAST	33	27	22	25	766	775	875	948	25	21	20	24
FAR EAST	480	662	460	580	381	399	451	418	183	264	207	242
CENTR PLANND	395	404	369	385	430	366	320	369	170	148	118	144
ASIAN CPE	184	174	169	169	487	403	426	426	90	70	72	72
E EUR+USSR	211	230	200	220	381	339	231	325	80	78	46	72
DEV.PED ALL	230	240	210	230	420	348	246	335	97	83	52	77
DEV.PING ALL	1204	1353	1017	1112	621	564	568	551	748	763	578	613

TABLE
TABLEAU 31
CUADRO

SUNFLOWER SEED	GRAINES DE TOURNESOL	SEMILLA DE GIRASOL
AREA HARV SUP RECOLTEE SUP COSECHAD (1000 HA)	YIELD RENDEMENT RENDIMIENTO (KG/HA)	PRODUCTION PRODUCTION PRODUCCION (1000 MT)

	AREA HARV (1000 HA)				YIELD (KG/HA)				PRODUCTION (1000 MT)			
	1969-71	1975	1976	1977	1969-71	1975	1976	1977	1969-71	1975	1976	1977
WORLD	8384	8647	9146	9844	1177	1086	1107	1194	9868	9389	10127	11754
AFRICA	229	360	407	537	681	792	826	1053	156	285	337	566
ALGERIA	4	3	2	2F	307	360	262	267	1	1		
ANGOLA	11	13F	13F	15F	882	769	769	867	10	10F	10F	13F
BOTSWANA					32	50	50	50				
EGYPT		8	8	8F	364	1504	1547	1548		12	13	13F
KENYA	4	5F	5*	7F	718	1000	1000	846	3	5F	5F	6F
MALAWI	4	4F	5F	5F	719	775	1000	963	3	3*	5F	5F
MOROCCO	19	23	23	49	651	667	682	388	12	16	16	19
MOZAMBIQUE	8	10F	10F	10F	538	800	900	1000	4	8F	9F	10F
RHODESIA	5	5F	6F	6F	551	574	569	565	3	3F	3F	4F
SOUTH AFRICA	140	239*	288*	389*	757	874	885	1224	106	209	255	476
TANZANIA	30	28F	25F	25F	426	321	237	280	13	9F	6	7F
UGANDA	3	3F	3F	3F	400	400	400	400	1	1F	1F	1F
ZAMBIA	2	18	19	18F	300	431	679	667	1	8	13	12*
N C AMERICA	175	320	361	972	965	1225	1155	1370	169	392	417	1332
CANADA	48	25	20	67	809	1193	1188	1189	39	30	24	79
MEXICO	17	8	13*	16*	528	700	308	313	9	6	4*	5*
USA	110	287	328	889	1101	1243	1186	1403	121	357	389	1248
SOUTH AMERIC	1398	1110	1406	1310	736	722	846	725	1029	801	1189	950
ARGENTINA	1283	1005	1258	1233	739	728	862	730	949	732	1085	900
CHILE	20	13	22	10	1275	1340	1253	1482	26	18	27	15
URUGUAY	95	92	126	67*	572	558	614	515	55	51	77	34*
ASIA	488	582	583	626	1036	1058	1163	946	506	615	678	592
AFGHANISTAN	2	2F	2F	2F	1555	1864	1826	1826	2	4F	4F	4F
BANGLADESH					655	500	500	500				
BURMA	1	6	6F	6F	326	204	333	328		1	2F	2F
CHINA	55F	55F	56F	56F	1273	1273	1250	1250	70F	70F	70F	70F
IRAN	70	80*	75*	81F	551	438	400	411	39	35*	30*	33F
IRAQ	3	5F	8	11*	775	860	933	909	2	4*	7	10F
ISRAEL	4	6	8	8F	1081	1027	1095	1088	4	6	8	9
LEBANON	5	6F	7F	7F	470	450	446	443	2	3F	3	4F
SYRIA	1	3	4	4F	1718	1129	892	925	2	4	3	4F
TURKEY	347	418	418	450*	1104	1167	1316	1016	383	488	550	457*
EUROPE	1373	2020	1718	1700	1404	1084	1251	1394	1928	2190	2149	2370
ALBANIA	18	27F	28F	30F	950	752	709	671	18	20F	20F	20F
AUSTRIA	1				1985	2207	2300	2721	1	1	1	1
BULGARIA	277	238	226	230F	1697	1791	1600	1609	471	426	362	370F
CZECHOSLOVAK	2	4	5	9	1625	1428	1257	1388	4	5	7	13F
FRANCE	30	72	60	38*	1772	1532	1262	1842	53	110	76	70*
GREECE	2	2	2F	2F	1042	1500	1500	1500	2	3	3F	3F
HUNGARY	98	144	154	141	1243	1083	1222	1135	122	155	188	160*
ITALY	4	25	25	30*	2009	1797	2135	1993	9	46	54	59F
PORTUGAL	1	12	15	9*	767	621	628	578	1	8	9	5*
ROMANIA	562	511	521	514*	1370	1425	1534	1570	769	728	799	807
SPAIN	179	792	506	489	813	525	616	780	146	416	312	381
YUGOSLAVIA	199	194	175	208	1679	1402	1823	2313	334	272	319	481
OCEANIA	38	210	137	125	680	541	588	594	26	113	80	74
AUSTRALIA	38	210	137	125	680	541	588	594	26	113	80	74
USSR	4682	4045	4534	4574	1293	1234	1164	1283	6055	4993	5277	5870
DEV.PED M E	755	1864	1564	2254	1113	842	978	1281	840	1570	1531	2886
N AMERICA	158	312	348	956	1012	1239	1186	1388	160	386	413	1327
W EUROPE	415	1097	784	776	1312	779	988	1289	544	855	774	1000
OCEANIA	38	210	137	125	680	541	588	594	26	113	80	74
OTH DEV.PED	143	245	296	397	765	878	891	1221	110	215	263	485
DEV.PING M E	1934	1759	2057	2036	786	808	911	765	1520	1421	1875	1558
AFRICA	89	113	111	140	563	565	618	549	50	64	69	77
LAT AMERICA	1415	1118	1419	1326	733	721	841	720	1038	807	1193	955
NEAR EAST	428	522	522	564	1007	1052	1170	930	431	550	611	524
FAR EAST	1	6	6	6	342	205	334	328		1	2	2
CENTR PLANND	5695	5023	5524	5554	1318	1274	1217	1316	7508	6398	6722	7310
ASIAN CPE	55	55	56	56	1273	1273	1250	1250	70	70	70	70
E EUR+USSR	5640	4968	5468	5498	1319	1274	1217	1317	7438	6328	6652	7240
DEV.PED ALL	6395	6832	7033	7752	1294	1156	1164	1306	8278	7898	8183	10126
DEV.PING ALL	1989	1814	2113	2092	799	822	920	778	1590	1491	1945	1628

TABLE
TABLEAU 32
CUADRO

RAPESEED · GRAINES DE COLZA · SEMILLAS DE COLZA

	AREA HARV / SUP RECOLTEE / SUP COSECHAD — 1000 HA				YIELD / RENDEMENT / RENDIMIENTO — KG/HA				PRODUCTION / PRODUCTION / PRODUCCION — 1000 MT			
	1969-71	1975	1976	1977	1969-71	1975	1976	1977	1969-71	1975	1976	1977
WORLD	8480	10207	8834	9313	778	831	855	822	6598	8487	7550	7653
AFRICA	50	50	50	50	400	400	400	400	20	20	20	20
ETHIOPIA	50	50F	50F	50F	400	400	400	400	20	20F	20F	20F
N C AMERICA	1540	1755	726	1354	989	1000	1162	1316	1523	1755	844	1783
CANADA	1533	1748	720	1348	989	1000	1163	1318	1517	1749	837	1776
MEXICO	6	6*	6*	6*	895	1000	1000	1000	6	6*	6*	6*
USA	1	1F	1F	1F	1333	1333	1333	1333	1	1F	1F	1F
SOUTH AMERIC	53	46	60	54	1400	1347	1746	1528	74	62	105	83
ARGENTINA	2		1F	1F	872	750	571	571	2			
CHILE	50	45	60	54	1424	1350	1760	1540	72	61	105	83
ASIA	5772	7042	6638	6470	533	570	552	497	3076	4011	3665	3214
BANGLADESH	219	196	193	194	604	593	578	588	132	116	112	114
BURMA	1	1F	1F	1F	1000	1000	1000	1000	1	1F	1F	1F
CHINA	1902F	2683F	2603F	2603F	524	505	501	463	996F	1354F	1305F	1205F
INDIA	3122	3680	3339	3145	522	612	580	497	1629	2252	1936	1562
JAPAN	21	4	4	4F	1614	1659	1811	1351	34	7	7	5*
KOREA REP	27	25	23	23F	1166	1288	1413	1413	31	32	32	33F
PAKISTAN	476	451	470	496	523	549	569	582	249	248	267	289
TURKEY	4		4	4F	1153	1218	1439	1439	5		6	6F
EUROPE	1017	1286	1336	1352	1840	2035	2164	1866	1870	2618	2892	2523
AUSTRIA	4	2	2	2	2004	2229	2258	2303	8	6	4	5
BELGIUM-LUX	1			1*	2345	2176	1256	2167	1	1		1*
BULGARIA					860	174	1500	1500				
CZECHOSLOVAK	40	64	63	73F	1752	2059	2116	1646	70	131	134	120*
DENMARK	17	72	44	42F	1783	1819	1830	1690	30	131	81	71*
FINLAND	6	17	14	23	1497	1527	1616	1524	9	26	22	35
FRANCE	318	301	298	273*	1837	1619	1919	1465	585	487	572	400*
GERMAN DR	102	132	130	128*	1760	2757	2410	2461	180	363	312	315*
GERMANY FED	85	90	95	104	2252	2211	2336	2674	190	199	222	279
HUNGARY	30	46	52	50*	1539	1404	1283	1300	46	65	67	65*
ITALY	3	1	1	1F	1867	2235	2299	2237	5	3	2	2F
NETHERLANDS	8	14	12	11	2770	2595	2775	2736	22	37	34	30
NORWAY	4	5	6	5	1242	1279	1419	1848	6	6	9	10
POLAND	269	309	398	425*	1691	2350	2465	1647	455	726	980	700
ROMANIA	3	6*	6F	6F	898	2453	2500	2333	3	16*	15F	14F
SWEDEN	105	170	145	123	2068	1933	1921	2259	218	329	279	277
SWITZERLAND	9	10	11	11	2167	1912	2154	1882	19	18	24	22
UK	6	39	48	55	1842	1564	2313	2582	10	61	111	142
YUGOSLAVIA	7	7	11	19	1809	1849	2162	1895	13	14	24	36
OCEANIA	44	16	8	20	698	751	1039	750	31	12	8	15
AUSTRALIA	44	16	8	20	698	751	1039	750	31	12	8	15
USSR	4	13	16	13F	977	717	981	1208	4	9	16	16F
DEV.PED M E	2171	2498	1419	2042	1243	1235	1576	1521	2698	3085	2236	3106
N AMERICA	1534	1749	720	1348	989	1000	1163	1318	1517	1749	838	1777
W EUROPE	572	729	688	670	1951	1806	2013	1954	1116	1317	1384	1309
OCEANIA	44	16	8	20	698	751	1039	750	31	12	8	15
OTH DEV.PED	21	4	4	4	1614	1659	1811	1351	34	7	7	5
DEV.PING M E	3958	4456	4147	3973	542	614	599	532	2146	2737	2485	2113
AFRICA	50	50	50	50	400	400	400	400	20	20	20	20
LAT AMERICA	59	52	66	60	1346	1307	1679	1475	80	68	111	89
NEAR EAST	4		4	4	1153	1218	1439	1439	5		6	6
FAR EAST	3845	4354	4026	3859	531	608	583	518	2042	2648	2348	1998
CENTR PLANND	2351	3253	3268	3298	746	819	865	738	1754	2665	2828	2434
ASIAN CPE	1902	2683	2603	2603	524	505	501	463	996	1354	1305	1205
E EUR+USSR	449	570	665	695	1690	2299	2293	1770	758	1310	1524	1230
DEV.PED ALL	2620	3068	2084	2737	1319	1433	1805	1584	3456	4396	3760	4336
DEV.PING ALL	5860	7139	6750	6576	536	573	561	504	3142	4091	3790	3318

TABLE
TABLEAU 33
CUADRO

	SESAME SEED		GRAINES DE SESAME			SEMILLA DE SESAMO					
	AREA HARV SUP RECOLTEE SUP COSECHAD	1000 HA	YIELD RENDEMENT RENDIMIENTO		KG/HA			PRODUCTION PRODUCTION PRODUCCION	1000 MT		
	1969-71	1975	1976	1977	1969-71	1975	1976	1977	1969-71	1975	1976	1977
WORLD	6256	6336	6336	6485	324	301	293	301	2026	1907	1855	1952
AFRICA	1406	1761	1702	1643	362	294	295	304	509	517	502	499
ANGOLA	6	6F	6F	7F	361	333	333	357	2	2F	2F	3F
BENIN									2	1F	1F	1F
CAMEROON	16	10F	10F	10F	365	600	600	600	6	6F	6F	6F
CENT AFR EMP	49	44	47F	50F	263	281	284	286	13	12	13F	14F
CHAD	31	28	28F	28F	368	280	280	280	11	8	8F	8F
CONGO					333	399	398	396				
EGYPT	16	14	13	15F	1152	1226	999	1067	19	17	13	16F
ETHIOPIA	153	160F	160F	160F	533	438	406	438	81	70*	65*	70F
GUINEA	2	2F	2F	2F	169	100	100	100				
IVORY COAST	4	4F	4F	4F	500	500	500	500	2	2F	2F	2F
KENYA	8	8F	9F	9F	400	346	402	398	3	3F	4F	4F
MOZAMBIQUE	6	7F	7F	7F	524	429	429	429	3	3F	3F	3F
NIGERIA	205	225F	230F	230F	290	298	304	304	59	67F	70F	70F
SIERRA LEONE	2	2F	2F	2F	401	415	425	425	1	1F	1F	1F
SOMALIA	20	70F	70F	70F	340	357	314	317	7	25F	22*	22F
SUDAN	720	962	924	850F	355	247	254	259	256	238	235	220F
TANZANIA	43	38	23	26F	260	263	258	250	11	10F	6	7F
TOGO	6	7F	7F	7F	200	214	214	211	1	2F	2F	2F
UGANDA	84	122	108	113	252	320	350	350	21	39	38	40
UPPER VOLTA	23	40F	40F	40F	200	175	175	150	5	7F	7F	6F
ZAIRE	12	11F	11F	11F	453	455	482	500	5	5F	5F	6F
N C AMERICA	294	244	220	250	662	547	493	617	195	133	112	154
COSTA RICA					533	533	533	533				
DOMINICAN RP	1	1F	1F	1F	485	456	466	466				
EL SALVADOR	1	4	4	4*	1044	754	677	750	1	3	3	3F
GUATEMALA	7	11F	15*	15*	1019	1192	1101	1057	7	13*	16*	16F
HAITI					533	556	556	556				
HONDURAS	1	1F	2F	2F	746	929	938	938	1	1F	2F	2F
MEXICO	273	219	198	219	653	506	429	580	178	111	85	127
NICARAGUA	10	7	7	8	640	603	785	893	6	4	6	5
PANAMA					498	494	483	494				
USA	1	1F	1F	1F	533	706	706	706		1F	1F	1F
SOUTH AMERIC	206	186	179	189	617	490	494	567	127	91	88	107
BRAZIL	5	5F	5F	5F	577	556	556	556	3	3F	3F	3F
COLOMBIA	31	42	36	36	680	498	562	579	21	21	20	21
ECUADOR	2	3	3	2*	846	854	1157	1750	2	3	3	4*
PERU					712	800	800	800				
VENEZUELA	167	136	135	146*	603	477	459	548	101	65	62	80*
ASIA	4330	4135	4219	4395	275	281	272	270	1189	1161	1149	1188
AFGHANISTAN	42	43F	50	50F	778	941	1240	1240	33	40	62	62F
BANGLADESH	47	54	49	55F	614	532	562	545	29	29	28	30F
BURMA	707	656	741	766F	164	145	181	181	116	95	134	139F
CHINA	894F	952F	942F	942F	412	400	399	383	368F	381F	376F	361F
CYPRUS		1			439	303	304	333				
INDIA	2378	2170	2165	2300F	205	221	187	196	487	479	404	450*
INDONESIA	18	16F	16F	16F	226	313	313	313	4	5F	5F	5F
IRAN	7	4*	4*	4F	841	1000	1000	1000	6	4*	4*	4F
IRAQ	17	12	12	13F	787	651	585	530	13	8	7	7F
ISRAEL	1				569	536	476	500	1			
JAPAN	1	1F	1F	1F	611	600	600	600				
JORDAN	1				280	162	385	379				
KAMPUCHEA DM	13	9F	9F	8F	711	611	588	563	10	6F	5F	5F
KOREA REP	24	39	41	42F	444	488	541	538	10	19	22	23F
LEBANON					773	846	846	846				
PAKISTAN	32	28	30	29F	334	380	394	414	11	11	12	12F
SAUDI ARABIA	18	2	2F	2F	943	263	250	250	17		1F	1F
SRI LANKA	12	21	26	26F	812	397	445	453	10	8	11	12F
SYRIA	9	31	43	45F	488	433	446	444	4	14	19	20F
THAILAND	27	23F	25F	28*	752	773	824	893	20	17	21	25*
TURKEY	64	54	43	46*	622	611	584	457	40	33	25	21*
VIET NAM	7	7F	7F	7F	430	438	435	432	3	3F	3F	3F
YEMEN AR	8	9	10	10F	502	556	566	561	4	5	6	6F
YEMEN DEM	4	4	4	4F	854	886	902	884	3	4	4	4F
EUROPE	20	9	8	8	346	409	412	413	7	4	3	3
BULGARIA	3	2	3	3F	86	66	11	11				
GREECE	14	6	5	5F	355	500	600	600	5	3	3	3F
ITALY	2	1			660	779	701	688	1			
YUGOSLAVIA	1				504	667	500	500				
USSR	1				80	200	200	200				
DEV.PED M E	19	9	7	7	417	553	613	613	8	5	4	4
N AMERICA	1	1	1	1	533	706	706	706		1	1	1
W EUROPE	17	7	6	6	395	532	605	604	7	4	3	3
OTH DEV.PED	2	1	1	1	589	577	563	571	1			
DEV.PING M E	5318	5356	5368	5517	308	282	273	286	1637	1513	1467	1578
AFRICA	669	785	764	778	350	334	332	338	234	262	254	263
LAT AMERICA	499	429	406	437	644	522	493	595	321	224	200	260
NEAR EAST	906	1136	1105	1040	436	319	339	346	395	363	375	360
FAR EAST	3244	3007	3093	3262	212	221	206	213	686	664	637	695
CENTR PLANND	919	971	961	960	415	402	400	384	381	390	384	369
ASIAN CPE	915	968	958	957	416	403	401	385	381	390	384	369
E EUR+USSR	4	3	3	3	85	71	17	18				
DEV.PED ALL	23	11	10	10	358	441	446	447	8	5	4	4
DEV.PING ALL	6233	6324	6326	6475	324	301	293	301	2018	1902	1851	1947

TABLE
TABLEAU 34
CUADRO

	AREA HARV / SUP RECOLTEE / SUP COSECHAD (LINSEED) 1000 HA				YIELD / RENDEMENT / RENDIMIENTO (GRAINES DE LIN) KG/HA				PRODUCTION / PRODUCCION (LINAZA) 1000 MT			
	1969-71	1975	1976	1977	1969-71	1975	1976	1977	1969-71	1975	1976	1977
WORLD	6769	5827	5645	6014	514	441	440	482	3481	2568	2484	2898
AFRICA	140	131	127	127	577	617	605	606	81	81	77	77
EGYPT	11	23	20	20F	1103	1190	1212	1215	13	27	24	24F
ETHIOPIA	120	100F	100F	100F	518	500	500	500	62	50*	50*	50F
KENYA					1000	1000	1000	1000				
MOROCCO	5	2	1		907	742	500	500	5	2		
TUNISIA	3	6F	6F	6F	361	344	344	344	1	2	2F	2F
N C AMERICA	1965	1194	730	1159	795	726	669	895	1563	867	489	1037
CANADA	1001	567	324	575	832	785	855	1061	833	445	277	610
MEXICO	22	16	8	12	1427	1688	1625	1500	31	27	13	18
USA	942	611	399	572	742	646	498	715	699	395	199	409
SOUTH AMERIC	832	537	793	878	759	806	877	801	632	433	696	703
ARGENTINA	692	446	674	793	788	845	915	807	545	377	617	640
BRAZIL	34	20F	20F	20F	673	800	800	800	23	16F	16F	16F
CHILE	1	1*	1*	1*	1016	833	833	800	1	1*	1*	1*
URUGUAY	106	70	98	64	592	563	629	725	63	39	62	46
ASIA	1985	2251	2299	2103	256	285	294	242	508	642	676	509
AFGHANISTAN	55	50F	52F	50F	315	300	308	300	17	15F	16F	15F
BANGLADESH	15	14	14	15	495	507	526	506	7	7	7	7
CHINA	77F	90F	91F	90F	414	444	440	444	32F	40F	40F	40F
CYPRUS					408							
INDIA	1799	2071	2119	1925	236	272	282	224	424	564	598	431
IRAN	5	3*	3*	3F	1059	1000	1000	1000	5	3*	3*	3F
IRAQ	14	2	1	1	741	565	786	390	11	1	1	
JAPAN					500	500	500	500				
PAKISTAN	6	8	8	8	506	553	563	550	3	4	4	4
TURKEY	14	13	11	12F	635	592	627	625	9	8	7	8F
EUROPE	311	288	314	317	636	586	582	595	198	169	183	189
BELGIUM-LUX	12	9	9	9*	663	680	870	778	8	6	8	7*
BULGARIA	9	6	7	7F	345	284	203	232	3	2	1	2F
CZECHOSLOVAK	31	31	30	33	488	357	337	305	15	11	10	10F
DENMARK					811	1098	1038	1000				
FRANCE	41	48	63	56*	589	875	556	727	24	42	35	41*
GERMAN DR	11	2	1	1F	446	438	436	436	5	1	1	1F
GREECE					103	50	50	50				
HUNGARY	18	19	19	20F	1101	939	900	920	20	18	17	19F
ITALY	1	3	5	5F	1060	691	799	800	1	2	4	4F
NETHERLANDS	7	5	5	6	1370	900	1105	1136	10	5	6	7F
POLAND	100	79	87	87F	660	455	566	575	66	36	49	50F
PORTUGAL					403	400	189	471				
ROMANIA	78	83	85	90F	587	539	592	533	46	45	50	48F
SPAIN	1	2	2	1F	591	775	723	1000		1	1	1F
YUGOSLAVIA	2	1	1F	1F	43	33	50	50				
OCEANIA	42	23	21	51	892	1155	1279	870	37	26	26	44
AUSTRALIA	37	16	16	46	702	775	1051	749	26	12	17	34
NEW ZEALAND	5	7	5	5F	2322	2020	2021	2000	11	14	10	10F
USSR	1493	1403	1361	1380F	309	249	248	246	462	350*	337	340*
DEV.PED M E	2050	1268	828	1276	787	727	672	880	1613	922	556	1123
N AMERICA	1944	1178	722	1147	788	713	658	888	1532	840	476	1019
W EUROPE	65	67	85	78	675	832	637	765	44	56	54	60
OCEANIA	42	23	21	51	892	1155	1279	870	37	26	26	44
OTH DEV.PED					500	500	500	500				
DEV.PING M E	2901	2846	3136	3030	420	402	453	418	1219	1144	1422	1266
AFRICA	128	109	107	107	530	498	492	492	68	54	53	52
LAT AMERICA	854	553	801	890	776	832	884	810	662	460	709	721
NEAR EAST	99	91	87	86	547	594	585	584	54	54	51	50
FAR EAST	1820	2093	2141	1947	239	275	285	227	434	575	610	442
CENTR PLANND	1817	1713	1681	1709	357	293	301	298	648	503	505	509
ASIAN CPE	77	90	91	90	414	444	440	444	32	40	40	40
E EUR+USSR	1740	1623	1590	1619	354	285	293	290	616	463	465	469
DEV.PED ALL	3790	2892	2418	2895	588	479	423	550	2230	1384	1022	1592
DEV.PING ALL	2979	2936	3227	3120	420	403	453	419	1251	1184	1462	1306

TABLE
TABLEAU 35
CUADRO

SAFFLOWER SEED · GRAINES DE CARTHAME · SEMILLA DE CARTAMO

	AREA HARV / SUP RECOLTEE / SUP COSECHAD — 1000 HA				YIELD / RENDEMENT / RENDIMIENTO — KG/HA				PRODUCTION / PRODUCTION / PRODUCCION — 1000 MT			
	1969-71	1975	1976	1977	1969-71	1975	1976	1977	1969-71	1975	1976	1977
WORLD	997	1288	1126	1356	708	781	665	754	706	1007	749	1023
AFRICA	63	64	64	64	586	391	391	391	37	25	25	25
ETHIOPIA	62	64F	64F	64F	583	391	391	391	36	25F	25F	25F
MOROCCO	1				818				1			
N C AMERICA	286	458	285	499	1766	1555	1510	1433	504	712	430	715
MEXICO	194	363	185	399	1564	1466	1299	1316	303	532	240	525
USA	92	95F	100F	100F	2191	1895	1900	1900	202	180*	190F	190F
SOUTH AMERIC	1	4	8	8	706	849	780	789		3	6	6
ARGENTINA	1	4	8	8F	706	849	780	789		3	6	6F
ASIA	584	655	679	674	226	332	356	328	132	217	242	221
INDIA	582	648	674	668	223	327	354	325	130	212	238	217
IRAN	1				711				1			
ISRAEL		5	3F	4F	824	770	625	625		4	2	3
TURKEY	1	2	2	2F	814	675	727	667	1	1	2	1F
EUROPE	25	62	66	56	664	464	521	404	16	29	35	23
PORTUGAL	9	28	31	22*	813	451	474	265	7	12	15	6*
SPAIN	16	34	36	34F	579	475	560	494	9	16	20	17*
OCEANIA	24	40	15	44	400	457	476	637	10	18	7	28
AUSTRALIA	24	40	15	44	400	457	476	637	10	18	7	28
USSR	14	6	10	11F	421	333	460	467	6	2	5	5F
DEV.PED M E	141	201	184	204	1614	1145	1268	1192	228	231	233	243
N AMERICA	92	95	100	100	2191	1895	1900	1900	202	180	190	190
W EUROPE	25	62	66	56	664	464	521	404	16	29	35	23
OCEANIA	24	40	15	44	400	457	476	637	10	18	7	28
OTH DEV.PED		5	3	4		824	770	625	625		4	2	3
DEV.PING M E	842	1081	932	1141	561	716	548	679	472	774	511	774
AFRICA	63	64	64	64	586	391	391	391	37	25	25	25
LAT AMERICA	194	367	193	407	1561	1460	1279	1306	303	536	246	531
NEAR EAST	2	2	2	2	776	675	727	667	1	1	2	1
FAR EAST	582	648	674	668	223	327	354	325	130	212	238	217
CENTR PLANND	14	6	10	11	421	333	460	467	6	2	5	5
E EUR+USSR	14	6	10	11	421	333	460	467	6	2	5	5
DEV.PED ALL	155	207	194	215	1504	1121	1227	1156	234	233	238	248
DEV.PING ALL	842	1081	932	1141	561	716	548	679	472	774	511	774

TABLE
TABLEAU 36
CUADRO

SEED COTTON COTON A GRAINES ALGODON SIN DESMOTAR

	AREA HARV / SUP RECOLTEE / SUP COSECHAD (1000 HA)				YIELD / RENDEMENT / RENDIMIENTO (KG/HA)				PRODUCTION / PRODUCTION / PRODUCCION (1000 MT)			
	1969-71	1975	1976	1977	1969-71	1975	1976	1977	1969-71	1975	1976	1977
WORLD	32667	30827	31182	33159	1067	1163	1153	1259	34858	35846	35961	41757
AFRICA	4813	4187	4183	4474	789	776	741	789	3795	3248	3100	3528
ALGERIA	3	1			644	873	1332	1350	2	1	1	1F
ANGOLA	80	61*	61*	61*	1032	639	639	639	83	39F	39F	39F
BENIN	42	31	54	56F	860	753	840	872	36	23	46	49F
BOTSWANA	1	1F	1F	1F	2745	2727	2727	2727	3	3F	3F	3F
BURUNDI	9	9F	9F	9F	925	424	444	467	8	4	4F	4F
CAMEROON	103	73	60*	78*	559	676	797	577	58	49*	48*	45F
CENT AFR EMP	131	135*	135*	140*	404	248	304	326	53	33*	41*	46F
CHAD	303	336*	319*	340F	353	517	462	471	107	174*	147*	160*
EGYPT	668	565	524*	611	2126	1848	2049	1874	1421	1045	1074	1145
ETHIOPIA	97	110F	120F	120F	412	503	600	600	40	55	72*	72*
GHANA	1	13	5*	5*	540	672	800	800	1	9	4*	4*
IVORY COAST	40	59	65	65*	908	1037	1009	1108	37	61	66	72*
KENYA	74	69*	69*	81*	215	234	229	262	16	16	16	21*
MADAGASCAR	10	17	17*	18*	1919	2025	2035	2056	20	34	35	37*
MALAWI	46	38	41*	52*	447	466	434	466	21	18	18	24*
MALI	73	90	113	118F	894	1182	1182	1270	66	106	133	150*
MOROCCO	17	12	18	18*	1302	1312	1220	1222	22	15	22	22*
MOZAMBIQUE	360	243*	243*	243F	353	350	350	370	127	85F	85F	90F
NIGER	20	18	13	16F	468	622	563	631	9	11	7	10F
NIGERIA	405	500*	526*	567*	460	314	352	388	186	157F	185F	220F
RHODESIA	60	91*	89*	89*	2150	1582	1315	1112	129	144*	117*	99*
RWANDA					918	401	390	385				
SENEGAL	14	41	44	41F	1054	791	1032	1098	15	32	45	45F
SOMALIA	13	12*	12*	12*	310	330	330	275	4	4*	4*	3*
SOUTH AFRICA	81	89*	68*	90*	705	1192	831	1284	57	106	57	116
SUDAN	502	491	402	417	1369	1337	886	1436	687	656	357	599
SWAZILAND	12	20	11F	11F	616	865	1636	1636	8	17	18F	18F
TANZANIA	425	233*	374*	374F	497	583	552	472	211	136	206	177*
TOGO	36	18	8	8F	183	594	1213	875	7	11*	10*	7*
TUNISIA					1367							
UGANDA	923	594	563	591	279	193	240	229	258	115	135	135F
UPPER VOLTA	83	62	68	80F	402	497	745	725	33	31	51	58*
ZAIRE	169	151	144	155F	378	358	355	355	64	54	51	55F
ZAMBIA	9	4	6	6F	904	741	753	500	8	3	5	3*
N C AMERICA	5275	4186	4978	6192	1431	1471	1491	1632	7547	6159	7419	10107
ANTIGUA		1F	1F	1F	476	500	508	522				
COSTA RICA	1	3	7		1637	1535	1437	1840	2	1*	4*	13*
CUBA	4	3F	3F	3F	857	968	968	645	3	3F	3F	2F
DOMINICAN RP	3	3*	4*	8*	1092	1000	1000	1000	3	3*	4*	8*
EL SALVADOR	56	88	74	79	2394	2384	2247	2449	135	210	166	194*
GRENADA					350	450	450	450				
GUADELOUPE	1				665	600	615	615	1			
GUATEMALA	77	111*	81*	122*	2691	3003	3346	3067	207	333	271*	375*
HAITI	6	6F	6F	6F	324	500	500	500	2	3F	3F	3F
HONDURAS	6	8	10*	15F	2123	1765	1920	1907	12	15*	20*	29*
MEXICO	464	227	235	386*	2108	2397	2480	2461	979	544	583	950F
MONTSERRAT					1946	1750	1750	1750				
NICARAGUA	112	179	144	199	2097	1788	2064	1809	236	320	297	359
ST KITTS ETC					1947	1176	1176	1176				
ST VINCENT					838							
USA	4543	3560	4416	5366	1313	1328	1374	1523	5967	4727	6068	8174
SOUTH AMERIC	3560	3372	2944	3547	815	937	861	972	2901	3159	2534	3446
ARGENTINA	408	505	414	536	906	1071	1076	978	370	541	445	524
BOLIVIA	11	54	30*	40*	1802	1251	1282	1206	19	67*	38*	48*
BRAZIL	2644	2226	1902	2145	691	786	656	875	1828	1751	1247	1876
COLOMBIA	241	281	286	415*	1462	1428	1431	1088	352	401	409	452*
ECUADOR	13	30	27	26	1081	1000	909	1031	14	30	25	26
PARAGUAY	46	100	110	219*	660	996	978	1038	31	100	107	227*
PERU	150	97*	115*	101*	1638	1856	1739	1996	245	180*	200*	202*
URUGUAY	1	1	1*	1F	719	1420	1172	1170	1	2*	1	1F
VENEZUELA	46	78	61*	65*	895	1117	1033	1385	41	88	63*	90*
ASIA	15967	15853	15830	15596	840	931	889	973	13412	14759	14075	15179
AFGHANISTAN	54	112	130*	130*	1380	1429	1228	1274	75	160	159	165F
BANGLADESH	11	6	5	5F	519	395	396	396	6	2	2	2F
BURMA	162	171F	171F	171F	242	204	222	222	39	35	38F	38F
CHINA	4791F	4856F	4897F	4856F	1245	1473	1448	1447	5965F	7155F	7089F	7026F
CYPRUS					983		1538					
INDIA	7701	7460	7461	7001	424	466	414	502	3264	3480	3090	3513
INDONESIA	10	19	16	16F	698	839	875	938	7	16	14	15F
IRAN	336	290*	299*	300*	1312	1348	1525	1650	440	391*	456*	495*
IRAQ	33	26	25	26F	1304	1470	1338	1451	43	39	34	37F
ISRAEL	34	41	43	54	2819	3199	3253	3074	95	131	141	166
JORDAN					2233	6648	6000			2		2F
KAMPUCHEA DM	4		2*	5*	669	1000	1250	1200	3		3F	6F
KOREA DPR	15	15F	15F	15F	600	600	600	600	9	9F	9F	9F
KOREA REP	16	10	8	9F	835	921	881	933	13	9	7	8F
LAO	6	5F	5F	6F	1479	1260	1283	1304	9	6F	7F	7F
PAKISTAN	1817	1851	1865	1902	983	833	673	855	1786	1542	1255	1626
PHILIPPINES		1F	1F	2*	1200	516	597	555			1*	1*
SRI LANKA	1	2	2	2F	1255	3603	4012	4000	2	6	8	8F
SYRIA	266	208	182	182*	1468	1992	2249	2143	391	414	409	390*
THAILAND	62	61	71	62*	1494	1076	1142	1057	93	65	81	66*
TURKEY	618	670	581	800	1853	1863	2105	1938	1146	1248	1224	1550*
VIET NAM	9	9F	9F	10	680	667	667	625	6	6F	6F	6F
YEMEN AR	6	28	30	30F	850	962	915	903	5	27	27	27F

TABLE
TABLEAU 36
CUADRO

	SEED COTTON				COTON A GRAINES				ALGODON SIN DESMOTAR			
	AREA HARV SUP RECOLTEE SUP COSECHAD		1000 HA		YIELD RENDEMENT RENDIMIENTO		KG/HA		PRODUCTION PRODUCCION PRODUCCION		1000 MT	
	1969-71	1975	1976	1977	1969-71	1975	1976	1977	1969-71	1975	1976	1977
YEMEN DEM	14	12*	12*	12*	1055	1350	1183	1250	15	16F	14F	15F
EUROPE	334	267	269	324	1688	2143	1806	2042	564	571	486	662
ALBANIA	22	23F	23F	23F	668	867	867	867	15	20F	20F	20F
BULGARIA	42	26	25	26F	922	1194	130	1231	39	32	3	32F
GREECE	146	138	149*	183*	2338	2652	2168	2306	342	366	323*	422*
ITALY	6	5	3	3F	635	382	650	680	4	2	2	. 2F
ROMANIA		6F	7F	6F		1033	1071	1033		6F	8F	6F
SPAIN	106	62	56	78	1456	2245	2226	2238	155	139	125	175
YUGOSLAVIA	12	6	5	5	861	1024	925	943	10	6	5	5
OCEANIA	33	39	29	34	2159	2243	2247	2187	72	86	66	74
AUSTRALIA	33	39	29	34	2159	2243	2247	2187	72	86	66	74
USSR	2685	2924	2949	2992	2445	2689	2808	2928	6566	7864	8281	8760
DEV.PED M E	4962	3939	4770	5812	1351	1413	1422	1571	6703	5564	6786	9133
N AMERICA	4543	3560	4416	5366	1313	1328	1374	1523	5967	4727	6068	8174
W EUROPE	270	211	213	269	1890	2435	2133	2244	511	513	455	604
OCEANIA	33	39	29	34	2159	2243	2247	2187	72	86	66	74
OTH DEV.PED	115	130	111	144	1328	1826	1775	1956	152	237	198	282
DEV.PING M E	20137	19028	18484	19414	772	798	744	863	15552	15189	13756	16758
AFRICA	3562	3042	3189	3356	458	474	506	497	1630	1441	1613	1669
LAT AMERICA	4291	3999	3506	4374	1044	1148	1109	1230	4480	4591	3886	5380
NEAR EAST	2499	2403	2185	2507	1691	1663	1718	1765	4224	3997	3755	4425
FAR EAST	9786	9585	9605	9177	533	539	469	576	5218	5162	4503	5285
CENTR PLANND	7568	7860	7928	7933	1665	1920	1945	2000	12603	15092	15419	15866
ASIAN CPE	4819	4880	4923	4886	1242	1469	1444	1442	5983	7170	7107	7047
E EUR+USSR	2749	2980	3005	3047	2408	2659	2766	2894	6620	7922	8312	8818
DEV.PED ALL	7711	6919	7775	8860	1728	1949	1942	2026	13323	13486	15098	17951
DEV.PING ALL	24956	23909	23407	24299	863	935	891	980	21535	22360	20863	23806

TABLE
TABLEAU 37
CUADRO

	COTTONSEED GRAINES DE COTON SEMILLA DE ALGODON — PRODUCTION			1000 MT	OLIVES OLIVES ACEITUNAS — PRODUCTION			1000 MT	OLIVE OIL, TOTAL HUILE D OLIVE, TOTAL ACEITE DE OLIVA, TOTAL — PRODUCCION			1000 MT
	1969-71	1975	1976	1977	1969-71	1975	1976	1977	1969-71	1975	1976	1977
WORLD	22503	23131	23134	26688	7652	9861	8123	8149	1554	2041	1516	1598
AFRICA	2438	2075	1974	2191	1040	1456	981	1088	177	295	180	199
ALGERIA	1	1			150	221	123F	113F	22	37	17	15
ANGOLA	55	26*	26*	27F								
BENIN	24	15F	30F	32F								
BOTSWANA	2	2F	2F	2F								
BURUNDI	5	2	3F	3F								
CAMEROON	33	29*	28*	27F								
CENT AFR EMP	36	24*	31*	34*								
CHAD	65	105*	87*	95F								
EGYPT	901	663	678	710	7	8	7	8F				
ETHIOPIA	27	37	48*	48*								
GAMBIA				1								
GHANA		5	3F	3F								
IVORY COAST	21	36	35*	37F								
KENYA	11	11	10	14*								
LIBYA					57	95F	95F	100F	12	19*	20F	20F
MADAGASCAR	12	20	20	22								
MALAWI	14	12	12	16*								
MALI	39	62*	75*	78F								
MOROCCO	15	10	14	14F	329	312	306*	252*	41	44*	41*	32*
MOZAMBIQUE	84	57*	57*	60F								
NIGER	6	7*	4*	6F								
NIGERIA	125	104F	122F	146F								
RHODESIA	86	96*	78*	66*								
SENEGAL	9	20*	28*	28*								
SOMALIA	3	3*	3*	2*								
SOUTH AFRICA	38	71	38	77								
SUDAN	442	418	233	350*								
SWAZILAND	5	12F	12F	12F								
TANZANIA	135	87	132	113*								
TOGO	3	5	3	3F								
TUNISIA					498	820	450F	615F	103	196	102	132
UGANDA	173	77	91	91F								
UPPER VOLTA	21	20F	33F	35F								
ZAIRE	42	36*	34*	36F								
ZAMBIA	5	2*	3*	2*								
N C AMERICA	4729	3795	4600	6211	65	76	83	50	2	2	2	2
COSTA RICA	1	3*	8									
CUBA	2	2F	2F	2F								
DOMINICAN RP	2	2*	3*	5*								
EL SALVADOR	80	125	98	119	2	2F	2F	2F				
GUATEMALA	138	193	163	228								
HAITI	1	2F	2F	2F								
HONDURAS	7	9*	12*	18*								
MEXICO	623	345	369	590*	10	14	9*	9F	1	1F	1F	1F
NICARAGUA	130	197	184	221								
USA	3742	2919	3764	5018	54	61	73	39	1	1*	1*	1F
SOUTH AMERIC	1836	1986	1556	2116	94	111	101	110	18	18	13	18
ARGENTINA	220	314	259	308	70	89	79	87	17	17	12	17F
BOLIVIA	13	41	24*	30*								
BRAZIL	1197	1138	795*	1180*	1	1F	1F	1F				
CHILE					12	7	7	8*	1	1F	1F	1F
COLOMBIA	205	235	235	260*								
ECUADOR	9	19	14	15F								
PARAGUAY	20	65*	65*	143*								
PERU	148	120*	124*	125*	9	12F	11F	12F				
URUGUAY		1*			2	2F	3F	3F				
VENEZUELA	25	53	39*	55*								
ASIA	8848	9757	9267	9988	668	829	1469	870	129	162	251	120
AFGHANISTAN	49	107*	81*	87F	5	5F	5F	5F				
BANGLADESH	4	2	1	1F								
BURMA	25	22	23F	23F								
CHINA	3977F	4770F	4726F	4684F	2F	1F	1F	1F	1F	1F	1F	1F
CYPRUS					14	20F	17F	20F	2	3F	3F	3F
INDIA	2176	2320	2060	2342								
INDONESIA	5	10	10	10F								
IRAN	285	252*	296*	315*	10	13*	17F	17F	1	2*	1F	1F
IRAQ	29	26*	22*	24F	9	11*	11F	11F				
ISRAEL	58	83	88	103	15	22	20	9	1	3F	1	
JORDAN		1F	1F	16	5	23	23F	3	1*	6*	6F
KAMPUCHEA DM	2	2F	4F									
KOREA DPR	6	6F	6F	6F								
KOREA REP	9	6	4	6F								
LAO	6	4F	5F	5F								
LEBANON					48	35F	46F	46F	11	8*	11	11F
PAKISTAN	1190	1028	837	1084								
PHILIPPINES				1*								
SRI LANKA	1	4	5	6F								
SYRIA	241	273	254*	230*	110	157	233	238F	23	35	27	17F
THAILAND	62	44	54	44*								
TURKEY	705	768	760	980*	438	561	1097	500*	86	110	201	81*
VIET NAM	4	4F	4F	4F								
YEMEN AR	3	17	17*	18F								
YEMEN DEM	10	11F	11F	11*								
EUROPE	365	365	315	443	5782	7387	5486	6030	1228	1563	1070	1258
ALBANIA	10	13F	13F	13F	33	47F	47F	47F	4	6F	6F	6F

TABLE
TABLEAU 37
CUADRO

	COTTONSEED GRAINES DE COTON SEMILLA DE ALGODON PRODUCTION			1000 MT	OLIVES OLIVES ACEITUNAS PRODUCTION			1000 MT	OLIVE OIL, TOTAL HUILE D OLIVE, TOTAL ACEITE DE OLIVA, TOTAL PRODUCCICN			1000 MT
	1969-71	1975	1976	1977	1969-71	1975	1976	1977	1969-71	1975	1976	1977
BULGARIA	26	21	2	20F								
FRANCE					15	13	13F	13F	2	2*	2*	2F
GREECE	227	234*	210*	267*	876	1365	1350F	1370F	178	288	251	254*
ITALY	2	1	1	1F	2530	3228	1668	2550	553	696	325	595
PORTUGAL					422	350	257	265	70	58	43	40
ROMANIA		4F	5F	4F								
SPAIN	94	88	81	135	1897	2358	2139	1772	419	509	442	359
YUGOSLAVIA	7	4	3	3	10	26	12	13F	2	4	2	2F
OCEANIA	45	54	41	46	2	2	2	2				
AUSTRALIA	45	54	41	46	2	2	2F	2F				
USSR	4241	5100*	5383*	5694*								
DEV.PED M E	4213	3453	4225	5650	5820	7424	5533	6033	1227	1561	1066	1253
N AMERICA	3742	2919	3764	5018	54	61	73	39	1	1	1	1
W EUROPE	330	327	295	406	5749	7340	5439	5983	1224	1557	1064	1252
OCEANIA	45	54	41	46	2	2	2	2				
OTH DEV.PED	96	153	125	180	15	22	20	9	1	3	1	
DEV.PING M E	10024	9760	8769	10609	1797	2390	2541	2068	323	473	443	338
AFRICA	1057	923	1025	1054	976	1353	879	980	165	276	160	179
LAT AMERICA	2823	2862	2392	3309	106	127	112	120	19	19	14	19
NEAR EAST	2666	2535	2353	2725	715	910	1551	968	138	177	269	140
FAR EAST	3478	3440	2999	3521								
CENTR PLANND	8266	9918	10140	10429	35	48	48	49	5	7	7	7
ASIAN CPE	3989	4780	4738	4698	2	1	1	1	1	1	1	1
E EUR+USSR	4277	5138	5403	5731	33	47	47	47	4	6	6	6
DEV.PED ALL	8490	8591	9627	11381	5853	7471	5580	6080	1231	1567	1072	1259
DEV.PING ALL	14013	14540	13507	15307	1798	2391	2542	2069	323	474	444	339

TABLE
TABLEAU 38
CUADRO

	COCONUTS NOIX DE COCO COCOS PRODUCTION		1000 MT	COPRA COPRAH COPRA PRODUCTION		1000 MT	TUNG OIL HUILE D.ABRASIN ACEITE DE TUNG PRODUCCION		1000 MT
	1969-71	1975	1976	1977	1969-71	1975	1976	1977	1969-71	1975	1976	1977
WORLD	27986	30896	33925	32422	3915	4368	5053	4765	109	105	112	95
AFRICA	1483	1549	1542	1569	157	158	179	181	2	1	1	1
BENIN	20	20*	20*	20F	3	3F	3F	3F				
CAMEROON	2	2F	2F	2F								
CAPE VERDE	10	9F	10F	10F								
COMOROS	69	55F	57F	58F	6	3F	4F	4F				
EQ GUINEA	6	7F	7F	7F								
GHANA	257	311	300F	304F	14	17F	17F	17F				
GUIN BISSAU	31	25F	25F	23F	6	5F	5F	5F				
IVORY COAST	49	95*	100F	105F	7	12F	14F	14F				
KENYA	77	81F	82F	83F	6	5F	5F	5F				
LIBERIA	6	7F	7F	7F								
MADAGASCAR	19	22	24	24F	2	2F	3F	3F	1			
MALAWI									1	1	1F	1F
MAURITIUS	8	5F	5F	5F								
MOZAMBIQUE	400	400F	410F	420F	59	63*	83*	85F				
NIGERIA	86	90F	90F	90F	8	9F	9F	9F				
SAO TOME ETC	42	43F	43F	43F	5	5F	5F	5F				
SENEGAL	4	4F	4F	4F								
SEYCHELLES	42	35F	11F	11F	5	2F	2F	2F				
SIERRA LEONE	20	20F	20F	20F								
SOMALIA	1	1F	1F	1F								
TANZANIA	310	300F	307F	314F	31	27F	27F	27F				
TOGO	22	19F	19F	19F	4	3F	3F	3F				
N C AMERICA	1537	1452	1445	1470	202	194	185	185	1			
BARBADOS	1	2F	2F	2F								
BELIZE	2	2F	2F	2F								
COSTA RICA	17	25F	25F	25F	2	2F	2F	2F				
CUBA	11	14	8	9F								
DOMINICA	13	18F	18F	18F	2	2*	3F	3F				
DOMINICAN RP	57	64F	65F	67F	6	9F	9F	9F				
EL SALVADOR	30	42F	44F	44F	3	5*	5F	5*				
GRENADA	9	6F	6F	7F	1	1F	1F	1F				
GUADELOUPE	2	2	2	2F								
GUATEMALA	2	2F	2F	2F								
HAITI	24	29F	29F	29F								
HONDURAS	16	22F	22F	22F	2	3F	3F	3F				
JAMAICA	150	89	86	87F	17	6	6F	6F				
MARTINIQUE	2	2F	3F	2								
MEXICO	998	960*	960F	980F	146	145*	135*	135*				
NICARAGUA	2	1F	1F	1F								
PANAMA	28	25	25	25F	1	1F	1F	1F				
PUERTO RICO	9	9	9	9F	2	2F	2F	2F				
ST KITTS ETC	1	1F	1F	1F								
ST LUCIA	35	38F	39F	39F	6	6	7F	7F				
ST VINCENT	24	23F	20F	20F	3	2F	3F	3F				
TRINIDAD ETC	105	76F	78F	79F	13	9	9	9F				
USA									1			
SOUTH AMERIC	612	512	514	532	29	29	30	31	35	34	34	30
ARGENTINA									20	20	17	12*
BRAZIL	331	241	232	237	2	2F	2F	2F	1	1F	1F	1F
COLOMBIA	35	41	43	50								
ECUADOR	25	28	30	30F	4	4F	5F	5F				
GUYANA	56	31	32F	33F	6	5	5F	5F				
PARAGUAY									14	13	16	17F
PERU	13	12F	13F	13F								
SURINAM	6	4	5F	5F	1	1F	1F	1F				
VENEZUELA	147	154	160F	165F	16	18	18*	19*				
ASIA	22217	25194	28240	26621	3229	3683	4361	4062	70	69	76	63
BANGLADESH	66	62	62F	63F	2	2F	2F	2F				
BRUNEI	1	1F	1F	1F								
BURMA	67	71	77	81F								
CHINA	53F	54F	55F	56F					70F	69F	76F	63F
EAST TIMOR	16	16F			2	2F						
INDIA	4472	4331	4337	4347F	354	314*	320*	320*				
INDONESIA	6088	6942*	7150*	6882*	833	885*	949*	950*				
KAMPUCHEA DM	44	40F	42F	42F	8	7F	8F	8F				
MAL PENINSUL	825	800*	790F	800F	148	140	125F	110F				
MAL SABAH	126	90F	90F	90F	23	22F	22F	22F				
MAL SARAWAK	88	95F	100F	104F	15	21	19	19F				
MALDIVES	10	9F	9F	9F	2	1F	1F	2F				
PHILIPPINES	7601	9903F	12950F	11587F	1582	2020*	2697*	2400*				
SINGAPORE	10	7	7F	7F								
SRI LANKA	1938	1965	1771	1771F	209	203	151	160F				
THAILAND	713	677	670	650F	31	41F	43F	45F				
VIET NAM	100	130F	130F	132F	20	23F	23F	23F				
OCEANIA	2137	2189	2185	2230	297	304	299	306				
AMER SAMOA	9	10F	10F	10F	1	1F	1F	1F				
COCOS IS	2	3F	3F	3F								
COOK ISLANDS	12	11	11F	11F	1	1F	1F	1F				
FIJI	260	260F	265F	270F	30	24	27	29*				
FR POLYNESIA	178	164F	165F	165F	24	22	23F	23F				
GILBERT IS	62	74F	82F	83F	8	10F	11F	11F				
GUAM	22	26F	27F	28F	1	1F	1F	1F				
NAURU	2	2F	2F	2F								

TABLE
TABLEAU 38
CUADRO

	COCONUTS NOIX DE COCO COCOS PRODUCTION			1000 MT	COPRA COPRAH COPRA PRODUCTION			1000 MT	TUNG OIL HUILE D ABRASIN ACEITE DE TUNG PRODUCCION			1000 MT
	1969-71	1975	1976	1977	1969-71	1975	1976	1977	1969-71	1975	1976	1977
NEWCALEDONIA	21	18F	19F	19F	1	1F	1F	2F				
NEW HEBRIDES	247	260F	264F	268F	35	37*	37F	37F				
NIUE ISLAND	3	6F	6F	6F		1F	1F	1F				
PACIFIC IS	101	72F	72F	73F	12	8	8F	8F				
PAPUA N GUIN	741	755*	744*	757F	131	135	132*	132F				
SAMOA	193	208	195	215*	15	21	14	19*				
SOLOMON IS	184	183F	183F	183F	25	25F	24	24F				
TOKELAU	2	2F	2F	2F								
TONGA	88	125*	125F	125F	10	16F	17F	17F				
TUVALU	3	2*	2F	2F								
WALLIS ETC	9	10F	10F	10F	1	1F	1F	1F				
USSR									1	1	1	1F
DEV.PED M E									1			
N AMERICA									1			
DEV.PING M E	27789	30672	33698	32192	3887	4337	5022	4734	37	35	35	31
AFRICA	1483	1549	1542	1569	157	158	179	181	2	1	1	1
LAT AMERICA	2149	1964	1959	2002	232	223	215	216	35	34	34	30
FAR EAST	22021	24970	28013	26391	3201	3652	4330	4030				
OTH DV.PING	2137	2189	2185	2230	297	304	299	306				
CENTR PLANND	197	224	227	230	28	30	31	31	71	70	77	64
ASIAN CPE	197	224	227	230	28	30	31	31	70	69	76	63
E EUR+USSR									1	1	1	1
DEV.PED ALL									2	1	1	1
DEV.PING ALL	27986	30896	33925	32422	3915	4368	5053	4765	107	104	111	94

TABLE
TABLEAU 39
CUADRO

	PALM KERNELS / PALMISTES / ALMENDRAS DE PALMA PRODUCTION MT				PALM OIL / HUILE DE PALME / ACEITE DE PALMA PRODUCTION MT				HEMPSEED / CHENEVIS / CANAMON PRODUCCION MT			
	1969-71	1975	1976	1977	1969-71	1975	1976	1977	1969-71	1975	1976	1977
WORLD	1180258	1360749	1491043	1514805	1937301	3263199	3426015	3750937	36534	29210	27692	27554
AFRICA	730445	681040	702477	722690	1068581	1278000	1292600	1321200				
ANGOLA	15867	12000F	12000F	13000F	40333	40000*	40000F	42000F				
BENIN	68652	73000F	70000F	70000F	28349	22000F	35000F	35000F				
BURUNDI	1350	1800F	1800F	1800F	1000	1000F	1000F	1000F				
CAMEROON	40737	45000F	45000F	47000F	63400	79900	80000F	82000F				
CENT AFR EMP	1333	1500F	1500F	1500F	500	500F	500F	500F				
CONGO	2356	815*	600*	600F	7867	5400F	5300F	5400F				
EQ GUINEA	2000	2300*	2400F	2400F	4000	4400F	4500F	4600F				
GABON	256	285	250F	210*	2400	2500F	2500F	2600F				
GAMBIA	2104	990F	1727F	180F	2000	2000F	2000F	1900F				
GHANA	36000	34000*	32000*	30000F	19000	24000*	18000F	15000F				
GUINEA	14900	10000F	15000F	15000F	43533	40000F	35000F	37800F				
GUIN BISSAU	7808	7000F	7300F	7000F	4400	4600F	4800F	4800F				
IVORY COAST	19333	36300	37500	38000F	46467	153400	158000	185000*				
LIBERIA	14443	14000F	13000F	13000F	14233	23000*	25000*	27000F				
NIGERIA	292333	295000*	321000*	340000*	528333	640000F	655000F	660000F				
SAO TOME ETC	2283	2300F	2400F	2400F	856	1000F	1000F	1000F				
SENEGAL	9077	5000F	5000F	5000F	4633	5400F	5600F	5700F				
SIERRA LEONE	53667	52000*	51000*	50000F	48667	55000*	56000*	56000F				
TANZANIA	2440	7000F	7000F	8000F	1559	2400F	2400F	2700F				
TOGO	17450	6250F	12000F	12600F	5800	6500F	6000F	6200F				
ZAIRE	126057	74500	64000*	65000*	201250	165000	155000*	145000*				
N C AMERICA	21465	24920	39780	40315	31370	41100	41750	43600				
COSTA RICA	3433	7000F	7000F	7200F	13167	22000F	23000F	23500F				
HONDURAS	2183	2920*	2780*	3115*	5537	9100*	8750*	10100*				
MEXICO	15848	15000	30000*	30000F	12667	10000F	10000F	10000F				
SOUTH AMERIC	252322	273900	304150	266000	46752	64865	68307	86037	2733	1700	1600	1100
BRAZIL	218599	240000*	270000	230000	7166	7000*	7000F	7000F				
CHILE									2733	1700*	1600*	1100*
COLOMBIA	10833	17000F	14350F	17400F	26933	39200	39700	48800				
ECUADOR	5667	4000*	6200*	4000*	5815	14186	15288	22037*				
PARAGUAY	16223	11700*	12400*	13400F	6838	3854	4589	4800F				
PERU							400*	3400*				
SURINAM	1000	1200F	1200F	1200F		625*	1330*					
ASIA	176016	376089	438136	479800	790583	1824834	1990608	2268100	5455	5700	5000	5800
CHINA	26667F	38000F	40000F	42000F	108333F	150000F	160000F	162000F				
EAST TIMOR	1007	1110F			5035	5551F						
INDONESIA	48980	81826*	95000*	95000F	217900	397000	439000	450000F				
MAL PENINSUL	93515	232538	275000*	315000F	426404	1135000	1250000*	1485400*				
MAL SABAH	5481	20000*	25000*	25600F	30894	123000	124000	145000*				
MAL SARAWAK		615	936			3283	5508F	13000*				
PAKISTAN									2821	2900F	3000F	3000F
PHILIPPINES	367	2000F	2200*	2200F	2017	11000F	12100*	12300F				
TURKEY									2633	2800	2000	2800F
EUROPE									7013	5810	7092	6654
CZECHOSLOVAK									210	127	91	91F
FRANCE									393	200F	200F	200F
GERMAN DR									239			
HUNGARY									653	822	1658	1700F
ITALY									54	15	13	13F
POLAND									2000	1000	2000	1500F
PORTUGAL									20	20F	20F	20F
ROMANIA									1600	2000F	2000F	2000F
SPAIN									78	326	210	330F
YUGOSLAVIA									1767	1300	900	800F
OCEANIA	10	4800	6500	6000	16	54400	32750	32000				
PAPUA N GUIN	10	4500F	6000F	6000F	16	53950*	32000*	32000F				
SOLOMON IS		300F	500F			450F	750F					
USSR									21333	16000	14000	14000F
DEV.PED M E									2311	1861	1343	1363
W EUROPE									2311	1861	1343	1363
DEV.PING M E	1153591	1322749	1451043	1472805	1828968	3113199	3266015	3588937	8188	7400	6600	6900
AFRICA	730445	681040	702477	722690	1068581	1278000	1292600	1321200				
LAT AMERICA	273787	298820	343930	306315	78122	105965	110057	129637	2733	1700	1600	1100
NEAR EAST									2633	2800	2000	2800
FAR EAST	149349	338089	398136	437800	682250	1674834	1830608	2106100	2821	2900	3000	3000
OTH DV.PING	10	4800	6500	6000	16	54400	32750	32000				
CENTR PLANND	26667	38000	40000	42000	108333	150000	160000	162000	26035	19949	19749	19291
ASIAN CPE.	26667	38000	40000	42000	108333	150000	160000	162000				
E EUR+USSR									26035	19949	19749	19291
DEV.PED ALL									28346	21810	21092	20654
DEV.PING ALL	1180258	1360749	1491043	1514805	1937301	3263199	3426015	3750937	8188	7400	6600	6900

TABLE
TABLEAU 40
CUADRO

	VEGETABLES+MELONS,TOTAL LEGUMES ET MELONS,TOTAL HORTALIZAS+MELONES,TOTAL PRODUCTION 1000 MT				FRUIT EXCL MELONS,TOTAL FRUITS,EXCEPTE MELONS,T. FRUTAS,EXCEPTO MELONES,T PRODUCTION 1000 MT				TREENUTS,TOTAL FRUITS A COQUE,TOTAL NUECES (TODA CLASE),TCT. PRODUCCION 1000 MT			
	1969-71	1975	1976	1977	1969-71	1975	1976	1977	1969-71	1975	1976	1977
WORLD	271816	307021	312293	318906	228259	256670	259695	257068	3207	3614	3606	3595
AFRICA	16225	19585	20055	20814	27720	29925	30383	31608	467	505	491	523
ALGERIA	504	748	546	632F	1914	1494	1172	1361F	3	5	8	7F
ANGOLA	194	186F	191F	197F	414	372F	435F	462F	1	1F	1F	2F
BENIN	66	66F	76F	79F	45	49F	79F	80F				
BOTSWANA	12	14F	15F	15F	6	8F	8F	8F				
BURUNDI	117	124F	125F	127F	861	957	976F	995F				
CAMEROON	243	297F	303F	310F	1157	1213F	1220F	1248F	17	24F	24F	24F
CAPE VERDE	4	4F	5F	5F	13	12F	13F	13F				
CENT AFR EMP	36	38F	39F	40F	129	149F	152F	155F				
CHAD	47	50F	51F	51F	28	31F	31F	31F				
COMOROS	2	2F	2F	2F	90	96F	98F	100F				
CONGO	27	27F	27F	28F	57	69	71	72				
EGYPT	5330	6355	6683	6965	1570	2065	2053	2291	2	2F	2F	2F
EQ GUINEA					13	15F	15F	15F				
ETHIOPIA	351	393F	403F	414F	153	184F	190F	196F				
GABON	27	30*	31F	31F	96	97F	97F	97F				
GAMBIA	6	7F	7F	5F	3	3F	4F	3F				
GHANA	196	298	294	285	1832	1231	1068	1032	11	14F	11F	11F
GUINEA	94	105F	110F	111F	367	380F	396F	400F				
GUIN BISSAU	19	22F	23F	20F	37	41F	42F	38F	2	3F	3F	3F
IVORY COAST	266	323F	308F	345F	975	1238	1189	1269	60	59	61	61F
KENYA	282	308F	325F	336F	417	503	542	560F	16	16	21F	22F
LESOTHO	23	26F	25F	26F	19	21F	22F	22F				
LIBERIA	53	58F	60F	61F	97	107F	109F	111F	1	2F	2F	2F
LIBYA	208	387	412F	427F	90	120	123F	126F	4	5F	5F	6F
MADAGASCAR	240	273	276	280F	535	799	848	865	2	2F	2F	2F
MALAWI	150	182F	184F	186F	172	200F	209F	209F				
MALI	107	94F	95F	96F	9	9F	9F	10F				
MAURITANIA	4	3F	3F	3F	17	14F	15F	16F				
MAURITIUS	25	25	28F	31F	12	5	6F	9F				
MOROCCO	864	1213	1141	1180F	1314	1091	1211	1289F	13	9	11	11
MOZAMBIQUE	174	183F	186F	188F	310	279F	307F	317F	180	180F	200F	200F
NAMIBIA	18	21F	21F	22F	21	23F	24F	24F				
NIGER	60	100	107	73F	28	22F	22F	23F				
NIGERIA	2156	2410F	2550F	2718F	2369	2600F	2750F	2880F	27	30F	30F	30F
REUNION	6	12	11	12F	16	16	16	16F				
RHODESIA	98	115F	119F	123F	63	74F	77F	79F				
RWANDA	120	145F	150F	152F	1664	1754	1872	1954				
SAO TOME ETC	1	2F	2F	2F	3	2F	2F	2F				
SENEGAL	50	68	82	84F	42	59	62F	66F				
SEYCHELLES	1	1F	1F	1F	1	1F	1F	2F				
SIERRA LEONE	140	144F	145F	147F	361	402F	406F	411F	1	1F	2F	2F
SOMALIA	45	50F	51F	52F	156	127	172F	172F				
SOUTH AFRICA	1183	1468	1524	1582	2394	2789	2806	2866				
SUDAN	624	757	769	779F	509	695F	712F	717F				
SWAZILAND	9	11F	11F	11F	73	77	98F	99F				
TANZANIA	835	883	913	938F	1489	1940F	1989F	2041F	117	126	84	114
TOGO	50	55F	58F	59F	31	36F	37F	37F				
TUNISIA	599	848	895	903	345	464	478	495	10	24	24	26*
UGANDA	211	247	251	257	2994	3294F	3384F	3484F				
UPPER VOLTA	57	55F	60F	55F	36	32F	35F	30F				
ZAIRE	147	182	185	189F	2321	2607	2669	2748F				
ZAMBIA	148	174F	180F	184F	51	58	61F	62F				
N C AMERICA	25440	31007	29029	30623	35205	40966	39889	41079	354	486	473	575
ANTIGUA	1		1	1F								
BAHAMAS	3	3F	3F	3F	3	4F	4F	4F				
BARBADOS	3	6	6	6	2	2F	2F	3F				
BELIZE	6	5F	5F	5F	47	57	49	42				
BERMUDA	1	1	1	1	1	1	1					
CANADA	1285	1504	1454	1477F	639	718	639	640				
COSTA RICA	42	49	52	53F	1274	1313	1422	1426F				
CUBA	169	449	475	477F	343	563	574	588F				
DOMINICA	4	4F	5F	5F	62	46F	48F	52F				
DOMINICAN RP	121	148	155	158	1165	1240	1348	1293	1	1F	1F	1F
EL SALVADOR	83	97	90	102F	232	280	280	282F				
GRENADA	3	3F	4F	4F	25	20F	20F	20F				
GUADELOUPE	32	26	24	25F	144	174	164	164F				
GUATEMALA	199	234F	240F	244F	656	708F	743F	746F				
HAITI	206	226F	230F	235F	625	687F	692F	700F				
HONDURAS	51	70	66F	66F	1612	1044	1380	1543F				
JAMAICA	55	71	86	89F	427	329	345	350F				
MARTINIQUE	24	32	28	45	220	221	281	316				
MEXICO	2003	2591	2480	2480	4610	6742	5182	5713	22	28F	32*	33F
MONTSERRAT					1	2F	2F	2F				
NICARAGUA	34	43	44	46	170	330	334	338				
PANAMA	39	45	29	31	1192	1188	1199	1200				
PUERTO RICO	32	23	24	24	288	276	269	271				
ST LUCIA	1	1F	1F	1F	105	86	93F	97F				
ST VINCENT		1F	1F	1F	49	40	41	41F				
TRINIDAD ETC	15	31	31F	32F	80	58	71	73F				
USA	21030	25345	23496	25014	21233	24838	24706	25176	331	458	440	541
SOUTH AMERIC	8946	9858	10170	10274	29017	34370	36914	37460	115	121	104	105
ARGENTINA	2274	2408	2308	2084	5192	6054	6586	6614	6	8	9	9
BOLIVIA	278	341	332	339	495	622	641	654	8	12	12F	12F
BRAZIL	2880	3404	3623	3868	10514	14454	16312	16406	95	95	77	76
CHILE	1131	1011	1039	1068	1149	1337	1355	1395	3	5	5	5
COLOMBIA	815	963	1056	1072	2693	3216	3351	3729				
ECUADOR	348	335	370	368F	4591	4268	4222	4077				

TABLE
TABLEAU 40
CUADRO

	VEGETABLES+MELONS,TOTAL / LEGUMES ET MELONS,TOTAL / HORTALIZAS+MELONES,TOTAL — PRODUCTION — 1000 MT				FRUIT EXCL MELONS,TOTAL / FRUITS,EXCEPTE MELONS,T. / FRUTAS,EXCEPTO MELONES,T — PRODUCTION — 1000 MT				TREENUTS,TOTAL / FRUITS A COQUE,TOTAL / NUECES (TODA CLASE),TOT. — PRODUCCION — 1000 MT			
	1969-71	1975	1976	1977	1969-71	1975	1976	1977	1969-71	1975	1976	1977
FR GUIANA		1F	1F	1F	7	7F	7F	8F				
GUYANA	6	7	8	8F	44	35	35	36F				
PARAGUAY	151	211	213F	216F	611	565F	570	594F				
PERU	716	742	775	797F	1573	1674F	1629F	1650F	2	2F	2F	2F
SURINAM	3	3	3F	3F	63	60	59	59F				
URUGUAY	119	144F	148F	150F	296	305	318F	233				
VENEZUELA	223	289	293F	302F	1788	1772	1828	2006				
ASIA	138771	157910	165208	169030	53193	62233	62468	64304	1020	1263	1218	1261
AFGHANISTAN	605	630	634	673F	676	776	795	813F	26	26F	26F	26F
BAHRAIN	7	9F	10F	10F	18	19F	19F	19F				
BANGLADESH	1248	1094	1060	995	1481	1287	1294	1332				
BHUTAN	36	40F	41F	41F	10	11F	11F	11F				
BRUNEI	3	3F	3F	4F	5	6F	6F	6F				
BURMA	1540	1689	1717F	1746F	728	816F	823F	829F				
CHINA	57321F	65316	66879F	68903F	5111F	5493F	5642F	5711F	249F	280F	286F	292F
CYPRUS	123	118	121	142F	413	432	419	469F	5	5	6	5F
EAST TIMOR	21	23F			26	29F						
GAZA STRIP					160	190F	199F	207F				
HONG KONG	179	170	186	197	3	2	3	3				
INDIA	28359	33448	37222	37975F	15026	16252	16506	17198F	187	157*	163	166
INDONESIA	4652	5350F	5460F	5590F	3332	5300F	5500F	5600F				
IRAN	2589	3287	3380F	3489F	1735	2058	2113	2141F	69	88	113	115F
IRAQ	1876	1295	1491F	1469F	667	810	698	703F	3	4F	4F	4F
ISRAEL	600	762	715	756	1606	1852	1796	1788	3	5	6	6
JAPAN	14822	14525	14566	15090F	5560	6991	6409	6924	49	60	63F	67F
JORDAN	266	303	222	222F	70	41	43	44F	1	1	1	1F
KAMPUCHEA DM	467	460F	460F	450F	243	172F	176F	177F				
KOREA DPR	1457	1872F	1914F	1941F	222	304F	313F	322F	5	6F	6F	6F
KOREA REP	3938	4357	4701	4880F	417	648	619	645F	11	27	30	30F
KUWAIT	15	16	16F	17F	1	1	1F	1F				
LAO	135	154F	156F	159F	71	78F	79F	80F				
LEBANON	270	269F	279F	289F	564	701	713F	726F	5	7F	7F	8F
MACAU	1	1F	1F	1F	4	4F	4F	4F				
MAL PENINSUL	331	389F	398F	407F	883	870	895F	920F	6	6F	6F	6F
MAL SABAH	15	20F	20F	21	14	17F	18F	18F				
MAL SARAWAK	35	40F	40F	41F	29	33F	34F	35F				
MALDIVES	11	14F	15F	15F	5	6F	7F	7F	1	1F	1F	1F
MONGOLIA	13	21	25	28*	2	3F	3F	3F				
NEPAL	167	170F	175F	180F	115	124F	125F	127F				
OMAN	6	8F	8F	8F	65	79F	80F	81F				
PAKISTAN	1616	2080	1943	2043	1178	1472	1503	1543F	5	5F	5F	5F
PHILIPPINES	1175	1386	1381	1421F	1943	2768	2761	2744	8	9	10F	10F
QATAR	23	31F	33F	35F	2	2F	2F	2F				
SAUDI ARABIA	369	759	748F	756F	378	466	473	480F				
SINGAPORE	51	38	41	42	19	15	19	20				
SRI LANKA	135	252	249	253F	533	1025	1102	1109F				1F
SYRIA	923	2120	2151	2202F	370	518	579	582F	13	26	21	22F
THAILAND	2112	2638	2766	2847F	1664	2216F	2306F	2397F				
TURKEY	9002	10291	11425	11080F	6110	6380	6384	6419	373	551	464	491
VIET NAM	2138	2194F	2250F	2310F	1567	1768F	1790F	1852F	1	1F	1F	1F
YEMEN AR	70	168	183	195F	101	129	137	139F				
YEMEN DEM	49	101	123	109F	68	72	73	75F				
EUROPE	57889	60680	58450	60380	68711	71533	71566	63858	1051	1029	1109	917
ALBANIA	225	280F	357F	371F	117	124F	128F	132F	4	4F	4F	4F
AUSTRIA	546	622	465	587	959	1026	1040	898	10	11	4	4
BELGIUM-LUX	1130	1359	1132	1152	465	388	394	237				
BULGARIA	1876	1847	1937	2052F	2399	1997	2208	2316F	26	25	30	31F
CZECHOSLOVAK	1100	1113	829	850F	688	645	653	667F	18	21	10	10F
DENMARK	231	189	198	204F	157	157	137	164				
FINLAND	75	99	90	92	59	81	98	97				
FRANCE	7299	6720	6551	6589	15162	15761	15609	11234	83	73	58	43
GERMAN DR	818	876	884	926F	631	643	610	595	1	1F	1F	1F
GERMANY FED	2132	1781	1616	1851	4676	4369	4009	3496	14	13	9	6
GREECE	2521	3629	2759	3196	2974	3283	3328	3379	68	87	82	87F
HUNGARY	2685	2013	2072	2291F	2140	2142	2201	2509	20	26	25	26F
ICELAND	1	1F	1F	1F								
IRELAND	256	290	291F	292F	30	28	29	28				
ITALY	12030	12399	11838	12141	18919	19375	19690	18895	413	336	337	357
MALTA	44	53	56	58F	11	11	11	11F				
NETHERLANDS	2152	2318	2248	2463	669	529	546	484				
NORWAY	179	159	164	181	114	111	125	106				
POLAND	3730	4115	3773	3775	1107	1384	1865	1628				
PORTUGAL	1799	1851	1726	1822	1995	1949	1955	1697	64	51	65	52F
ROMANIA	2657	3175	4152	3932F	2279	2258	2856	2670F	34	26	30	32F
SPAIN	6491	8229	7915	7943	8364	10777	9825	8183	245	309	405	218
SWEDEN	201	227	260	266	233	254	251	253F	1	1*	1*	1F
SWITZERLAND	269	258	261	289	830	879	679	601	7	7	8	5
UK	3725	3568	3117	3366F	768	553	626	523	4	4F	4F	4F
YUGOSLAVIA	3716	3509	3753	3692F	2966	2809	2693	2853	40	33	36	36F
OCEANIA	1484	1498	1546	1589	3259	3408	3249	3310	4	5	6	6
AMER SAMOA					3	4F	4F	4F				
AUSTRALIA	913	878	905	931F	2045	2067	1864	1945	1	1	2	2F
COOK ISLANDS	1	2F	2F	2F	13	13	16	14				
FIJI	11	13F	13F	14F	12	12F	12F	12F				
FR POLYNESIA	1	1F	1F	1F	3	3F	3F	3F				
GILBERT IS	3	4F	4F	4F	3	4F	4F	4F				
GUAM	1				1							
NEW CALEDONIA	3	4	4F	4F	9	8	8F	8F				
NEW HEBRIDES	4	5F	6F	6F	3	4F	5F	5F				

TABLE
TABLEAU
CUADRO 40

	VEGETABLES+MELONS,TOTAL LEGUMES ET MELONS,TOTAL HORTALIZAS+MELONES,TOTAL PRODUCTION 1000 MT				FRUIT EXCL MELONS,TOTAL FRUITS,EXCEPTE MELONS,T. FRUTAS,EXCEPTO MELONES,T PRODUCTION 1000 MT				TREENUTS,TOTAL FRUITS A COQUE,TOTAL NUECES (TODA CLASE),TOT. PRODUCCION 1000 MT			
	1969-71	1975	1976	1977	1969-71	1975	1976	1977	1969-71	1975	1976	1977
NEW ZEALAND	334	352	367	379F	227	256	266	227				
NIUE ISLAND						1F	1F	1F				
PACIFIC IS	2	3F	3F	3F	3	3F	3F	3F				
PAPUA N GUIN	203	226	230F	235F	868	963F	981F	999F	3	3F	4F	4F
SAMOA					49	48	62	63				
SOLOMON IS	3	4F	4F	4F	9	10F	10F	10F				
TONGA	5	6F	6F	6F	12	10F	11F	11F				
USSR	23061	26483	27836	26196	11155	14235	15226	15449	156	205F	205F	209F
DEV.PED M E	84966	92096	87472	91412	93054	101853	99532	92710	1332	1450	1520	1429
N AMERICA	22315	26849	24949	26491	21872	25556	25345	25817	331	458	440	541
W EUROPE	44799	47262	44446	46184	59351	62341	61046	53142	948	926	1010	813
OCEANIA	1247	1230	1272	1310	2272	2324	2129	2172	2	1	2	2
OTH DEV.PED	16605	16755	16805	17427	9559	11632	11011	11579	52	65	69	73
DEV.PING M E	89305	105160	111453	113470	107544	123651	126495	130130	1322	1569	1488	1554
AFRICA	8880	10619	10668	11061	23157	24255	24688	25608	461	458	484	516
LAT AMERICA	12071	14015	14249	14406	42349	49780	51457	52722	138	150	137	138
NEAR EAST	22354	26903	28685	28866	13565	15554	15615	16035	501	714	650	678
FAR EAST	45762	53355	57576	58858	27485	32977	33614	34626	218	205	214	218
OTH DV.PING	238	268	274	280	988	1085	1120	1139	3	3	4	4
CENTR PLANND	97545	109765	113368	114025	27661	31167	33669	34229	553	594	598	612
ASIAN CPE	61394	69863	71529	73633	7146	7739	7923	8064	255	287	293	299
E EUR+USSR	36151	39902	41840	40392	20515	23427	25746	26165	298	308	304	312
DEV.PED ALL	121117	131997	129312	131804	113565	125280	125278	118875	1630	1758	1825	1742
DEV.PING ALL	150700	175024	182981	187102	114690	131390	134417	138194	1577	1856	1781	1853

TABLE
TABLEAU **41**
CUADRO

	CABBAGES AREA HARV SUP RECOLTEE SUP COSECHAD — 1000 HA				CHOUX YIELD RENDEMENT RENDIMIENTO — KG/HA				COLES PRODUCTION PRODUCTION PRODUCCICN — 1000 MT			
	1969-71	1975	1976	1977	1969-71	1975	1976	1977	1969-71	1975	1976	1977
WORLD	1160	1188	1187	1203	17608	17645	18263	18495	20427	20961	21671	22247
AFRICA	20	23	23	24	21980	23063	23794	23745	437	540	559	568
EGYPT	12	14	14	15F	23318	22474	23034	22807	271	321	329	331F
ETHIOPIA	2	2F	2F	3F	15412	15417	15833	16000	31	37F	38F	40F
LIBYA					17409	14091	13600	13600				
MADAGASCAR				1F	14123	12030	12030	11961	3	6	6	6F
MALAWI	1	1F	1F	1F	11495	17000	17000	18000	11	17F	17F	18F
MAURITIUS					22979	23581	25000	25333	2	3	4F	4F
SOUTH AFRICA	3	3F	3F	3F	36222	47419	50323	50000	109	147	156	160F
TUNISIA					17062	14167	14400	14400	4	3F	4F	4F
ZAIRE	2	2F	2F	2F	3003	3086	3111	3222	5	5F	6F	6F
N C AMERICA	78	83	83	83	17677	18719	18424	18983	1384	1545	1522	1578
ANTIGUA					4000	3333	4000	3857				
BARBADOS					10057	6527	6557	6333	1	1	1	1F
BERMUDA					16143	15000	18167	18333				
CANADA	3	4	4	4F	23336	22579	25005	25000	70	93	104	105F
COSTA RICA	1	1F	1F	1F	7709	7778	8333	8211	6	7F	8F	8F
CUBA	1	2F	2F	2F	14525	15429	18156	18500	15	23	33	33F
DOMINICAN RP					10666	10938	10909	10882	3	4F	4F	4F
EL SALVADOR					8000	9000	9333	8571	2	3F	3F	3F
GUADELOUPE					15018	15000	20000	17391	4	3	4	4F
GUATEMALA	1	1F	1F	1F	16006	15455	15455	15455	14	17F	17F	17F
HAITI	1	1F	1F	1F	10088	10000	10000	10000	7	8F	9F	9F
HONDURAS	1	1	1F	1F	7133	7961	8462	8358	5	5	6F	6F
JAMAICA	1	1F	2F	2F	6151	7356	7709	7700	7	8	15	15F
MARTINIQUE					14106	14333	14333	22667	4	4*	4*	7*
NICARAGUA	5	6F	6F	6F	1656	1673	1667	1667	8	9F	10F	10F
PANAMA					20043	20150	20333	19900	3	3	1	1
PUERTO RICO					9259	16085	15892	15476	3	2	2	3
TRINIDAD ETC					26522	25224	25769	26154	2	7	7F	7F
USA	63	65	64	64	19407	20723	20336	21141	1229	1348	1296	1346
SOUTH AMERIC	16	17	18	18	24943	24455	27052	26551	409	421	475	478
BOLIVIA	2	2	2F	2F	6000	5877	6087	6042	12	13	14F	15F
CHILE	1	1F	1F	1F	15000	12500	13000	12000	15	13*	13*	12*
COLOMBIA	7	8	8	8F	34592	35952	40098	39759	241	286	329	330F
ECUADOR	2	2	2	2F	37805	23527	29705	28824	84	43	50	49F
GUYANA					5700	6479	6550	6316		1	1*	1F
PERU	3	3F	3F	3F	13063	16033	15938	15853	42	48F	51F	54F
SURINAM					7692	7000	7000	7000				
URUGUAY					5129	5814	5909	5778	2	3F	3F	3F
VENEZUELA		1	1F	1F	25233	26913	27885	28113	12	14	15F	15F
ASIA	687	731	729	745	14836	14973	15710	15575	10193	10943	11448	11599
BANGLADESH	6	6	6	6	8432	7617	7562	7521	51	44	45	43
CHINA	329F	368F	368F	373F	11885	12874	13360	13598	3910F	4744F	4911F	5075F
CYPRUS					23021	18182	20455	20455	4	4F	5F	5F
HONG KONG	6	6	7	7	11733	11557	11420	10789	73	71	77	80
INDIA	65	74F	75F	77F	5846	5811	5880	5894	380	430F	441F	452F
IRAQ	1	1	1F	1F	13716	8072	8500	8523	15	9	10F	12F
ISRAEL	1	1	1	1F	22790	28467	32961	33548	15	21	25	26F
JAPAN	117	106	105	107F	32061	33302	34438	35586	3753	3530	3616	3801F
JORDAN					15984	14545	12308	12308	2	2	2	2F
KOREA DPR	19	22F	23F	23F	11261	13409	13244	13333	208	295F	298F	300F
KOREA REP	73	70	65	70F	12386	12592	15690	15503	909	880	1023	1079F
LEBANON	1	1F	1F	1F	18194	15214	16143	17071	25	21F	23F	24F
PHILIPPINES	6	8	8	8F	7150	6625	6726	6741	44	53	54	55F
SAUDI ARABIA					4333	6352	6250	6250	1	1	1F	1F
SRI LANKA	2	3	3	3F	6194	4055	5500	5538	14	12	14	14F
SYRIA	1	2	3	4F	15145	21286	21100	20000	19	45	63	70F
THAILAND	32	34F	34F	35F	5432	3521	5441	5362	174	118	185F	185F
TURKEY	24	26F	27F	28F	22223	23462	22222	22143	543	610	600	620F
VIET NAM	2	2F	2F	2F	25263	24651	25000	24783	48	53F	55F	57F
EUROPE	354	330	330	329	22295	22463	22913	23169	7892	7409	7561	7617
AUSTRIA	3	4	4	4	37201	35795	33692	38069	101	142	120	138
BELGIUM-LUX	1	1F	1F	1F	47893	59560	44950	47900	52	60	45	48F
BULGARIA	3	3	3	4F	56896	53071	51487	51000	168	175	180	184F
CZECHOSLOVAK	9	10	9	9F	30867	30095	25224	25532	268	287	237	240F
DENMARK	2	1	2	2F	28302	24302	28703	28250	55	32	44	45F
FINLAND		1	1	1F	23927	28316	21413	25000	12	16	14	15
FRANCE	17	15F	12	12F	23324	21379	21480	21483	406	310F	249	249F
GERMAN DR	12	12	14	14F	26319	24977	26436	26500	320	305	359	371F
GERMANY FED	15	12	11	12	35210	39230	37043	44415	531	465	423	553
GREECE	8	8F	8F	8F	16651	17051	17813	17750	127	133F	143F	142F
HUNGARY	14	11*	11F	12F	15760	17586	19112	18333	220	197	214	220F
ICELAND					15500	17000	17000	17000				
IRELAND	5	5F	5F	5F	27076	27451	27059	27235	130	140F	138F	139F
ITALY	40	30	29	28	20208	20630	20457	21080	801	625	600	599
MALTA					24326	26167	24684	25316	2	2	2F	2F
NETHERLANDS	9	9	9	9F	26568	24964	26200	26783	229	225	241	246F
NORWAY	1	1	1	2	30201	29372	28158	25839	44	39	38	51
POLAND	63	68	66	67F	23126	22990	22045	22388	1459	1563	1455	1500F
PORTUGAL	7	8F	8F	8F	17094	17722	18125	17722	120	140F	145F	140F
ROMANIA	24	22	25	25F	22652	23960	36652	34008	541	520	905	840F
SPAIN	27	23	23	20*	26306	24044	23652	22500	718	546	544	450
SWEDEN	1	1	1	1F	39230	42034	42034	42069	30	37	37	37F
SWITZERLAND	1	1	1	1	47040	43000	43692	43692	26	28	28	28

TABLE
TABLEAU 41
CUADRO

	CABBAGES				CHOUX				COLES			
	AREA HARV 1000 HA SUP RECOLTEE SUP COSECHAD				YIELD KG/HA RENDEMENT RENDIMIENTO				PRODUCTION 1000 MT PRODUCTION PRODUCCION			
	1969-71	1975	1976	1977	1969-71	1975	1976	1977	1969-71	1975	1976	1977
UK	48	38	38	38F	19424	19335	16493	18158	926	733	632	690F
YUGOSLAVIA	45	47	48	47	13462	14660	16021	14681	605	689	769	690F
OCEANIA	4	4	4	4	25160	25348	24905	25405	111	103	105	107
AUSTRALIA	3	3	3F	3F	27244	26985	26714	27179	71	74	75F	76F
NEW ZEALAND	2	1	1F	1F	22129	21972	21286	21857	40	29	30F	31F
DEV.PED M E	421	387	383	381	24251	24819	24865	25746	10202	9603	9513	9806
N AMERICA	66	69	68	68	19584	20834	20623	21380	1299	1441	1400	1451
W EUROPE	229	204	202	198	21443	21393	20885	21521	4915	4362	4211	4262
OCEANIA	4	4	4	4	25160	25348	24905	25405	111	103	105	107
OTH DEV.PED	121	110	109	111	32113	33667	34880	35988	3877	3698	3797	3987
DEV.PING M E	265	282	283	293	11617	11397	12504	12460	3082	3219	3543	3655
AFRICA	5	6	6	6	10861	11952	12104	12448	57	72	74	77
LAT AMERICA	28	31	32	33	17419	17168	18491	18195	494	525	597	606
NEAR EAST	41	45	47	49	21825	22314	21775	21603	885	1013	1033	1064
FAR EAST	191	200	198	205	8610	8028	9313	9326	1645	1609	1840	1908
OTH DV.PING					16143	15000	18167	18333				
CENTR PLANND	474	519	521	529	15066	15695	16545	16619	7143	8139	8614	8786
ASIAN CPE	349	393	392	398	11925	12968	13419	13648	4166	5052	5264	5432
E EUR+USSR	125	126	128	131	23860	24195	26097	25666	2977	3047	3350	3355
DEV.PED ALL	545	513	511	512	24162	24666	25174	25726	13179	12651	12863	13161
DEV.PING ALL	615	675	676	691	11792	12311	13035	13144	7248	8310	8807	9086

TABLE
TABLEAU 42
CUADRO

	ARTICHOKES / AREA HARV SUP RECOLTEE SUP COSECHAD (1000 HA)				ARTICHAUTS / YIELD RENDEMENT RENDIMIENTO (KG/HA)				ALCACHOFAS / PRODUCTION PRODUCTION PRODUCCION (1000 MT)			
	1969-71	1975	1976	1977	1969-71	1975	1976	1977	1969-71	1975	1976	1977
WORLD	129	119	117	119	9891	11091	10941	10766	1271	1315	1280	1276
AFRICA	11	11	12	13	7770	7254	7726	7839	83	83	96	98
ALGERIA	5	4	5	5F	8168	6280	7262	7298	38	27	34	34F
EGYPT			1	1F	26190	33333	25000	25000	11	14	21	21F
MOROCCO	4	4F	5F	5F	5983	6279	6222	6444	23	27F	28F	29F
TUNISIA	2	2*	2F	3F	6238	6176	5417	5600	11	15	13	14*
N C AMERICA	4	4	4	4	7403	8065	8522	7399	32	33	37	32
USA	4	4	4	4	7403	8065	8522	7399	32	33	37	32
SOUTH AMERIC	6	6	4	5	15727	16483	14909	17398	87	100	66	94
ARGENTINA	4	4	3	4F	18695	19017	17730	20000	68	83	50	80F
CHILE	2	2F	1F	1F	10002	10000	10000	10000	17	15*	14*	12*
PERU					10668	8889	8889	8889	2	2F	2F	2F
ASIA	2	1	1	1	9709	14214	12278	12295	17	16	17	17
CYPRUS	1	1F	1F	1F	9825	8929	8929	9091	5	5F	5F	5F
ISRAEL	1				6549	11875	5778	6667	6	3	3	3F
LEBANON					12091	11961	11961	11961	1	1F	1F	1F
TURKEY					18954	26333	28333	26667	5	8	9	8F
EUROPE	106	96	94	95	9905	11300	11272	10910	1052	1083	1065	1035
FRANCE	19	18F	16	16F	7825	6317	6211	6389	146	114	97	100F
GREECE	5	5F	5F	5F	9646	13191	13617	13617	44	62	64	64F
ITALY	63	54	53	53	10785	12658	12848	12144	681	680	677	640
SPAIN	20	19	22	22	9125	11731	10558	10601	181	226	227	231
DEV.PED M E	111	100	99	100	9781	11168	11128	10737	1090	1119	1104	1071
N AMERICA	4	4	4	4	7403	8065	8522	7399	32	33	37	32
W EUROPE	106	96	94	95	9905	11300	11272	10910	1052	1083	1065	1035
OTH DEV.PED	1				6549	11875	5778	6667	6	3	3	3
DEV.PING M E	17	18	18	19	10602	10671	9902	10919	181	196	176	206
AFRICA	10	11	12	12	7016	6257	6479	6607	72	69	75	77
LAT AMERICA	6	6	4	5	15727	16483	14909	17398	87	100	66	94
NEAR EAST	1	1	2	2	17163	20669	20051	19879	22	28	35	35
DEV.PED ALL	111	100	99	100	9781	11168	11128	10737	1090	1119	1104	1071
DEV.PING ALL	17	18	18	19	10602	10671	9902	10919	181	196	176	206

TABLE
TABLEAU 43
CUADRO

	TOMATOES				TOMATES				TOMATES			
	AREA HARV SUP RECOLTEE SUP COSECHAD		1000 HA		YIELD RENDEMENT RENDIMIENTO		KG/HA		PRODUCTION PRODUCTION PRODUCCION		1000 MT	
	1969-71	1975	1976	1977	1969-71	1975	1976	1977	1969-71	1975	1976	1977
WORLD	1867	2104	2116	2200	18009	20580	19853	20465	33625	43307	42002	45023
AFRICA	236	302	305	333	12948	14184	13603	13797	3061	4284	4152	4601
ALGERIA	9	12	12	12F	10408	11533	8780	11333	92	135	106	136F
ANGOLA	2	2F	2F	2F	1334	1250	1250	1500	3	3F	3F	3F
BENIN	3	2F	2F	2F	3202	2500	2500	2800	11	5F	5F	6F
CAMEROON	7	8F	8F	8F	1727	1750	1744	1738	13	14F	14F	15F
CAPE VERDE					1250	1250	1250	1250				
EGYPT	102	137	129	150F	15524	15433	15968	16000	1580	2107	2066	2400*
ETHIOPIA	3	4F	4F	4F	11325	12571	12500	12778	37	44F	45F	46F
GHANA	15	19	21*	22*	4958	4737	4619	4545	73	90	97*	100*
IVORY COAST	1	1F	1F	2F	10000	9692	9429	9267	10	13F	13F	14F
KENYA					15051	15000	17647	17568	4	5F	6F	7F
LIBYA	8	16F	16F	17F	16607	11613	12000	12364	134	180*	192F	204F
MADAGASCAR	1	2	2F	2F	6508	6408	6250	6250	6	10	10F	10F
MALAWI	2	3F	3F	3F	8432	9259	9286	9286	20	25F	26F	26F
MALI	2	2F	2F	2F	3995	3733	3800	3867	6	6F	6F	6F
MAURITIUS	1	1	1F	1F	10533	9559	12963	13636	7	6	7F	8F
MOROCCO	12	11	15	15F	19759	49250	30494	31333	246	542	450	470F
MOZAMBIQUE	1	1F	1F	1F	9373	9231	10000	10000	12	12F	13F	13F
NIGER		1	1F	1F	8120	13400	12346	13415	3	12	10F	11F
NIGERIA	22	25F	27F	30F	9985	9200	8704	8333	220	230F	235F	250F
REUNION					18333	19535	19535	19816	2	4	4	4F
SENEGAL	1	2F	2F	2F	9996	9833	15000	15000	8	18	30*	30F
SIERRA LEONE	1	1F	1F	1F	7002	6923	6923	7000	9	9F	9F	9F
SOUTH AFRICA	8	12F	12F	13F	29303	27500	27000	25385	225	330	324	330F
SUDAN	9	12F	12F	12F	11935	11667	11833	11626	109	140F	142F	143F
SWAZILAND					12273	13333	13333	13333	4	4F	4F	4F
TANZANIA	1	2F	2F	2F	7839	8000	8056	7895	12	14F	15F	15F
TOGO	4	4F	4F	4F	664	674	674	682	3	3F	3F	3F
TUNISIA	13	16*	17F	18F	12618	16099	14706	15000	160	260	250	270*
UGANDA	1	1F	1F	1F	6683	6563	6566	6700	5	6F	7F	7F
UPPER VOLTA					9113	8667	8667	8000	1	1F	1F	1F
ZAIRE	4	5F	5F	5F	6899	7447	7500	7592	30	35F	36F	37F
ZAMBIA	2	2F	2F	2F	11098	11500	12000	12250	19	23F	24F	25F
N C AMERICA	275	326	284	300	25684	32534	30125	32375	7074	10599	8565	9714
ANTIGUA					3438	3429	3250	3226				
BARBADOS					5280	4933	4589	4681				
BERMUDA					11125	17000	17000	17000				
CANADA	11	12	12	12F	30537	33809	36822	36904	350	397	424	424F
COSTA RICA	2	2F	2F	2F	7431	7381	7273	7174	13	16F	16F	17F
CUBA	11	25F	25F	25F	6013	7368	7772	7812	64	184	194	195F
DOMINICAN RP	2	3F	3F	3F	29559	31852	32143	31724	62	86F	90F	92F
EL SALVADOR	1	1	1	2F	14324	15734	11440	14000	18	19	12	21F
GUADELOUPE					13472	12000	10000	10000	3	1	2	2F
GUATEMALA	11	10F	10F	10F	6358	7400	7500	7600	71	74F	75F	76F
HAITI					22500	15714	16667	18182	3	3F	4F	4F
HONDURAS	1	2	1F	1F	2813	9753	11111	10870	3	15	10F	10F
JAMAICA	1	1F	1F	1F	8975	8505	9205	9000	7	9	12	13F
MARTINIQUE		1	1	1	9938	8750	6250	2000	5	7	5	2
MEXICO	68	59	48	49	13231	17796	16813	17469	901	1056	807	856
MONTSERRAT					8100	8333	6833	6667				
NICARAGUA		1	1	1	41566	42509	42542	42506	17	24	25	26
PANAMA	2	3F	2F	2F	11884	10823	9485	10000	24	30	19	20F
PUERTO RICO	1				8912	11545	11560	14060	12	3	3	3
TRINIDAD ETC		1	1F	1F	11337	15970	15806	15625	5	10	10F	10F
USA	162	206	177	191	34053	42123	38791	41691	5516	8665	6857	7943
SOUTH AMERIC	113	114	115	116	16110	19505	21158	21442	1814	2230	2427	2487
ARGENTINA	23	27	29	25F	18007	18264	18236	18000	418	486	533	450*
BOLIVIA	5	5	5	5	10066	11732	8000	7600	50	63	41	38
BRAZIL	45	46	46	51	16944	22528	25557	25567	762	1047	1168	1291
CHILE	10	6	7	7	19966	24999	24999	25001	199	156	164	172
COLOMBIA	12	10	8	9F	12080	18400	24320	24706	146	177	206	210F
ECUADOR	1	2	3	3F	21120	16221	16366	16346	25	37	43	43F
GUYANA					4626	5993	6200	6176	2	2*	2*	2F
PARAGUAY	3	3F	3F	3F	18463	19074	18929	18690	47	52F	53F	54F
PERU	5	6F	6F	6F	11446	12727	13745	14764	60	70F	76F	81F
SURINAM					6519	6500	6500	6393	1			
URUGUAY	3	3F	3F	4F	10341	12121	12353	12286	30	40F	42F	43F
VENEZUELA	5	6	5F	5F	14779	17025	20000	19692	76	101	100F	102F
ASIA	591	655	678	689	12812	14737	15114	15330	7574	9655	10249	10556
BAHRAIN					23107	26000	27000	28000	2	3F	3F	3F
BANGLADESH	9	8	8	8	8841	7706	7510	6858	75	61	58	53
CHINA	228F	249F	253F	258F	11313	12919	13034	13418	2584F	3210F	3299F	3466F
CYPRUS	2	2F	2F	2F	15910	14375	14286	14286	25	23F	25F	25F
HONG KONG					17753	16706	18442	16706	1	1	1	1
INDIA	65	72F	73F	75F	9538	9444	9521	9544	620	680F	695F	712F
INDONESIA	52	59F	61F	62F	6019	6356	6364	6371	310	375F	385F	395F
IRAN	21	22F	22F	23F	8571	10909	11364	11130	180	240F	250F	256F
IRAQ	34	29	38F	32F	8855	9032	9211	9063	305	258	350F	290F
ISRAEL	5	7	7	7F	36129	38329	34362	35714	165	278	237	250F
JAPAN	20	19	18	19F	41347	53895	49669	51898	811	1024	899	965F
JORDAN	16	10	9	9F	9113	14006	9966	9989	142	145	88	88F
KOREA DPR	2	2F	3F	3F	13397	13333	13200	13462	22	32F	33F	35F
KOREA REP	3	4	3	4F	15872	19619	18072	16405	54	77	61	61F
KUWAIT					12424	12821	12800	12593	3	3	3F	3F
LAO	2	2F	2F	2F	2924	3000	3042	3083	7	7F	7F	7F
LEBANON	5	5F	5F	5F	13062	13333	13265	13208	68	60F	65F	70F

TABLE
TABLEAU 43
CUADRO

	AREA HARV / SUP RECOLTEE / SUP COSECHAD (1000 HA)				YIELD / RENDEMENT / RENDIMIENTO (KG/HA)				PRODUCTION / PRODUCTION / PRODUCCION (1000 MT)			
	1969-71	1975	1976	1977	1969-71	1975	1976	1977	1969-71	1975	1976	1977
MAL PENINSUL	5	5F	6F	6F	5000	4981	4964	4948	23	27F	28F	29F
MAL SABAH					14500	15455	16364	17273				
PHILIPPINES	17	19	21	20F	5651	7211	7283	7500	97	137	153	150F
SAUDI ARABIA	9	21	21F	22F	10750	14442	14524	14091	93	301	305F	310F
SRI LANKA	4	6	6F	6F	1860	3419	3569	3567	7	21	21	21F
SYRIA	18	27	32	35F	11895	14060	16348	15714	211	375	517	550F
THAILAND	5	5F	5F	6F	2889	2843	2736	2727	13	15F	15F	15F
TURKEY	72	82F	84F	86F	24506	28049	32738	32558	1756	2300	2750	2800F
EUROPE	441	468	444	472	24848	27187	26383	27345	10951	12714	11727	12915
AUSTRIA					50803	64317	54185	67884	22	31	18	24
BELGIUM-LUX	1	1	1	1F	101810	167497	182803	175250	90	135	149	140*
BULGARIA	24	28	29	30F	29272	21721	26850	26667	711	599	785	800F
CZECHOSLOVAK	4	4F	4	4F	26147	26379	20699	21024	105	97	84	86F
DENMARK					172942	195972	213381	207273	21	21	22	23F
FINLAND					115828	129801	127922	126667	11	20	20	19
FRANCE	18	23F	23F	23F	30363	27773	24913	27174	535	639	573	625
GERMAN DR	1	1	2	2F	19466	24566	22100	22125	18	33	34	35F
GERMANY FED	1	1	1	1	47207	54202	55157	54852	31	32	33	33
GREECE	33	42	32	36*	29852	38738	32031	40138	976	1627	1025	1457*
HUNGARY	19	17	17F	20F	19095	18002	23252	22500	359	313	395	450F
ICELAND					132667	100000	103333	106667				
IRELAND					74639	75000	76471	77941	19	26	26F	27F
ITALY	127	113	100	108	28108	31030	29705	30630	3571	3512	2969	3304
MALTA	1	1	1	1F	15606	26962	28617	26429	11	16	18	19F
NETHERLANDS	3	2	2	2F	111903	143584	152349	157542	365	347	360	378F
NORWAY					132242	166667	177193	185565	9	10	10	11
POLAND	27	30	32	31F	12609	13930	11859	11290	339	418	380	350F
PORTUGAL	23	25F	15	27*	32689	33608	42309	29259	764	840*	631	790*
ROMANIA	57	57*	75*	70F	13757	16946	19637	19514	780	968*	1473*	1366F
SPAIN	66	81	68	73	25436	30604	30469	31744	1687	2488	2078	2330
SWEDEN					103928	110000	111818	114545	7	12	12	13F
SWITZERLAND	1				36055	54386	64164	65000	22	16	19	20
UK	2	3	3F	3F	74217	57596	57742	56774	167	181	179	176F
YUGOSLAVIA	33	37	39	39	10148	9054	11198	11282	332	335	433	440F
OCEANIA	9	9	9	10	25882	25057	25853	26080	236	235	245	250
AUSTRALIA	8	8	8	8F	23123	20481	21375	21630	177	162	171	175F
COOK ISLANDS					3486	10909	11538	11538				
FIJI					7526	6875	6970	7059	1	2F	2F	2F
FR POLYNESIA					24111	28000	29000	30000				
GUAM					19167	1714	1333	1091				
NEW ZEALAND	1	1	1F	1F	48556	70658	71546	71837	56	68	69F	70F
PAPUA N GUIN					3111	3767	3836	3836				
TONGA					20146	21212	21739	22222	1	1F	2F	2F
USSR	202	230F	279	280F	14453	15609	16600	16071	2915	3590F	4637	4500F
DEV.PED M E	523	595	520	565	30462	35639	33772	35354	15940	21210	17558	19985
N AMERICA	173	217	188	202	33821	41675	38671	41418	5866	9061	7281	8367
W EUROPE	309	331	286	316	27946	31117	30025	31139	8640	10287	8576	9827
OCEANIA	9	9	9	9	26425	25948	26800	27048	233	231	240	246
OTH DEV.PED	32	38	37	39	37699	42665	39462	40034	1201	1632	1460	1545
DEV.PING M E	781	891	902	937	12623	14404	14772	14886	9853	12837	13324	13950
AFRICA	110	126	136	142	9228	12119	10513	10756	1012	1527	1428	1524
LAT AMERICA	215	223	211	214	14083	16920	17609	17913	3022	3768	3710	3833
NEAR EAST	295	361	370	393	15614	16998	18267	18172	4608	6136	6755	7142
FAR EAST	161	181	185	188	7508	7745	7700	7692	1208	1402	1425	1445
OTH DV.PING			1	1	9826	9000	9208	9372	3	4	5	5
CENTR PLANND	563	618	694	698	13903	14984	16027	15895	7832	9260	11120	11088
ASIAN CPE	230	251	256	261	11328	12923	13035	13418	2607	3242	3332	3501
E EUR+USSR	333	367	438	437	15682	16393	17773	17375	5226	6018	7788	7586
DEV.PED ALL	857	962	958	1002	24711	28297	26455	27518	21165	27227	25346	27573
DEV.PING ALL	1011	1142	1158	1198	12328	14079	14389	14566	12459	16079	16656	17450

TABLE
TABLEAU 44
CUADRO

| | CAULIFLOWER CHOUX-FLEURS COLIFLORES | | | | | | | | | | | |
| | AREA HARV SUP RECOLTEE SUP COSECHAD (1000 HA) | | | | YIELD RENDEMENT RENDIMIENTC (KG/HA) | | | | PRODUCTION PRODUCTION PRODUCCICN (1000 MT) | | | |
	1969-71	1975	1976	1977	1969-71	1975	1976	1977	1969-71	1975	1976	1977
WORLD	295	313	311	320	13180	12831	12835	12659	3885	4014	3998	4047
AFRICA	5	5	6	6	22597	21611	22627	23090	105	110	126	134
EGYPT	3	3	4	4F	21918	20833	21958	21949	74	70	83	86F
LIBYA					19905	20625	20625	20588				
MADAGASCAR					20109	11269	11538	11538	1	1	2F	2F
MAURITIUS					11457	20959	20000	22500		1	2F	2F
SOUTH AFRICA	1	1F	1F	1F	26333	25385	26923	28571	26	33	35	40F
TUNISIA					17895	18000	17308	17037	3	5F	5F	5F
N C AMERICA	12	14	15	15	10994	10718	10555	10801	131	155	158	165
CANADA	2	1	1	1F	12241	12782	13706	13538	19	18	17	18F
JAMAICA					5900	7375	6800	6800				
TRINIDAD ETC					9051	8806	8750	8701		1F	1F	1F
USA	10	13	14	14	10822	10501	10278	10560	111	136	140	147
SOUTH AMERIC	4	4	4	4	19232	18576	20310	20274	74	71	84	86
CHILE	1	1	1	1	5221	10002	9993	9993	6	5	6	6
COLOMBIA	1	1	2	2F	24671	19808	25000	25000	34	26	38	35F
ECUADOR	1	1F	1F	1F	26984	27841	27778	27778	17	25F	25F	25F
PERU	1	1F	1F	1F	14283	13895	13333	13200	15	13F	13F	13F
VENEZUELA					11636	13805	13889	14444	2	2	3F	3F
ASIA	140	157	160	163	8788	9365	9441	9448	1233	1473	1513	1541
BANGLADESH	6	5	6	6	7375	6951	6883	6676	42	37	38	37
CHINA	47F	52F	53F	54F	10803	11878	11924	11889	511F	619F	632F	642F
CYPRUS					19561	18750	21212	21212	4	3F	4F	4F
INDIA	75	84F	85F	87F	7133	7083	7176	7204	535	595F	610F	624F
IRAQ	1	1	1F		10521	7817	7910	8028	6	5	5F	6F
ISRAEL	1	1			10669	14831	16087	18889	7	9	7	9F
JAPAN	3	4	4	4F	19303	18509	17900	18114	50	72	75	80F
JORDAN	1				14847	16512	16486	16486	9	7	6	6F
LEBANON		1F	1F	1F	5875	2400	2222	2069	2	1F	1F	1F
SYRIA	1	3	3	3F	13559	19520	19200	18667	14	49	58	56F
THAILAND	3	4F	4F	4F	5940	6571	6571	6571	20	23F	23F	23F
TURKEY	2	3F	3F	3F	11128	15000	14516	14839	27	45	45	46F
VIET NAM		1F	1F	1F	12790	12698	12656	12615	6	8F	8F	8F
EUROPE	130	128	123	127	17119	16437	16454	15869	2224	2109	2021	2023
BELGIUM-LUX	1	1	1	1F	70433	46141	39658	39155	44	32	28	28F
BULGARIA					105280	174324	139581	139785	18	12	13	13F
CZECHOSLOVAK	3	4	3	4F	20299	19738	18655	18429	66	70	64	65F
DENMARK	1	1	1	1F	10758	10922	11887	11728	10	8	9	10F
FINLAND					10872	9762	7714	9667	2	3	3	3
FRANCE	32	40F	36	40F	12560	11499	13245	11000	407	460	471	440*
GERMAN DR	3	4	5	5F	23595	24085	24231	23922	76	102	122	122F
GERMANY FED	4	4	4	4	21411	21149	15929	21661	80	79	70	77
GREECE	3	3F	3F	3F	13721	11000	11500	11500	37	33F	35F	35F
HUNGARY	2	2	2F	2F	9964	11441	11599	11111	19	19	20	20F
ICELAND					18250	16200	16600	17000				
IRELAND					20000	20000	20000	20000	2	2F	2F	2F
ITALY	38	29	29	28	18828	19929	19623	19576	721	584	564	554
MALTA					16060	18581	16277	16500	4	4	3	3F
NETHERLANDS	4	5	5	5F	12839	11015	11516	11538	53	60	60	60F
NORWAY	1	1	1	1F	12832	12075	11612	11552	7	6	7	7F
POLAND	12	10	11	10F	11828	13130	12927	13500	138	131	142	135F
SPAIN	8	8	8	8	24983	23585	24075	22679	205	193	193	184
SWEDEN		1	1	1F	10952	8726	8726	8767	4	6	6	6F
SWITZERLAND					31671	26190	23810	26190	11	11	10	11
UK	17	15	14	15F	18956	18994	14613	16667	321	293	200	250F
OCEANIA	4	4	4	4	28381	25115	25474	25921	118	95	97	99
AUSTRALIA	3	3	3F	3F	32385	27212	27538	27962	88	71	72F	73F
NEW ZEALAND	1	1	1F	1F	20995	20583	21000	21500	31	25	25F	26F
DEV.PED M E	130	133	126	132	17232	16100	16100	15574	2239	2138	2031	2060
N AMERICA	12	14	15	15	11009	10729	10567	10815	130	154	157	165
W EUROPE	110	109	102	107	17386	16306	16346	15595	1907	1775	1660	1669
OCEANIA	4	4	4	4	28381	25115	25474	25921	118	95	97	99
OTH DEV.PED	4	6	6	6	19715	19680	19731	20512	82	114	117	128
DEV.PING M E	97	108	110	112	8386	8483	8745	8751	812	914	965	983
AFRICA					17439	16298	16129	16458	5	7	8	8
LAT AMERICA	4	4	4	4	19062	18386	20079	20036	74	72	84	86
NEAR EAST	9	11	12	12	15821	17021	17355	17265	136	180	202	204
FAR EAST	84	93	94	96	7102	7056	7137	7150	597	655	671	684
CENTR PLANND	68	72	75	75	12263	13311	13364	13371	834	962	1001	1005
ASIAN CPE	48	53	54	55	10824	11888	11932	11898	518	627	640	650
E EUR+USSR	20	19	21	20	15672	17165	16973	17299	317	334	361	355
DEV.PED ALL	150	152	147	153	17022	16236	16226	15805	2555	2473	2392	2414
DEV.PING ALL	145	161	164	167	9192	9602	9787	9781	1330	1542	1605	1633

TABLE
TABLEAU 45
CUADRO

PUMPKINS, SQUASH, GOURDS   COURGES ET AUTRES CUCURB   CALABAZAS,TODAS CLASES

	AREA HARV / SUP RECOLTEE / SUP COSECHAD (1000 HA)				YIELD / RENDEMENT / RENDIMIENTO (KG/HA)				PRODUCTION / PRODUCTION / PRODUCCION (1000 MT)			
	1969-71	1975	1976	1977	1969-71	1975	1976	1977	1969-71	1975	1976	1977
WORLD	1318	1130	1064	1014	4611	5468	5748	5865	6075	6179	6115	5949
AFRICA	49	57	59	60	14595	14528	14878	15052	713	829	876	905
CENT AFR EMP	1	1F	1F	1F	7971	9067	9467	9867	6	7F	7F	7F
EGYPT	19	19	21	22F	17905	17966	18364	18370	346	350	378	397F
LIBYA		1F	1F	1F	13170	9000	9000	9000	4	9F	9F	9F
MAURITIUS					16986	17154	16364	18182	1	2	2F	2F
MOROCCO	5	5F	6F	6F	14030	13962	13393	13571	65	74	75F	76F
MOZAMBIQUE					2286	2429	2429	2429	1	1F	1F	1F
RWANDA	8	10F	10F	10F	6016	6000	6000	6020	49	60F	60F	61F
SOUTH AFRICA	8	12F	12F	12F	16401	15667	17000	17333	137	188	204	208
SUDAN	3	3F	3F	3F	19811	21154	20741	20357	50	55F	56F	57F
TUNISIA	3	3F	3F	3F	11161	14000	13125	13750	29	42F	42F	44F
ZAIRE	2	3F	3F	3F	13984	16800	16346	16500	26	42F	43F	43F
N C AMERICA	5	11	11	11	12333	12378	12077	12233	65	131	128	129
ANTIGUA					3800	4450	4440	4615				
CUBA	2	4F	4F	4F	15031	17668	17667	17667	24	74	74F	74F
DOMINICAN RP	1	1F	1F	1F	17833	16667	16667	15714	11	10	10	11*
JAMAICA	2	3F	2F	2F	9048	9415	10148	10045	14	26	21	22F
MARTINIQUE					19877	20000	20000	20000	1	1	1	1
PUERTO RICO	1	3	3	3F	10398	5771	5051	5367	13	15	16	16
ST VINCENT					10429	10000	10000	10000				
TRINIDAD ETC					10008	11892	11892	11892	1	4F	4F	4F
USA					15667	13944	13889	13889				
SOUTH AMERIC	56	69	66	60	12090	10837	10738	9705	681	752	711	580
ARGENTINA	27	35	31	25	13258	11765	11704	9429	353	410	364	231
CHILE	5	5F	5F	5F	23778	24420	25000	25000	125	122*	125*	125*
COLOMBIA	2	2	2	2F	15064	15366	15366	15388	30	32	32	32F
ECUADOR	10	14F	14F	14F	7143	6884	6786	6786	70	95F	95F	95F
PERU	5	6F	6F	6F	13427	10818	10875	10930	73	60F	61F	62F
URUGUAY	7	8F	9F	9F	4207	4146	4118	4118	31	34F	35F	35F
ASIA	163	197	176	177	11288	10144	11515	11375	1835	2002	2022	2019
BANGLADESH	16	13	13	13	8126	7256	7182	6876	128	96	93	89
CHINA	65F	94F	72F	73F	10292	8285	11087	11011	665F	779F	794F	799F
CYPRUS					18377	16000	16800	17200	4	4F	4F	4F
HONG KONG					19978	17865	17832	17862	6	5	8	8
IRAQ	7	5F	5F	5F	9812	8283	8510	8712	68	38F	42F	45F
ISRAEL	1	1	1	1F	15100	21867	23671	24667	16	16	19	19F
JAPAN	18	14	14	14F	16881	17444	16667	16667	301	248	230	230F
JORDAN	1	1	3	3F	11008	11232	6222	6222	12	16	17	17F
KOREA DPR	4	5F	5F	5F	10000	9800	10000	10000	40	49F	50F	50F
KOREA REP	9	11	11	11F	12078	13153	14955	14761	105	148	164	167F
LEBANON	1	1F	1F	1F	8976	10000	10000	10000	9	9F	9F	9F
MAL PENINSUL					2981	2969	2941	3056	1	1F	1F	1F
SAUDI ARABIA	2	1	1F	1F	12471	8126	8000	8000	19	7	8F	8F
SRI LANKA	7	12	10	10F	2247	4466	4429	4417	16	53	46	46F
SYRIA	5	12	13	13F	9342	13778	11992	12109	46	161	154	155F
THAILAND	8	8F	8F	9F	8750	8537	8333	8140	70	70F	70F	70F
TURKEY	19	19F	19F	19F	17167	16324	17027	16324	328	302	315	302F
EUROPE	1035	786	743	697	2602	3033	3091	3206	2694	2385	2297	2233
BULGARIA	3	2F	2F	2F	38831	38550	38550	38550	121	77F	77F	77F
DENMARK					21050	16235	28263	27500		1		1F
FRANCE	3	3F	3F	3F	28507	27692	27692	27692	93	72F	72F	72F
GREECE	5	5F	5F	5F	14574	15769	15962	15962	71	82F	83F	83F
HUNGARY	2	2F	2F	2F	17066	18729	18009	18421	30	33	29	35F
ITALY	13	13	13	13	23586	24714	24345	24350	308	331	322	327
MALTA					8116	14309	16052	16294	1	3	3	3F
ROMANIA	233	167	173	160F	1531	2975	2621	2813	357	497	452	450F
SPAIN	4	4	5	4F	21939	20591	22894	20476	78	91	108	86F
YUGOSLAVIA	773	590	541	507	2117	2036	2128	2170	1636	1201	1151	1100F
OCEANIA	9	9	9	10	9338	8742	8574	8464	87	79	81	82
AUSTRALIA	8	8	8F	8F	8541	8029	7963	7952	71	63	65F	66F
NEW ZEALAND	1	1F	1F	1F	16000	13583	12538	11643	16	16F	16F	16F
PAPUA N GUIN					5222	2000	2000	2000				
DEV.PED M E	834	652	603	569	3269	3545	3769	3885	2727	2311	2272	2210
N AMERICA					15667	13944	13889	13889				
W EUROPE	798	616	567	533	2742	2889	3067	3137	2187	1779	1739	1671
OCEANIA	9	9	9	10	9346	8765	8596	8485	87	79	81	82
OTH DEV.PED	27	27	27	27	16666	16776	17025	17194	453	452	453	457
DEV.PING M E	177	208	208	204	12052	11675	11726	11418	2136	2434	2441	2327
AFRICA	18	22	23	23	9602	10336	10139	10302	176	227	229	234
LAT AMERICA	62	80	77	70	12110	11041	10922	10085	746	883	839	709
NEAR EAST	57	61	65	67	15551	15510	15171	15029	888	951	951	1003
FAR EAST	40	45	43	44	8116	8255	8794	8654	326	373	381	381
OTH DV.PING					5222	2000	2000	2000				
CENTR PLANND	306	270	253	242	3957	5318	5546	5845	1212	1434	1402	1412
ASIAN CPE	69	99	77	78	10275	8362	11016	10946	705	828	844	849
E EUR+USSR	238	171	176	164	2133	3552	3168	3430	507	606	558	562
DEV.PED ALL	1072	822	779	733	3017	3547	3633	3783	3234	2917	2831	2772
DEV.PING ALL	246	307	285	281	11556	10609	11535	11288	2841	3262	3285	3176

TABLE
TABLEAU **46**
CUADRO

CUCUMBERS AND GHERKINS CONCOMBRES ET CORNICHONS PEPINOS Y PEPINILLOS

	AREA HARV / SUP RECOLTEE / SUP COSECHAD — 1000 HA				YIELD / RENDEMENT / RENDIMIENTO — KG/HA				PRODUCTION / PRODUCTION / PRODUCCION — 1000 MT			
	1969-71	1975	1976	1977	1969-71	1975	1976	1977	1969-71	1975	1976	1977
WORLD	474	530	516	522	13815	14965	14664	14934	6544	7930	7570	7803
AFRICA	13	18	17	17	13689	14601	15271	15203	184	259	259	259
EGYPT	12	16	15	15F	13863	14506	15238	15238	165	226	224	224F
LIBYA					7500	7333	7333	7333		1F	1F	1F
MADAGASCAR					9889	10125	10000	10244	1	1	1F	1F
MAURITIUS					16231	15179	15455	16667	1	2	2F	2F
SOUTH AFRICA	1	1F	1F	1F	14667	20000	20000	18333	11	20	22	22
TUNISIA	1	1F	1F	1F	10205	12027	12000	12267	5	9F	9F	9F
ZAIRE					4167	4167	4167	4167				
N C AMERICA	79	99	80	79	10106	9651	11613	11955	797	953	926	945
ANTIGUA					4769	5158	4889	4947				
CANADA	4	4	4	4F	16954	18782	18193	18574	71	71	67	72F
CUBA	1	3F	3F	3F	11805	13259	13032	12844	13	40	40F	41F
DOMINICAN RP					36625	37143	40000	35000	1	1F	1F	1F
GUADELOUPE					20222	20000	12000	12000			1	1F
JAMAICA	1		1F	1F	8180	9534	9223	9167	4	4	6	6F
MARTINIQUE					9978	10000	10000	36000	1	3	3	2
PANAMA					4120	4265	4214	4800				
TRINIDAD ETC					9123	10108	10000	10000	2	2	2F	2F
USA	73	91	72	71	9691	9139	11239	11549	705	831	805	819
SOUTH AMERIC	2	2	2	2	17354	17839	18093	18017	42	41	43	44
CHILE	1	1F	1F	1F	25806	25824	26087	25914	23	24F	24F	24F
COLOMBIA					14595	15000	15000	15217	3	2	3	4
PERU	1	1F	1F	1F	10130	9400	9400	9400	7	5F	5F	5F
VENEZUELA	1	1	1F	1F	14619	14263	14865	14868	9	11	11F	11F
ASIA	272	295	303	308	13770	14287	14093	14229	3750	4208	4275	4386
BANGLADESH	3	2	2	2	6425	5400	5351	5296	17	12	13	12
CHINA	163F	183F	187F	190F	11545	11733	11827	11907	1884F	2151F	2212F	2264F
CYPRUS	1	1F	1F	1F	18857	18571	17143	18571	12	13F	12F	13F
HONG KONG					24138	19588	19600	19104	2	2	2	3
IRAQ	14	14F	15F	15F	9251	6866	6986	7100	127	98F	102F	107F
ISRAEL	3	2	2	3F	17113	21437	20412	22000	48	53	50	55F
JAPAN	32	26	26	26F	30903	39046	38521	40508	976	1023	990	1045F
JORDAN	1	1	2	2F	5858	15547	8243	8063	7	21	13	13F
KOREA DPR	2	3F	3F	3F	9728	11034	11000	11333	24	32F	33F	34F
KOREA REP	8	9	9	9F	12319	13076	15039	15110	94	121	130	138F
LEBANON	3	3F	3F	3F	10548	9400	9800	10200	28	28F	29F	31F
MAL PENINSUL	1	1F	1F	1F	1991	2063	2121	2174	1	1F	1F	2F
SINGAPORE					27814				7			
SRI LANKA	2	3	3	3F	1959	4092	3885	3900	5	13	12	12F
SYRIA	10	13	18	18F	6150	13664	10339	10275	61	183	183	187F
THAILAND	6	7F	8F	8F	10446	9444	9067	8718	62	68F	68F	68F
TURKEY	25	24F	25F	25F	15990	16167	17000	16200	394	388	425	405F
EUROPE	106	116	113	114	16648	21265	18234	18820	1758	2457	2053	2154
AUSTRIA	2	2	2	2	21320	21952	16507	26603	45	43	31	58
BELGIUM-LUX					130250	154790	159801	164333	29	66	48	49F
BULGARIA	4	5	5	5F	12105	33792	36233	36000	52	165	171	180F
CZECHOSLOVAK	9	9	9	9F	13878	16300	8749	9172	122	142	76	80F
DENMARK	1				38058	40535	47317	50000	22	16	17	18F
FINLAND	1	1	1	1F	23260	39117	29303	29000	14	23	24	23
FRANCE	4	4F	4F	4F	20155	22343	23486	24629	81	78F	82F	86F
GERMAN DR	4	4	4	4F	14536	26510	22042	22050	59	59	80	88F
GERMANY FED	2	2	1	1	24905	34348	31950	36093	57	56	46	54
GREECE	4	3F	3F	3F	25657	33333	36300	36667	96	100F	109F	110F
HUNGARY	14	12	12F	11F	8375	12562	9572	13182	115	146	112	145F
ICELAND					120500	120000	120000	120000				
ITALY	5	5	5	5	19059	21496	22208	22175	103	110	113	111
NETHERLANDS	2	2	1	2F	159886	210903	217385	228733	300	327	323	343F
NORWAY					48305	68142	80392	86000	6	8	8	9F
POLAND	30	34	34	36F	12269	14326	8462	8333	366	487	288	300F
ROMANIA	15	22F	20F	20F	6799	12727	10750	10021	103	280F	215F	155F
SPAIN	4	7	7	7F	20015	27704	30712	29091	85	197	203	192F
SWEDEN	1	1F	1F	1F	21446	16300	17608	18308	23	21	23	24F
SWITZERLAND					148543	141667	133333	128205	7	5	4	5
UK					212827	234500	200000	210000	34	47	40F	42F
YUGOSLAVIA	3	4F	4F	4F	11765	11000	10895	10795	40	41F	41F	42F
OCEANIA	1	1	1	1	12562	11646	11587	11855	13	13	14	15
AUSTRALIA	1	1	1	1F	11337	11294	11761	12091	10	11	13	13F
NEW ZEALAND					23056	16473	10164	10000	2	1	1	1F
DEV.PED M E	143	156	136	136	19351	20175	22554	23531	2766	3150	3061	3194
N AMERICA	77	95	75	75	10085	9525	11580	11927	775	902	873	891
W EUROPE	30	31	30	30	31537	37157	37215	38537	943	1139	1112	1165
OCEANIA	1	1	1	1	12562	11646	11587	11855	13	13	14	15
OTH DEV.PED	35	30	29	30	29457	36938	36319	38037	1035	1096	1062	1122
DEV.PING M E	89	103	108	109	11802	12460	12278	12089	1055	1279	1324	1323
AFRICA	1	1	1	1	10324	11733	11750	12144	7	12	12	12
LAT AMERICA	4	6	7	7	14448	14494	14236	14443	64	92	96	97
NEAR EAST	65	72	77	78	12273	13235	12780	12509	796	958	989	980
FAR EAST	19	23	23	23	9683	9487	9999	9998	188	218	226	234
CENTR PLANND	241	271	273	277	11283	12912	11682	11850	2723	3500	3186	3286
ASIAN CPE	166	186	190	193	11518	11722	11814	11898	1908	2183	2245	2298
E EUR+USSR	76	85	83	84	10770	15524	11378	11739	815	1318	941	988
DEV.PED ALL	219	241	218	220	16379	18537	18321	19016	3581	4468	4002	4182
DEV.PING ALL	255	289	298	303	11617	11984	11982	11967	2963	3462	3568	3621

TABLE
TABLEAU 47
CUADRO

EGGPLANTS — AUBERGINES — BERENJENAS

	AREA HARV / SUP RECOLTEE / SUP COSECHAD (1000 HA)				YIELD / RENDEMENT / RENDIMIENTO (KG/HA)				PRODUCTION / PRODUCTION / PRODUCCION (1000 MT)			
	1969-71	1975	1976	1977	1969-71	1975	1976	1977	1969-71	1975	1976	1977
WORLD	271	293	298	302	12593	12723	12549	12910	3416	3727	3746	3898
AFRICA	18	25	25	25	17657	13626	14099	14311	316	339	352	363
EGYPT	10	10	11	11F	22631	20915	21905	22110	225	217	230	241F
GHANA	4	10	10F	10F	2917	2800	2700	2600	12	28	27F	26F
IVORY COAST	1	1F	1F	1F	12102	12615	12923	13308	15	16F	17F	17F
LIBYA					9926	10143	9733	10000		1F	1F	1F
MADAGASCAR					13514	13545	14455	14561	1	1	1	1F
MAURITIUS					15573	13250	16667	25000	3	1	1F	2F
SUDAN	2	3F	3F	3F	24727	25000	25333	25333	60	75F	76F	76F
N C AMERICA	3	4	4	4	14058	18544	19409	19261	46	78	81	80
ANTIGUA					5000	5222	5556	5700				
CUBA					21471	15075	15663	16279	2	1	1F	1F
DOMINICAN RP					20955	20667	20000	20000	3	3F	3F	3F
GUADELOUPE					17000	25000	50000	50000		3	8	8F
HAITI					18000	10000	10000	10000	1	1F	1F	1F
JAMAICA					5348	5467	4867	4867				
MARTINIQUE					19725	22222	20000	30000	2	4	2	2
MEXICO	2	2F	2F	2F	9785	15000	15000	14857	15	30F	30F	31F
TRINIDAD ETC					22320	20178	21176	21176	2	4	4F	4F
USA	1	1	1	1	17167	23205	23273	23098	21	32	32	29
SOUTH AMERIC					13297	14118	14000	14146	4	6	6	6
COLOMBIA					14574	15000	15000	15000	2	3	3	3F
VENEZUELA					12095	13223	13000	13333	2	3	3F	3F
ASIA	229	240	247	250	11209	11444	11259	11651	2565	2752	2777	2909
CHINA	120F	132F	137F	139F	7910	8163	8113	8864	946F	1077F	1111F	1231F
CYPRUS					17048	16667	15882	15882	2	3F	3F	3F
HONG KONG					27273	25250	24675	24691	2	1	2	2
IRAQ	10	5F	6F	6F	12623	14370	14864	15281	125	78F	88F	98F
ISRAEL	1	1	1	1F	27301	39310	39167	41667	21	23	24	25F
JAPAN	28	23	22	22F	26118	29445	28624	29545	719	668	624	650F
JORDAN	3	2	3	3F	12023	18249	15678	15852	32	40	43	43F
KOREA DPR	1	2F	2F	2F	11182	11875	12500	11765	14	19F	20F	20F
KOREA REP	3	3	3	3F	11763	12235	13833	14036	31	33	37	39F
LEBANON	1	1F	2F	2F	12152	15000	14000	13125	18	21F	21F	21F
PHILIPPINES	17	17F	16	16F	3757	5294	5079	5000	64	90F	82	80F
SAUDI ARABIA	1	4	4F	4F	9000	10012	10000	10000	9	39	40F	40F
SYRIA	5	7	7	7F	11667	18182	14465	14569	54	120	103	105F
THAILAND	10	10F	10F	11F	5201	5437	5385	5333	50	56F	56F	56F
TURKEY	31	34F	35F	35F	15587	14265	15000	14171	482	485	525	496F
EUROPE	21	23	22	22	23326	24041	23774	24177	482	552	530	541
BULGARIA	1	1	1	1F	28837	21773	21786	22727	25	27	23	25F
FRANCE	1	1F	1	1F	23350	23917	24353	25000	31	29F	23	24F
GREECE	3	3F	3F	3F	16021	18235	18824	18824	55	62	64	64F
ITALY	12	13	13	13	25281	26252	25457	25617	311	332	322	336
SPAIN	3	5	4	4F	21963	22867	23095	24211	59	103	97	92F
DEV.PED M E	49	46	45	45	24694	26909	26362	27031	1218	1249	1186	1220
N AMERICA	1	1	1	1	17167	23205	23273	23098	21	32	32	29
W EUROPE	20	22	21	21	23083	24168	23875	24252	457	525	506	516
OTH DEV.PED	28	23	22	23	26149	29691	28906	29867	740	691	648	675
DEV.PING M E	100	112	114	115	12102	12126	12337	12179	1213	1356	1406	1402
AFRICA	5	11	11	11	5496	4021	3994	3997	30	46	46	46
LAT AMERICA	2	3	3	3	12288	15985	17045	17138	28	52	54	56
NEAR EAST	63	67	70	71	15960	16065	16129	15806	1009	1078	1129	1123
FAR EAST	29	30	29	29	4992	5998	6031	6035	146	180	177	177
CENTR PLANND	122	135	140	142	8094	8330	8269	9006	985	1122	1154	1276
ASIAN CPE	121	134	139	141	7944	8208	8164	8899	960	1096	1131	1251
E EUR+USSR	1	1	1	1	28837	21773	21786	22727	25	27	23	25
DEV.PED ALL	50	48	46	46	24765	26778	26254	26929	1243	1276	1209	1245
DEV.PING ALL	221	245	252	256	9830	9994	10047	10376	2173	2451	2536	2653

TABLE
TABLEAU 48
CUADRO

CHILLIES+PEPPERS, GREEN POIVRONS PIMIENTOS FRESCOS

	AREA HARV / SUP RECOLTEE / SUP COSECHAD 1000 HA				YIELD / RENDEMENT / RENDIMIENTO KG/HA				PRODUCTION / PRODUCTION / PRODUCCION 1000 MT			
	1969-71	1975	1976	1977	1969-71	1975	1976	1977	1969-71	1975	1976	1977
WORLD	558	641	684	696	8257	8292	8057	8195	4608	5315	5509	5703
AFRICA	97	106	110	114	7637	7613	7727	7918	738	809	849	900
EGYPT	7	8	8	9F	17069	15789	16429	16562	127	126	138	147F
IVORY COAST	2	2F	2F	2F	8034	7909	7739	7625	16	17F	18F	18F
LIBYA	1	1F	1F	1F	5373	5000	5000	5000	3	4F	4F	4F
MAURITIUS					3085	3396	4167	4286	1			
NIGERIA	60	62F	64F	65F	8061	8871	8906	9077	484	550F	570F	590F
SUDAN					7862	8444	8656	8696	3	4F	4F	4F
TUNISIA	23	29F	30F	32F	4013	3207	3333	3750	91	93	100	120
ZAIRE	3	4F	4F	4F	4010	3897	3950	3951	13	15F	16F	16F
N C AMERICA	61	63	65	65	7287	8759	9701	9605	445	554	627	628
ANTIGUA					4167	4333	4200	4600				
CANADA									3	5	5	5F
CUBA	1	1F	1F	1F	21474	24290	24909	24909	12	27	27F	27F
DOMINICAN RP		1F	1F	1F	21878	23000	24000	23774	11	12F	12F	13F
JAMAICA					2951	3269	3288	3250	1	1	1	1F
MEXICO	39	40	40	40	5816	6825	8475	8325	226	273	339	333
MONTSERRAT					3125	800	800	800				
PANAMA					15691	1854	2075	2000	1	1	1	1F
PUERTO RICO	1				5818	7836	7560	7667	4	3	2	2F
ST VINCENT					6000	6000	6000	6600				
TRINIDAD ETC					7138	8049	7976	7907		1F	1F	1F
USA	20	20	22	23	9309	11333	10948	10872	186	232	238	245
SOUTH AMERIC	20	16	17	16	6360	6853	7104	6828	128	107	123	107
ARGENTINA	10	6	7	5F	6799	7683	7836	7692	67	46	57	40*
BOLIVIA	2	2	2F	2F	2000	2024	2048	2000	4	4	4F	4F
CHILE	2	2F	2F	2F	10000	10345	10539	10286	20	21F	22F	22F
COLOMBIA			1	1F	14516	15000	15000	15200	5	6	8	8F
ECUADOR	1	1F	1	1F	3143	3750	5389	5455	2	3F	3	3F
PERU	4	3F	3F	4F	4863	5067	4879	4722	19	15F	16F	17F
URUGUAY	1	1F	1F	1F	6067	6462	6324	6286	4	4F	4F	4F
VENEZUELA	1	1	1F	1F	9384	10560	11000	10976	7	8	9F	9F
ASIA	249	323	351	358	6341	5817	5596	5883	1581	1878	1962	2107
CHINA	109F	121F	127F	129F	8600	8754	8676	9471	933F	1061F	1101F	1222F
CYPRUS					10167	18815	16926	16667	1	1		1F
IRAQ	2	1F	1F	1F	5927	7750	7692	7643	11	9F	10F	11F
ISRAEL	1	1	2	2F	22136	21901	18855	18421	31	31	34	35F
JAPAN	4	4	4	4F	27806	34460	35084	38286	122	147	147	161F
JORDAN	1	1F	1	1F	8508	3765	5809	5753	5	3F	4	4F
KOREA DPR	10	13F	13F	14F	2326	2462	2538	2444	23	32F	33F	33F
KOREA REP	37	53	54	57F	1730	1610	1645	1661	63	85	90	95F
PAKISTAN	33	34	51	52F	1220	1535	1542	1541	41	52	79	80F
SRI LANKA	21	55F	57F	58F	1306	851	864	876	27	47F	49F	51F
SYRIA	2	3	3	3F	6225	10412	10813	10667	14	35	35	35F
TURKEY	29	36F	37F	37F	10512	10417	10270	10270	309	375	380	380F
EUROPE	131	133	141	143	13103	14793	13777	13710	1716	1967	1949	1962
AUSTRIA	1	1	1	1	18631	21473	13395	19963	19	25	13	22
BULGARIA	16	18	17	17F	15351	14796	14618	15294	249	263	253	260F
CZECHOSLOVAK	2	3	3	3F	14438	19669	11977	12185	32	49	32	33F
FRANCE	1	2F	2F	2F	12710	13400	13375	13353	15	20F	21F	23F
GREECE	2	4F	4F	4F	11935	10350	10600	10600	29	41F	42F	42F
HUNGARY	14	13	14F	14F	11271	13052	10338	13571	161	175	141	190F
ITALY	19	21	20	20	21872	23249	23619	23784	421	483	481	483
ROMANIA	20	13	21	21F	7007	11328	8146	7619	143	148	173	160F
SPAIN	23	28	30	30	16455	16606	16566	14487	384	460	492	435
YUGOSLAVIA	31	31	30F	31F	8585	9710	10000	10048	263	301	300F	315F
DEV.PED M E	104	112	114	117	14223	15548	15505	15079	1474	1746	1774	1764
N AMERICA	20	20	22	23	9476	11586	11159	11085	189	237	243	249
W EUROPE	78	86	87	88	14531	15448	15581	14921	1131	1331	1350	1319
OTH DEV.PED	6	6	6	6	26420	31320	30226	32098	154	178	181	196
DEV.PING M E	283	348	375	382	5631	5294	5345	5348	1592	1841	2003	2041
AFRICA	88	97	100	104	6856	6956	7013	7191	605	676	704	745
LAT AMERICA	61	59	60	59	6267	7261	8426	8292	383	425	506	485
NEAR EAST	42	51	52	53	11116	11001	11093	11148	472	557	575	586
FAR EAST	91	141	162	167	1445	1297	1341	1351	131	184	218	226
CENTR PLANND	172	181	195	197	8980	9552	8899	9623	1542	1729	1733	1898
ASIAN CPE	119	134	140	143	8069	8144	8106	8806	957	1093	1134	1255
E EUR+USSR	53	47	55	55	11011	13587	10924	11753	585	636	599	643
DEV.PED ALL	157	159	169	172	13134	14971	14021	14019	2059	2382	2373	2407
DEV.PING ALL	401	482	515	524	6351	6088	6096	6288	2548	2934	3137	3296